Gravitation and Astrophysics

On the Occasion of the 90th Year of General Relativity

Proceedings of the VII Asia-Pacific International Conference

Gravitation and Astrophysics

On the Occasion of the 90th Year of General Relativity

Proceedings of the VII Asia-Pacific International Conference

National Central University, Taiwan
23 – 26 November 2005

editors

James M Nester, Chiang-Mei Chen
National Central University, Taiwan

Jong-Ping Hsu
University of Massachusetts Dartmouth, USA

World Scientific

NEW JERSEY • LONDON • SINGAPORE • BEIJING • SHANGHAI • HONG KONG • TAIPEI • CHENNAI

Published by

World Scientific Publishing Co. Pte. Ltd.

5 Toh Tuck Link, Singapore 596224

USA office: 27 Warren Street, Suite 401-402, Hackensack, NJ 07601

UK office: 57 Shelton Street, Covent Garden, London WC2H 9HE

British Library Cataloguing-in-Publication Data
A catalogue record for this book is available from the British Library.

GRAVITATION AND ASTROPHYSICS
On the Occasion of the 90th Year of General Relativity
Proceedings of the VII Asia-Pacific International Conference

ISBN-13 978-981-256-950-9
ISBN-10 981-256-950-2

Printed in Singapore by World Scientific Printers (S) Pte Ltd

Preface

2005 was an auspicious year: not only was it an International Year of Physics commemorating the centennial of Einstein's great achievements of 1905---it also was an anniversary of Einstein's development of his gravity theory, General Relativity, and of Hilbert's discovery of the general-coordinate-invariant action for gravity at the same time. The final form of the field equations for this theory, Einstein's greatest achievement, was submitted to the Prussian Academy of Sciences on 25 November, 1915. By the way, this year was also the jubilee of Utiyama's 1955 paper, which followed Yang-Mills' work and proposed the gauge-invariant interpretation of all interactions, a work which paved the way for far-reaching research on gauge theories of gravity and quantum gravity.

On the 90 anniversary of the birth of general relativity, at National Central University (NCU), Jhongli, Taiwan, 38 local participants were joined by 47 distinguished guests from Canada, China, France, Japan, Korea, Russia, and the USA to participate in the seventh in a series of International Conferences on Gravitation and Astrophysics (ICGA7). This series of conferences has been aimed especially to serve the growing needs of the workers in this research field in the Asia-Pacific region. Previous ones, with a growing number of local, regional and international participants, had been held in Seoul, Korea (1993), Hsinchu, Taiwan (1995), Tokyo, Japan (1997), Beijing, China (1999), Moscow, Russia (2001), and Seoul, Korea (2003).

Over 50 papers were presented at the ICGA7 conference. This volume includes almost all of these reports. We regret that due to circumstances beyond the control of the editors we were not able to include 7 reports (interested readers can get some idea of the missing presentations from works of these individuals which are available elsewhere). We do believe that the depth and breadth of the reports included here well reflect the quality of the conference and the development of this field in this Asia-Pacific region.

According to our standards and goals, as well as the feedback received from participants, the ICGA7 meeting turned out to be quite successful. Some factors which contributed to this included good support, nice weather and a good venue, along with the great effort and thoughtful attention to numerous details of the many local workers. NCU President Chuan Sheng Liu (a plasma physicist) came and spoke about the development of research in physics in general and relativity in particular at NCU, in Taiwan and in the Asia-Pacific region. We were especially pleased that a good number of people (about 13) from mainland China managed to cross the Taiwan strait and attend the meeting. This signified a growing scientific communication and exchange across the Taiwan strait. The various funding agencies, the Taiwan National Science Council, Ministry of Education, and National Center for Theoretical Sciences, the Asia Pacific Center for Theoretical Physics (Korea), the National Central University and the University of Massachusetts Dartmouth Foundation were quite helpful and provided enough support. We note that the ICGA7 meeting happened towards the end of a year packed with meetings that had kept the science historians so busy that, regretfully, none were in the end able to make it to ICGA7.

We believe that this ICGA series of conferences fills a big need and serves an important purpose for the researchers in this field in this region, especially since most of the other gravity meetings are held very far away from this Asia-Pacific neighborhood, and therefore only a limited number of the researchers from here are able to attend. We believe that this series of meetings will continue and that ICGA7 will play its role in carrying on the interchange of scientific and cultural ideas between East and West.

August, 2006

James M. Nester
Chiang-Mei Chen
Jong-Ping Hsu

CONTENTS

Experimental Tests of Gravity

PROGRESS IN TESTING NEWTONIAN INVERSE SQUARE LAW

Sheng-Gun Guan, Liang-Cheng Tu, Jun Luo*

Department of Physics, Huazhong University of Science and Technology, Wuhan 430074, P.R.China.

E-mail: junluo@mail.hust.edu.cn

Motivated by extra dimensions theories that predict new effects, we are testing the gravitational inverse square law at distance down to 100μm using a torsion balance and gold plate attractor. We tent to improve previous short-range forces constraints by up to a factor of 10 and search for deviations from Newtonian physics, predicted by ADD theory.

String theory is the most promising approach to the long-sought unified description of the four forces of nature and the elementary particles [1], but direct evidence supporting it is lacking. One generic prediction of the theory is the existence of extra dimensions in addition to our familiar 3-dimensional space. These extra dimensions had been thought to be extremely tiny (of order the Planck length $\sim 10^{-33}$cm), until in recent years the idea of large extra dimensions was proposed to address the "hierarchy problem"-a problem associated with the factor of the 10^{17} huge difference between the Planck scale and the weak interaction [2−3]. A special example is the ADD theory [2], which assumes that the fundamental energy scale of gravity is the same as that of the Standard Model, about 1 TeV. The apparent weakness of gravity is then interpreted as a consequence of the fact that gravitons are free to propagate in all spatial dimensions, while the particles and fields of the Standard Model are confined to a 3+1-dimensional 'brane'. Thus, if one assumes there are n large extra dimensions of equal size R^* and the fundamental Planck mass is M*=1 TeV, one can conclude [2−5]

$$R^* = \frac{\hbar c}{M^* c^2}\left(\frac{M_p}{M^*}\right)^{2/n}, \tag{1}$$

with the usual Planck mass $M_P=1.2\times 10^{16}$ TeV. According to above equation, the scenario with $n=1$ is ruled out by solar-system observations. If there are 2 extra dimensions, must be about 0.2 mm, which is interest for the laboratory experiments. The gravitational inverse square law will smoothly change from a $1/r^2$ form for $r \gg R^*$ to a $1/r^{2+n}$ form for $r \ll R^*$.

For the experimentally relevant case where $r \sim R^*$, the gravitational potential is usually parametrized as a Yukawa modification to the Newtonian potential :

$$V = -\frac{G_4 m}{r}(1 + \alpha e^{-r/\lambda}), \tag{2}$$

with strength α and range λ. G_4 is the four-dimensional Newtonian constant and m is the mass. The experiments can bound on the possible values of α and λ, which are represented in an α-λ diagram. Although the gravitational interaction has been tested with high precision for separations greater than 1 cm, very little is known about gravity at range below 1 mm (or λ <1 mm). Some other considerations, such as the dilaton and moduli exchange in string theories [6−9], also suggested the Yukawa potential and predicted that α will be large as 10^5 for Yukawa ranges $\lambda \sim$0.1 mm. While the simplest scenario with 2 large extra dimensions in ADD theory predicts α=3 or 4 for compactification on a 2-sphere or 2-torus, respectively [10−11].

Up to now, ISL holds to high precision on the scale of the solar system by the astronomical observations. The experimental limit at range of 1 cm to several km was determined by searching for a "fifth force' in the past two decades [12]. Due to the interests in searching for the "fifth force", in particular the possible compact extra dimensions, the sensitivity of experiments searching for deviations from Newtonian gravity at short distances has been improved by many orders of magnitude in the past decades. Adelberger and his colleagues at the University of Washington tested the transverse force between two disks with ten equally space holes using a torsion pendulum, and the sensitive range is 100μm to 300μm [13]. In 2003, Long *et al* at the University of Colorado tested the force between two flat plates using a torsion oscillator working at 1000 Hz, corresponding a sensitive range of 100μm and below[14]. Chiaverini *et al* at Stanford University used a Micro-cantilever modulated at its resonated frequency of 300Hz to test the force between the two flat plates, and the sensitive range is 20μm and below [15]. Figure 4 shows the current experimental constraints[13−17] of the inverse square law of gravitation together with the theoretical predictions in (α,λ) parameter space.

In this letter, we reported a scheme to test the gravitational $1/r^2$ law using planar plates separated by a gap of 100μm.

A schematic of the apparatus used in our measurement is shown in Figure 1. A 66 cm-long, 25-μm-diameter tungsten fiber, hanging from an x-

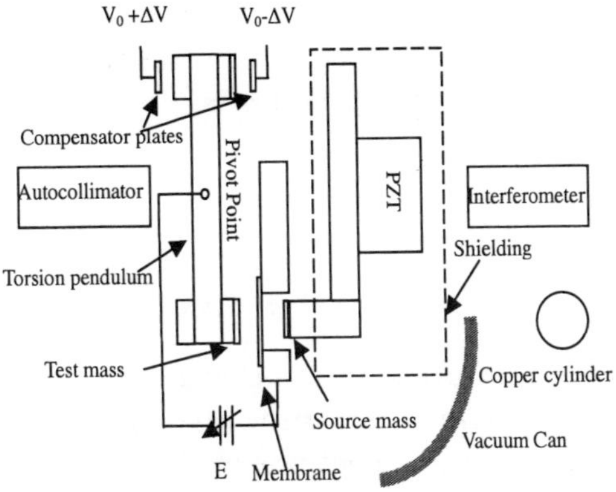

Fig. 1. Schematic of the apparatus. The vacuum vessel dimensions are 60 cm diam by 250 cm tall. The PZT was used to drive the source mass gently closer to test mass;. The six-freedom platform was not shown here.

y-z-θ stage, supported a torsion pendulum with gold coat. The test mass and source mass comprise a $20\times20\times0.2$ mm^3 gold plate, and a $18\times17\times0.2$ mm^3 gold plate. As shown in Fig. 1, the source mass and an electrostatic shields(a glass plate of 50μm in thickness, coated with gold) were mounted on one arm of the torsion pendulum, while the source mass and the electrostatic shields were placed on a micro positioning assembly which was formed by a six-freedom platform and a piezoelectric stack translators (PZT). The six-freedom platform, with the resolution of 1μrad for the circle goniometer and 0.05μm for the linear displacement, is used to adjust the parallelism and position of the source mass and the electrostatic shields relative to the test mass, the PZT was just used to drive source mass while a Michelson interferometer with the linear displacement sensitivity of 0.3 μm, detected the gap variance between the source mass and test mass. A vacuum of order 10^{-7}torr was maintained (to eliminate viscous effects and effectively decrease the impact of air current) by an ion pump;

A feedback system was used to keep the torsion pendulum angle fixed; as shown in Fig. 1, two "compensator plates" form a capacitor with respect to the pendulum body. An autocollimator (angular detective sensitivity of 0.05 μrad) was used to determine whether the torsion pendulum is in parallel position relative to the source mass; any unparallel was detected by the autocollimator which provides an error signal to an integral-plus-proportional feedback circuit. A dc correction voltage was applied to the compensators, as required to keep the torsion pendulum angle fixed. In addition, a constant dc voltage of $V_0 = 5.0$ V was applied to the compensators in order to linearize the effect of the small correction voltage δV[the force on the pendulum due to one compensator is:

$$F = (V^2/2)dC/dX \approx (V_0^2 \pm 2V_0)dC/dX, \quad (3)$$

and the net force from both is:

$$F = 2V_0\delta V dC/dX, \quad (4)$$

where dC/dX is the magnitude of the change in compensator-pendulum capacitance as the gap size x is varied. Since the feedback only affects the torsional mode, a strong magnet was used to overdamp all vibrational modes of the pendulum system. The angular fluctuations were consistent with the expected thermal noise[18];

$$\Delta\theta_{\text{rms}} = (KT/a)^{-1/2} \approx 0.7\mu\text{rad} \quad (5)$$

Where a $=8.6\times10^{-9}$Nm/rad is the torsion constant for the tungsten fiber , so our torsion pendulum can be sensitive to the moment of 6×10^{-15}Nm. A micro-rotation stage allowed turning the fiber to set δV=0. Before experiment, the fiber was annealed, with the pendulum hanging in the vacuum, by keeping it about 70°c over 24 hours; the drift was less than 1μrad/h.

The gravitation force was measured by simply stepping the voltage applied to the PZT up and down, and at each step, measuring the restoring force, implied by a change in δV, required to keep the pendulum angle fixed. The maximum displacement is 200 μm; the relative displacement was measured to 0.3 μm accuracy by use of a laser interferometer.

The system calibration was obtained through gravitation measurements based on the variation of a known gravitational force. A copper cylinder was placed in a fixed place around pendulum; the invariable gravitational force was determined by measuring the torsion pendulum deflexion angle. The calibration was determined by the change in δV from the feedback circuit to balance the restoring force, and is 1.2×10^{-11} Nm/V, with 1% accuracy.

With the gold plates separated but externally shorted together, there was an apparent shockingly large potential of 240mV; there are several decade separate electrical connections in this loop and a potential this large is consistent with what is expected for the various metallic contacts. This potential was easily canceled by setting an applied voltage between the plates to give a minimum δV; this applied voltage was taken as "zero" in regard to the calibration.

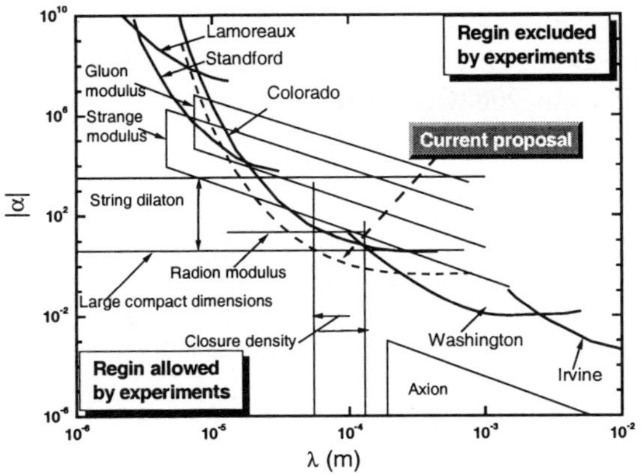

Fig. 2. 95% confidence upper limits on $1/r^2$-law violating interactions of the form given by E.q. 2. The region excluded by previous work [13-17]. Constraints from previous experiments and the theoretical predictions are adapted from Ref. [2], except for 2 large extra dimensions prediction which is from Ref. [10, 11].

The PZT give very accurate and reproducible relative changes in the plate separation; the absolute separation between test mass and membrane was determined by measuring the angle between the pendulum and membrane as a function of separation; the absolute separation between membrane and source mass was determined by a contact measurement.

The uncertainty in absolute distance between source mass and test mass was normally less than 5μm. Each up/down sweep cycle was measured repetitiously. The balance voltage was the measured gravitational torque change signal in the experiment. The quantity measured in the experiment was:

$$\Delta = \tau_{\text{up}} - \tau_{\text{down}} \tag{6}$$

Where τ_{up} is the total torque produced by all source masses after PZT driving and τ_{down} is the total torque produced by them before PZT driving. The Newtonian value of $\Delta(\Delta_{\text{thy}})$ was calculated by numerical integrations. Numerical integration over all source masses at between 0.1 mm to 0.2 mm separate.

Δ_{exp} was found by directly measuring the torque difference $\tau_{\text{up}} - \tau_{\text{down}}$. The discrepancy between the two results is $\Delta = \Delta_{\text{exp}} - \Delta_{\text{thy}}$, If the discrepancy is produced by a Yukawa potential, the resulting contraints on new physics of the form given in Eq. 2 are shown in Fig. 2; scenarios with $\alpha \geq 1$ are excluded at 95% confidence for λ=100μm.

This work was supported by the National Natural Science Foundation of China under Grant 10121503, and by the National Basic Research Program of China under Grant 2003CB716300.

References

1. B. Greene,. The Elegant Universe: Superstrings, Hidden Dimensions, and the Quest for the Ultimate Theory (Norton, New York, 1999).
2. N. Arkani-Hamed, S. Dimopoulos and G. Dvali, Phys. Lett.B429,263(1998)
3. N. Arkani-Hamed, S. Dimopoulos, G. Dvali and N. Kaloper, Phys. Rev. Lett. 84,586(2000)
4. C. D. Hoyle, U. Schmidt, B. R.Heckel, E. G.Adelberger, J. H.Gundlach, D. J. Kapner, H. E Swanson, Phys.Rev.Let. 86,1418 (2001).
5. C. D. Hoyle, D. J. Kapner, B. R.Heckel, etal, Phys. Rev. D70,042004(2004)
6. S. Dimopoulos and G. Guidice, Phys. Lett. B379,105(1996)
7. I. Antoniadis, S. Dimopoulos, G. Dvali, Nucl. Phys. B 516, 70 (1998).
8. D.B. Kaplan, M.B. Wise, J. High Energy Phys. 8, 37 (2000).
9. J.C. Long, H.W. Chan, A.B. Churnside, Nature,421, 922 (2003)
10. E.G. Floratos, G.K. Leontaris, Phys. Lett. B465,95(1999)
11. A. Kehagias, K. Sfetsos, Phys. Lett. B472,39(2000)
12. E. Fischbach D.E. Krause, V.M. Mostepanenko, M. Novello,, Phys. Rev. D 64, 075010 (2001).
13. C. D. Hoyle, U. Schmidt, B. R.Heckel, E. G.Adelberger, J. H.Gundlach, D. J. Kapner, H. E Swanson, Phys.Rev.Let. 86,1418 (2001).
14. J. C. Long, H. W Chan., A. B Churnside, E. A Gulbis, M. C. M.Varney, J. C Price, Nature, 421, 922 (2003)
15. J. Chiaverlini S. J. Smullin, A. A. Geraci, D.M.Weld, and A. Kapitulnik, Phys.Rev.Let. 90, 151101 (2003)
16. R. Spero,J. K. Hosking,R. Newman,J. Pellam and J. Schultz : Phys.Rev.Lett.,44(1980) 1654.
17. K. Lamoreaux : Phys.Rev.Lett. 44 (1997)1654.
18. G. Ising, Philos. Mag. 1, 827 (1926).

Null Test of the Inverse-Square Law at 100-Micrometer Distance

HO JUNG PAIK*, VIOLETA A. PRIETO, AND M. VOL MOODY
Department of Physics, University of Maryland,
College Park, MD 20742-4111, U.S.A.
**hpaik@physics.umd.edu*

In string theories, extra dimensions must be compactified. The possibility that gravity can have large radii of compactification leads to a violation of the inverse square law at submillimeter distances. We are preparing a null test of Newton's law with a resolution of one part in 10^3 at $100\,\mu$m, which will probe the extra dimensions down to $10\,\mu$m. The experiment will be cooled to 2 K. To minimize Newtonian errors, a near null source in the form of a circular disk of large diameter-to-thickness ratio is employed. Two test masses, also disk-shaped, are suspended on the two sides of the source mass at a distance of 150 mm. The signal is detected by a superconducting differential accelerometer. We discuss the design and principle of this experiment and report the progress.

1. Objective of Research

The objective of this experiment is to test Newton's inverse-square ($1/r^2$) law to better than one part in 10^3 at a 100-μm range. Figure 1 shows the existing limits and the sensitivity (2σ) of this experiment for the $1/r^2$ law at $\lambda \leq 1$ mm, where the total potential is written as

$$V(r) = -G\frac{m_1 m_2}{r}\left[1 + \alpha \exp\left(-\frac{r}{\lambda}\right)\right]. \qquad (1)$$

The line and the shaded region represent violations predicted by higher-dimensional string theory and the axion theory, respectively. The expected resolution of this experiment represents an improvement by over three orders of magnitude beyond the limits obtained by Chiaverini *et al.*[1], Long *et al.*[2], and Hoyle *et al.*[3]. Our experiment will be marginally sensitive to detect the axion with highest allowed coupling and will test a string theory prediction for the radius of compactification larger than 10 μm.

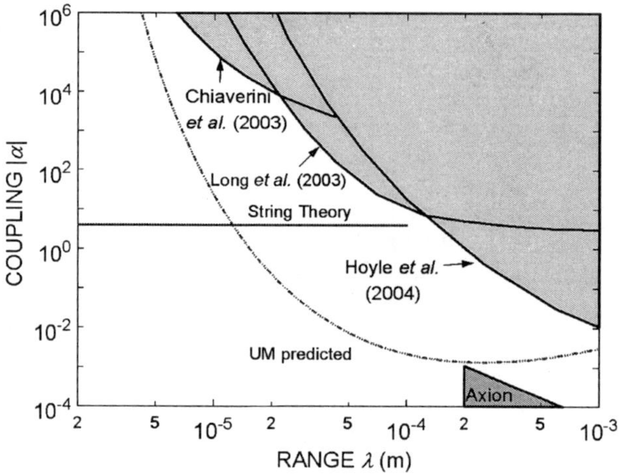

Fig. 1. Sensitivity of ISLES versus the existing limits.

1.1. *Test of string theories*

String theory is defined in terms of a fundamental scale M_*. If there are n compact dimensions with radii $R_1, R_2, \ldots, R_n$, Gauss's law implies that the Planck mass M_{Pl} is related to M_* by

$$M_{Pl}^2 \approx M_*^{2+n} R_1 R_2 \ldots R_n. \tag{2}$$

As we probe distances shorter than one of the radii R_i, a new dimension opens up and changes the r dependence of the gravitational force law. For $r > R_i$, the deviation is found to be Yukawa-type. In particular, when extra dimensions are compactified on an n-torus, the strength of the potential is $\alpha = 2n$.[4,5] The string theory derived deviation shown in Figure 1 corresponds to $n = 2$.

One theoretically well-motivated value for M_* is 1 TeV, which solves the gauge hierarchy problem. For two large dimensions of similar size, one obtains $R_1 \approx R_2 \approx 1$ mm.[6] Cosmological and astrophysical constraints give a bound $M_* > 100$ TeV,[7,8] which corresponds to $R_1 \approx R_2 < 1\,\mu$m. While this is beyond the reach of our experiment, there are cosmological assumptions going into these bounds.

1.2. *Search for the axion*

The Standard Model of particle physics successfully accounts for all existing particle data; however, it has one serious blemish: the strong CP problem.

Peccei and Quinn[9] developed an attractive resolution of this problem. One ramification of their theory is the existence of a new light-mass boson, the *axion*.[10,11] The axion mediates a short-range mass-mass interaction. The upper bound $\theta \leq 3 \times 10^{-10}$ corresponds to a violation of the $1/r^2$ law at the level of $|\alpha| \approx 10^{-3}$ at $\lambda = 200$ μm, which is within the reach of ISLES. The axion could also solve a major open question in astrophysics: the composition of dark matter. The axion is one of the strongest candidates for the cold dark matter.[12]

2. Principle and Design of the Experiment

To maximize the masses that can be brought to 100 μm from each other, flat disk geometry is used for both the source and test masses. An infinite plane slab is a Newtonian null source. We approximate this by using a circular disk of large diameter-to-thickness ratio. Figure 2 shows the configuration of the source and test masses with associated coils. Two disk-shaped test masses are suspended on two sides of the source and are coupled magnetically to form a differential accelerometer.

As the source mass is driven at frequency $f_S/2$ along the symmetry

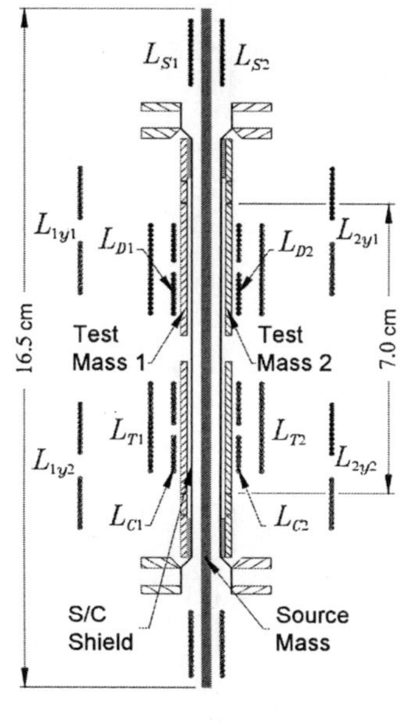

Fig. 2. Configuration of the source and test masses with associated coils.

axis, the first-order Newtonian fields arising from the finite diameter of the source mass are canceled upon differential measurement, leaving only a second-order error at f_S. By symmetry, the Yukawa signal also appears at f_S. The second harmonic detection, combined with the common-mode rejection of the detector, reduces vibration coupling by over 200 dB.

2.1. *Overview of the apparatus*

Figure 3 shows an expanded cross section of the experiment. To eliminate differential contraction and provide good electromagnetic shielding, the entire housing is fabricated from niobium (Nb). The source mass is made out of tantalum (Ta), which closely matches Nb in thermal contraction. It is suspended by cantilever springs and driven magnetically. The test masses are also made out of Ta and suspended by cantilever springs. A thin Nb sheet provides electrostatic and magnetic shielding between the source and each test mass.

The experiment is suspended from the top of the cryostat via three rub-

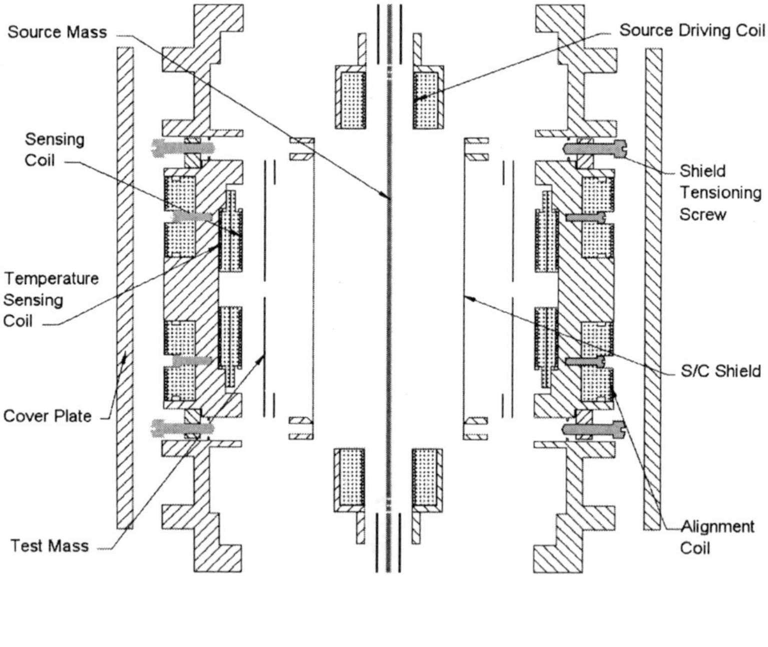

Fig. 3. Expanded cross section of the experiment.

ber tubes. Voice-coil transducers, incorporated into each vertical leg of the suspension, are used to shake the instrument for balance and calibration. By varying the magnitude and phase of the current through the coils, vertical acceleration or tilt in any direction is applied to the instrument. The tilt is sensed with a two-dimensional optical lever consisting of a laser, a beam splitter, an x-y photodiode, and a planar mirror mounted at the top of the instrument.

The cryostat has a cold plate and copper can, which will be cooled to below 2 K by pumping on liquid helium through a capillary.

2.2. *Source and test masses*

The source is a disk 1.65 mm thick by 165 mm in diameter, with mass 590 g. The source mass, cantilever springs, and rim are machined out of a single plate of Ta. Ta is chosen for its high density (16.6 g cm^{-3} compared to 8.57 g cm^{-3} for Nb), which increases the signal, and its relatively high critical field ($H_c = 0.070$ T at 2 K). The test masses are identical Ta disks 250 μm thick by 70 mm in diameter. Their dynamic mass is $m = 16.7$ g. The mechanical resonance frequency of the test mass is 11 Hz.

The equilibrium spacing between the source and each test mass is 150 μm. They are shielded from each other by means of a 12.5-μm thick Nb shield. The source mass is driven magnetically by sending AC currents to the two source coils. The design allows a source displacement of ± 87 μm.

The differential acceleration signals expected from the Yukawa force

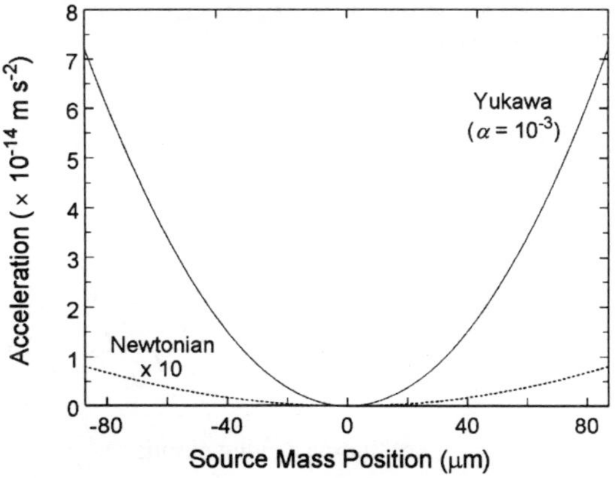

Fig. 4. Newtonian and Yukawa signals versus source position.

with $|\alpha| = 10^{-3}$ and $\lambda = 200\,\mu$m are plotted in Figure 4 as a function of the source mass position. The small Newtonian term arising from the finite source mass diameter is also shown. The source mass looks like an "infinite plane slab" to the test mass due to its proximity. The Yukawa signal is almost purely second harmonic to the source motion. Its rms amplitude, corresponding to a $\pm 87\,\mu$m displacement, is 2.6×10^{-14} m s^{-2}.

2.3. *Superconducting circuits*

Figure 5(a) is the differential-mode (DM) sensing circuit. A persistent current is stored through the loop comprising the *parallel* combination of L_{D1} and L_{D2} and the transformer primary. Another current is stored in the loop comprised of the *series* combination of L_{D1} and L_{D2}, which in turn tunes the ratio I_{D2}/I_{D1}. For redundancy, the CM circuit (not shown) is designed to be physically identical but with a different current configuration.

Figure 5(b) is the temperature sensing circuit. The coils L_{T1} and L_{T2} are mounted directly on the Nb housing (see Figure 2), making them sensitive only to temperature variation. The output of this circuit is used to compensate the temperature sensitivity of the differential accelerometer.

Figure 5(c) shows the superconducting circuit for the source. A large

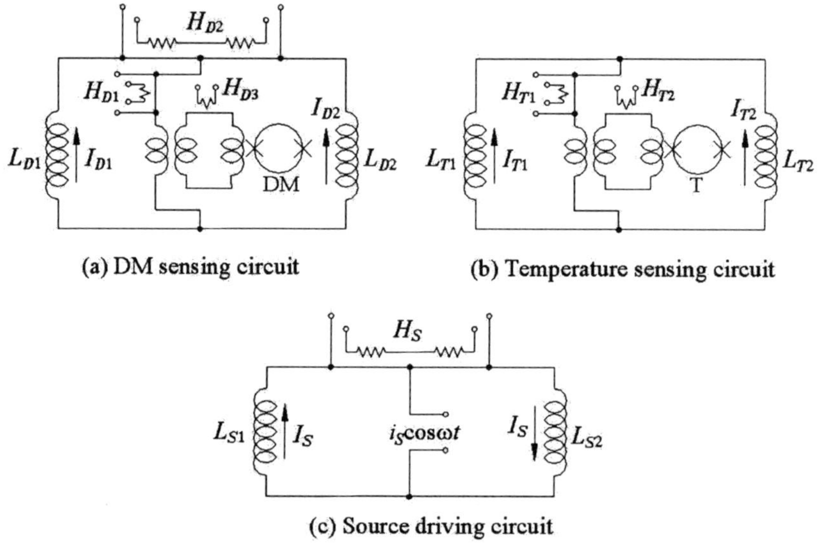

Fig. 5. Superconducting circuits for the detector and source.

persistent current, $I_S \approx 5$ A, is stored in the main loop. The source is then driven by sending a small current $i_S \cos \omega t$ across the loop. This reduces the magnetic crosstalk between the source and the detector.

3. Experimental Errors

3.1. *Metrology errors*

For the thin disk geometry of the source, errors in the thickness are dominant over density inhomogeneity. Due to the cylindrical symmetry of the test masses, linear taper of the source produces a second-order error and azimuthal asymmetry is averaged out. Radial thickness variation is the dominant error source. Due to the null nature of the source, test mass metrology is unimportant, except for the suspension spring. The source mass is lapped to meet the required dimensional tolerances and is stress relieved.

3.2. *Intrinsic noise of the detector*

The intrinsic power spectral density of a superconducting differential accelerometer can be written[13,14] as

$$S_a(f) = \frac{8}{m}\left[\frac{k_B T \omega_D}{Q_D} + \frac{\omega_D^2}{2\eta\beta} E_A(f)\right]. \tag{3}$$

where m is the mass of each test mass, $\omega_D = 2\pi f_D$ and Q_D are the DM (angular) resonance frequency and quality factor, β is the electromechanical energy coupling coefficient, η is the electrical energy coupling coefficient of the SQUID, and $E_A(f)$ is the input energy resolution of the SQUID. For our experiment, we find $S_a^{1/2}(f) = 8.4 \times 10^{-12}$ m s^{-2} Hz$^{-1/2}$ at $f = 0.1$ Hz. We have chosen $f_S = 0.1$ Hz since the seismic noise dips at that frequency and the SQUID noise is close to the white noise level.

3.3. *Other noise*

Modulation of the penetration depth of a superconductor with temperature gives rise to temperature sensitivity in a superconducting accelerometer.[13] Extrapolating from the low-frequency noise spectrum measured during the initial cooldown of the experiment, we find a temperature-induced noise of 2.4×10^{-12} m s^{-2} Hz$^{-1/2}$ at $f = 0.1$ Hz. This will be reduced by a factor of 10 by the temperature compensation.

Table 1. Error budget (1σ) of the experiment.

Error Source	**Error** ($\times 10^{-15}$ m s^{-2})
Metrology	1.4
Random (10^6-s integration)	
Intrinsic	8.4
Temperature	2.4
Seismic	0.5
Source Dynamic	17.4
Gravity Noise	< 0.1
Magnetic Coupling	< 0.1
Electrostatic Coupling	< 0.1
Total	19.5

The displacement of the source induces a platform tilt of 1.2×10^{-6} rad at $f_S/2$. The tilt modulates the Earth's gravity and produces an unacceptably high level of linear acceleration as well as angular acceleration. We will cancel the source driven tilt by a factor of 10^2 with a feedback loop. The tilt is measured with the optical tilt sensor and the signal is fed back to the voice-coil actuators to null the tilt signal.

Magnetic cross talk between the source and detector is an important error source. The entire housing is machined out of Nb and a Nb shield is provided between the source and each test mass. Further rejection by the frequency discrimination provides a comfortable margin. The Casimir force[15] is not important in our experiment since gaps between conducting planes are $> 10\,\mu$m.[16]

3.4. *Expected resolution*

Table 1 combines all the errors. To reduce the random noise to the levels listed, a 10^6-s integration was assumed. By equating the total error with the expected Yukawa signal, we compute the minimum detectable $|\alpha|$. Figure 1 shows the 2σ error plotted as a function of λ. The best resolution of our experiment is $|\alpha| \approx 1 \times 10^{-3}$ at $\lambda = 150\,\mu$m. The experiment will probe extra dimensions down to $10\,\mu$m when two extra dimensions are compactified on a square torus.

4. Progress of the Experiment

4.1. *Construction of the apparatus*

Two 1.75-mm thick Ta sheets were "two face grounded" commercially to improve the surface flatness. A measurement with dial indicators showed the surface to be flat to 10 μm, which is an acceptable tolerance for the initial experiment. A source mass was produced from one of these sheets by wire EDM and then was heat-treated to relieve stress. Four test masses were constructed, also with wire EDM, from a 0.25-mm thick Ta sheet.

Two Nb housings that hold the source mass, test masses, and various coils together were machined. Superconducting shields for the test masses were fabricated by diffusion-bonding 12.5-μm Nb sheets to Nb rings. All the superconducting coils were wound and tested.

The entire apparatus was assembled and integrated with the cryostat. Figure 6 shows the instrument suspended from the cryostat insert. The exposed coil and a like one on the opposite side (not shown in Figure 3) are used to cancel the recoil of the platform in response to source mass motion.

Fig. 6. The apparatus integrated with the cryostat.

4.2. *Test of the apparatus*

The apparatus has been cooled down several times and several mechanical problems have been discovered and corrected. One of the source coils partially came off from the Macor coil form upon cooling, touching the source mass. To reduce the driving current and thus the magnetic cross talk, the source coils had been wound with five layers of Nb-Ti coils. The resulting thick layer of epoxy cracked due to differential contraction. The test masses also touched the Nb shields due to stresses put on them by the stretching of the shields.

We have now rewound all the failed superconducting coils. The source coils have been rewound using two layers of wire. We have also inserted capacitor plates to monitor the source position *in situ* relative to the superconducting shields. We are now reassembling the apparatus, with increased spacing between the test masses and the shields, to assure that the test masses remain free at low temperature.

The resonance frequencies of the Nb shields were measured to be ~ 1 kHz, with Q's in excess of 10^5. This result is very encouraging since the modulated Casimir force from the source mass will be sufficiently attenuated by the stiff shields.

The He^4 cold plate was tested while monitoring the temperature sensing circuit. The cold plate reached a steady operating temperature of 1.6 K. The temperature remained stable for approximately 10 hr.

Acknowledgments

This work is partially supported by NASA under grant NAG32874 and by NSF under grant PHY0244966.

References

1. J. Chiaverini, *et al.*, *Phys. Rev. Lett.* **90**, 151101 (2003).
2. J. C. Long, *et al.*, *Nature* **421**, 922 (2003).
3. C. D. Hoyle, *et al.*, *Phys. Rev. D*, **70**, 042004 (2004).
4. E. G. Floratos and G. K. Leontaris, *Phys. Lett. B*, **B465**, 95 (1999).
5. A. Kehagias and K. Sfetsos, *Phys. Lett. B*, **B472**, 39 (2000).
6. N. Arkani-Hamed, S. Dimopoulos and G. Dvali, *Phys. Rev. D* **59**, 086004(1999).
7. S. Cullen and M. Perelstein, *Phys. Rev. Lett.* **83**, 268(1999).
8. L. J. Hall and D. Smith, *Phys. Rev. D* **60**, 085008(1999).
9. R. D. Peccei and H. Quinn, *Phys. Rev. Lett.* **38**, 1440 (1977).
10. S. Weinberg, *Phys. Rev. Lett.* **40**, 223 (1978).

11. F. Wilczek, *Phys. Rev. Lett.* **40**, 279 (1978).
12. M. S. Turner, *Phys. Rep.* **197**, 67 (1990).
13. H. A. Chan and H. J. Paik, *Phys. Rev. D* **35**, 3551 (1987).
14. M. V. Moody, E. R. Canavan and H. J. Paik, Rev. Sci. Instrum. **73**, 3957 (2002).
15. H. B. G. Casimir, *Proc. K. Ned. Akad. Wet.* **51**, 834 (793).
16. S. K. Lamoreaux, *Phys. Rev. Lett.* **78**, 5 (1997).

Numerical Relativity

A New Generalized Harmonic Evolution System

Lee Lindblom[1], Mark A. Scheel[1], Lawrence E. Kidder[2], Robert Owen[1], and Oliver Rinne[1]

[1] *Theoretical Astrophysics 130-33, California Institute of Technology Pasadena, CA 91125, USA*

[2] *Center for Radiophysics and Space Research, Cornell University Ithaca, NY 14853 USA*

This note describes recent work on finding a formulation of the Einstein equations suitable for constructing stable numerical evolutions. The formulation described here specifies the coordinate degrees of freedom with a generalized harmonic gauge source function rather than with the usual lapse and shift. This type of formulation appears to have played a critical role in the very impressive binary black hole evolutions performed recently by Pretorius. This note analyzes why this type of formulation is so effective for numerical work, describes a recent extension of the system that makes it possible to construct boundary conditions (including constraint-preserving boundary conditions), and describes numerical tests that demonstrate the effectiveness of the new equations and boundary conditions.

Two properties have made harmonic or generalized harmonic (GH) coordinates an important tool throughout the history of general relativity theory. The first property is well known: this method of specifying the coordinates transforms the principal parts of the Einstein equations into a manifestly hyperbolic form, in which each component of the metric is acted on by the standard second-order wave operator. The second property is not as widely appreciated: this method of specifying coordinates fundamentally transforms the constraints of the theory. This new form of the constraints makes it possible to modify the evolution equations in a way that prevents small constraint violations from growing during numerical evolutions—without changing the physical solutions of the system and without changing the fundamental hyperbolic structure of the equations. The purpose of this note is to explore these important properties and to describe how the GH evolution system has been extended in a way that makes it very useful for numerical computations.

Coordinates are fixed in the generalized harmonic (GH) method by specifying a gauge source function H_a, defined as the action of the scalar-wave operator on the coordinate functions x^a:

$$H_a \equiv \psi_{ab}\nabla^c\nabla_a x^b = -\psi^{bc}\Gamma_{abc} \equiv -\Gamma_a, \tag{1}$$

where ψ_{ab} is the spacetime metric and Γ_{abc} is the usual Christoffel symbol. The coordinates are fixed in this approach by requiring that $\Gamma_a = -H_a$, where $H_a = H_a(x, \psi)$ is a prescribed function of the coordinates x^a and the metric ψ_{ab}. The choice $H_a = 0$ corresponds to standard harmonic coordinates; the existence of solutions to the inhomogeneous wave equation, Eq. (1), implies the existence of such coordinates more generally. Choosing the coordinates in this way has two important consequences. The first is well known: the vacuum Einstein equations have a simple manifestly hyperbolic structure when expressed in GH coordinates. The Ricci curvature tensor can be written as

$$R_{ab} = -\frac{1}{2}\psi^{cd}\partial_c\partial_d\psi_{ab} + \nabla_{(a}\Gamma_{b)} + \psi^{cd}\psi^{ef}\left(\partial_e\psi_{ca}\partial_f\psi_{db} - \Gamma_{ace}\Gamma_{bdf}\right), \tag{2}$$

in any coordinate system, where $\nabla_a\Gamma_b \equiv \partial_a\Gamma_b - \psi^{cd}\Gamma_{cab}\Gamma_d$. In GH coordinates, $\Gamma_a = -H_a$, so the only second-derivative term remaining in the Ricci tensor is $\psi^{cd}\partial_c\partial_d\psi_{ab}$. Therefore, in GH coordinates the vacuum Einstein equations, $R_{ab} = 0$, form a manifestly hyperbolic system,

$$\psi^{cd}\partial_c\partial_d\psi_{ab} = -2\nabla_{(a}H_{b)} + 2\,\psi^{cd}\psi^{ef}\left(\partial_e\psi_{ca}\partial_f\psi_{db} - \Gamma_{ace}\Gamma_{bdf}\right), \tag{3}$$

for any choice of guage source function H_a.[1]

The second consequence of using GH coordinates is less widely appreciated: The constraints of the system are profoundly transformed. The vacuum Einstein equations, Eq. (3), can also be written in the more covariant form

$$0 = R_{ab} - \nabla_{(a}\mathcal{C}_{b)}, \tag{4}$$

where $\mathcal{C}_a = H_a + \Gamma_a$. The condition $\mathcal{C}_a = 0$ is the primary constraint of this system, while the standard Hamiltonian and momentum constraints $\mathcal{M}_a = G_{ab}t^b$ (where t^a is the unit normal to a Cauchy surface) are determined by the derivatives of $\mathcal{C}_a$: $\mathcal{M}_a = t^b(\nabla_{(a}\mathcal{C}_{b)} - \frac{1}{2}\psi_{ab}\nabla^c\mathcal{C}_c)$. This means that the primary constraints depend on the first but not the second derivatives of the metric.

Adding multiples of the constraints to the Einstein equations is known to have a significant effect on the growth rates of constraint violating solutions.[2] However, adding multiples of the Hamiltonian and momentum constraints has not been found to be very effective in controlling the growth of constraint violating solutions. This is because the addition of those constraints changes the principal part of the equations, so constraints can be added only in very restricted ways consistent with the hyperbolic structure of the equations. In contrast, arbitrary multiples of the gauge constraint $\mathcal{C}_a$ can be added to the system, Eq. (4), without effecting the hyperbolic structure at all. Pretorius,[3] based on the suggestion of Gundlach, et al.,[4] used a modified evolution system that included the following additional gauge constraint terms designed to suppress the growth of the constraints:

$$0 = R_{ab} - \nabla_{(a}\mathcal{C}_{b)} + \gamma_0\left[t_{(a}\mathcal{C}_{b)} - \frac{1}{2}\psi_{ab}\, t^c\mathcal{C}_c\right]. \tag{5}$$

The Bianchi identities then imply that $\mathcal{C}_a$ satisfies the damped wave equation,

$$0 = \nabla^c\nabla_c\mathcal{C}_a - 2\gamma_0\nabla^b[\,t_{(b}\mathcal{C}_{a)}] + \mathcal{C}^b\nabla_{(a}\mathcal{C}_{b)} - \frac{1}{2}\gamma_0\, t_a\mathcal{C}^b\mathcal{C}_b, \tag{6}$$

which exponentially suppresses all small short-wavelength constraint violations when the parameter γ_0 is positive.[4] This constraint suppressing feature of the modified generalized harmonic system, Eq. (5), contributed significantly to the success of Pretorius' impressive binary black-hole evolutions.[3,5]

We have recently extended the modified generalized harmonic evolution system, Eq. (5), to a first-order symmetric-hyperbolic form. (See Ref. [6] for the details.) The vacuum Einstein system expressed in this new GH first-order form is given by

$$\partial_t\psi_{ab} - (1+\gamma_1)N^k\partial_k\psi_{ab} = -N\Pi_{ab} - \gamma_1 N^i\Phi_{iab}, \tag{7}$$

$$\begin{aligned}\partial_t\Pi_{ab} &- N^k\partial_k\Pi_{ab} + Ng^{ki}\partial_k\Phi_{iab} - \gamma_1\gamma_2 N^k\partial_k\psi_{ab}\\ &= 2N\psi^{cd}\left(g^{ij}\Phi_{ica}\Phi_{jdb} - \Pi_{ca}\Pi_{db} - \psi^{ef}\Gamma_{ace}\Gamma_{bdf}\right)\\ &\quad -2N\nabla_{(a}H_{b)} - \frac{1}{2}Nt^ct^d\Pi_{cd}\Pi_{ab} - Nt^c\Pi_{ci}g^{ij}\Phi_{jab}\\ &\quad +N\gamma_0\left[2\delta^c{}_{(a}t_{b)} - \psi_{ab}t^c\right](H_c+\Gamma_c) - \gamma_1\gamma_2 N^i\Phi_{iab},\end{aligned} \tag{8}$$

$$\begin{aligned}\partial_t\Phi_{iab} &- N^k\partial_k\Phi_{iab} + N\partial_i\Pi_{ab} - N\gamma_2\partial_i\psi_{ab}\\ &= \frac{1}{2}Nt^ct^d\Phi_{icd}\Pi_{ab} + Ng^{jk}t^c\Phi_{ijc}\Phi_{kab} - N\gamma_2\Phi_{iab},\end{aligned} \tag{9}$$

where the dynamical field Π_{ab} is defined by Eq. (7), and Φ_{iab} is defined by $\Phi_{iab} = \partial_i\psi_{ab}$. We use the lapse N, shift N^i, and spatial metric g_{ij} (the standard functions of ψ_{ab}) to simplify the principal parts of Eqs. (7)–(9). The terms on the right sides of Eqs. (7)–(9) are algebraic functions of the dynamical fields. The connection terms Γ_{cab} appearing on the right side of Eq. (8) are computed using the standard definition of Γ_{abc}, with the partial derivatives of ψ_{ab} determined from the dynamical fields by

$$\partial_t\psi_{ab} = -N\Pi_{ab} + N^i\Phi_{iab}, \tag{10}$$

$$\partial_i\psi_{ab} = \Phi_{iab}. \tag{11}$$

The parameter γ_0 that appears in these expressions is the one used by Pretorius in Eq. (5). The parameter γ_1 was introduced to control the characteristic speed of the field ψ_{ab}. And the parameter γ_2 was introduced to suppress the growth of the new constraint $\mathcal{C}_{kab} = \partial_k\psi_{ab} - \Phi_{iab}$ that arises in this first-order form of the equations. Choosing the parameter $\gamma_0 > 0$ causes the constraint $\mathcal{C}_a$ to be exponentially suppressed via Eq. (6). Choosing the parameter $\gamma_1 = -1$ makes the new system linearly degenerate, so shocks do not form from smooth initial data.[7] Choosing the parameter $\gamma_2 > 0$ in this new system causes the constraint $\mathcal{C}_{iab}$ to be exponentially suppressed,[8] because the modified Eq. (9) implies an evolution equation for $\mathcal{C}_{iab}$ having the form $\partial_t\mathcal{C}_{iab} - N^k\partial_k\mathcal{C}_{iab} \simeq -\gamma_2 N\mathcal{C}_{iab}$.

Boundary conditions for hyperbolic evolution systems are applied to the characteristic fields of those systems. The characteristic fields for the new GH evolution system, Eqs. (7)–(9), are given by

$$u^{\hat{0}}_{ab} = \psi_{ab}, \tag{12}$$

$$u^{\hat{1}\pm}_{ab} = \Pi_{ab} \pm n^i\Phi_{iab} - \gamma_2\psi_{ab}, \tag{13}$$

$$u^{\hat{2}}_{iab} = P_i{}^k\Phi_{kab}, \tag{14}$$

where n_i is the outgoing unit normal at a point on the boundary, and $P_i{}^k = \delta_i{}^k - n_i n^k$. The characteristic fields $u^{\hat{0}}_{ab}$ have coordinate characteristic speed $-(1+\gamma_1)n_k N^k$, the fields $u^{\hat{1}\pm}_{ab}$ have speed $-n_k N^k \pm N$, and the fields $u^{\hat{2}}_{iab}$ have speed $-n_k N^k$. Characteristic fields with negative characteristic speeds propagate into the computational domain, so boundary conditions must be imposed on each characteristic field that has a negative characteristic speed. The simplest boundary condition that enforces the physical idea of no incoming waves sets each incoming characteristic speed to zero at the boundary. A similar condition, which we often find useful, freezes each incoming characteristic field to its initial value. We have also derived a set of rather more complicated constraint preserving and physical boundary conditions for this system (see Ref. [6]).

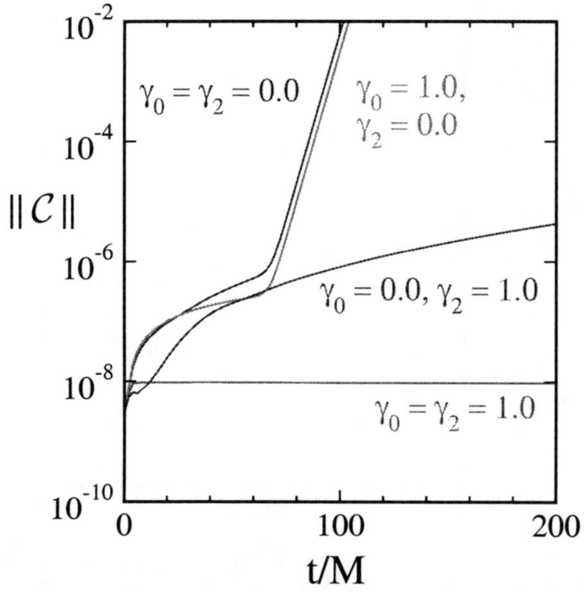

Fig. 1. Evolution of Schwarzschild initial data using different values of the constraint damping parameters γ_0 and γ_2.

The well-posedness of the initial-boundary value problem can be analyzed using the Fourier-Laplace technique[9] for the complicated physical and constraint preserving boundary conditions that we use. We have analyzed the well-posedness of this system for high-frequency perturbations of any spacetime in any GH gauge. Applying the Fourier-Laplace technique to this case yields a necessary (but not sufficient) condition for well-posedness, the so-called determinant condition;[9] failure to satisfy this condition would mean the system admits exponentially growing solutions with arbitrarily large growth rates. We have verified that this determinant condition is satisfied for the GH system using the combined set of physical and constraint preserving boundary conditions that we use.

We tested this new evolution system by evolving initial data for a Schwarzschild black hole. In these evolutions we "freeze" the values of the incoming characteristic fields on the boundaries. We performed these numerical evolutions using spectral methods as described in Ref. [10] for a range of numerical resolutions specified by N_r (the highest order radial basis function) and L_{max} (the highest order spherical harmonic). Figure 1 shows the time dependence of the constraint norm $||\mathcal{C}||$ for several val-

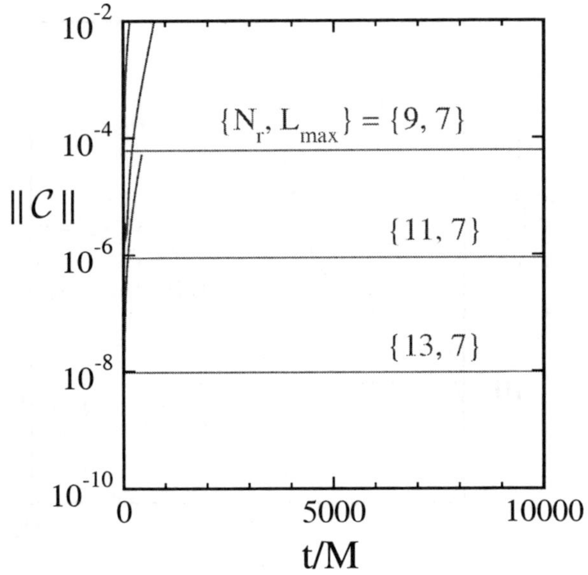

Fig. 2. Evolution of Schwarzschild initial data with $\gamma_0 = \gamma_2 = 1$ show stability and convergence for several numerical resolutions.

ues of the constraint damping parameters γ_0 and γ_2. These tests show that without constraint damping the extended evolution system is extremely unstable. But Figure 2 illustrates that with constraint damping, $\gamma_0 = \gamma_2 = 1$, the evolutions of the Schwarzschild spacetime are completely stable up to $t = 10,000M$ (and forever, we presume). These tests illustrate that both the γ_0 and the γ_2 constraint damping terms are essential.

We also tested our new constraint-preserving boundary conditions by evolving a black hole perturbed by an incoming gravitational wave (GW) pulse. We perturb Schwarzschild initial data by injecting a GW pulse through the outer boundary of the computational domain with time profile $f(t) = \mathcal{A} e^{-(t-t_p)^2/w^2}$ and $\mathcal{A} = 10^{-3}$, $t_p = 60M$, and $w = 10M$. Figure 3 shows the evolution of $||\mathcal{C}||$ using constraint-preserving boundary conditions (dashed curves) and simple freezing boundary conditions (solid curves). These results illustrate that the new boundary conditions are effective in preventing the influx of constraint violations. Figure 4 illustrates the time dependence of the Weyl tensor component $|\Psi_4|$ averaged over the outer boundary of the computational domain. The dashed curve (using constraint-preserving boundary conditions) shows black-hole quasi-normal

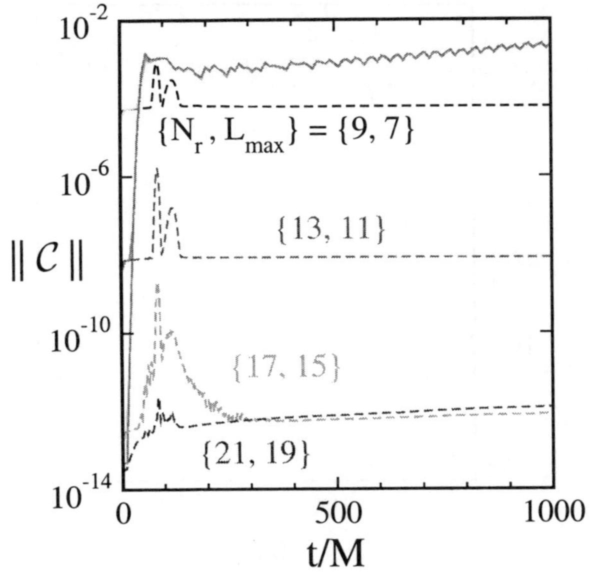

Fig. 3. Evolution of perturbed Schwarzschild spacetime. Solid curves use boundary conditions that freeze all the incoming characteristic fields, while dashed curves use constraint preserving and physical boundary conditions.

oscillations with the correct complex frequency, while the solid curve (using freezing boundary conditions) is completely unphysical. These results show that proper constraint preserving boundary conditions are essential if accurate gravitational waveforms are needed.

This work was supported in part by a grant from the Sherman Fairchild Foundation to Caltech and Cornell, by NSF grants PHY-0099568, PHY-0244906 and NASA grants NAG5-12834 and NNG05GG52G at Caltech, and by NSF grants PHY-0312072, PHY-0354631, and NASA grant NNG05GG51G at Cornell.

References

1. H. Friedrich. On the hyperbolicity of Einstein's and other gauge field equations. *Commun. Math. Phys.*, 100:525–543, 1985.
2. M. A. Scheel, L. E. Kidder, L. Lindblom, H. Pfeiffer, and S. A. Teukolsky, Toward stable 3D numerical evolutions of black-hole spacetimes. *Phys. Rev. D*, 66:124005, 2002.
3. F. Pretorius. Evolution of binary black hole spacetimes. *Phys. Rev. Lett.*, 95:121101, 2005.
4. C. Gundlach, G. Calabrese, I. Hinder, and J. M. Martin-Garcia. Constraint

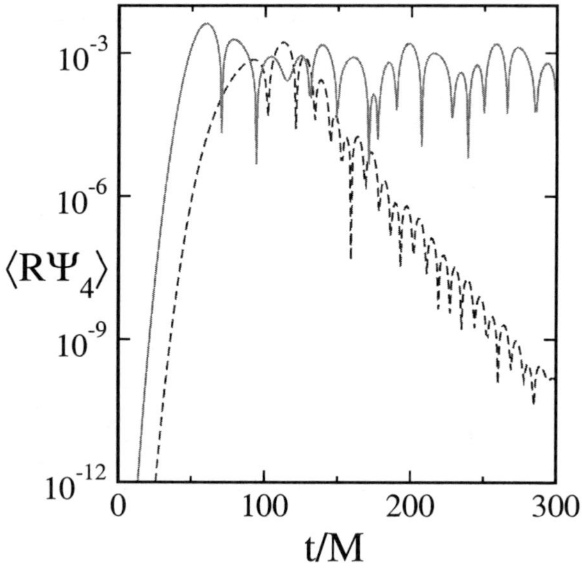

Fig. 4. Evolution of the Weyl curvature component $|\Psi_4|$ in a perturbed Schwarzschild spacetime. Solid curves use boundary conditions that freeze all the incoming characteristic fields, while dashed curves use constraint preserving and physical boundary conditions.

damping in the Z4 formulation and harmonic gauge. *Class. Quantum Grav.*, 22:3767–3774, 2005.

5. F. Pretorius. Numerical relativity using a generalized harmonic decomposition. *Class. Quantum Grav.*, 22:425–452, 2005.
6. L. Lindblom, M. A. Scheel, L. E. Kidder, R. Owen, and O. Rinne, A New Generalized Harmonic Evolution System. *Class. Quantum Grav.*, submitted, 2006; gr-qc/0512093.
7. T. P. Liu. Development of singularities in the nonlinear waves for quasilinear hyperbolic partial differential equations. *J. Diff. Equations*, 33(1):92–111, 1979.
8. M. Holst, L. Lindblom, R. Owen, H. P. Pfeiffer, M. A. Scheel, and L. E. Kidder. Optimal constraint projection for hyperbolic evolution systems. *Phys. Rev. D*, 70:084017, 2004.
9. B. Gustafsson, H.-O. Kreiss, and J. Oliger. *Time Dependent Problems and Difference Methods.* Pure and Applied Mathematics. Wiley, New York, 1995.
10. L. E. Kidder, L. Lindblom, M. A. Scheel, L. T. Buchman, and H. P. Pfeiffer. Boundary conditions for the Einstein evolution system. *Phys. Rev. D*, 71:064020, 2005.

Numerical Relativity beyond I^+

Charles W. Misner

Department of Physics, University of Maryland
College Park, Maryland 20742-4111; and
Albert-Einstein-Institut, Max-Planck-Institut für Gravitationsphysik
Am Mühlenberg 1, D-14476 Potsdam, Germany
E-mail: misner@umd.edu

James R. van Meter

Laboratory for Gravitational Astrophysics
NASA Goddard Space Flight Center, Greenbelt, Maryland 20771
E-mail: vanmeter@milkyway.gsfc.nasa.gov

David R. Fiske

Intelligent Systems Division, Decisive Analytics Corporation
1235 South Clark Street, Arlington, Virigina 22202; and
Laboratory for Gravitational Astrophysics
E-mail: david.fiske@gmail.com

This is a study of the behavior of wave equations in conformally compactified spacetimes suited to the use of computational boundaries beyond Scri+. There light cones may be adjusted for computational convenience and/or Scri+ may be converted to a spacelike hypersurface just outside a de Sitter horizon. Our preliminary numerical implementation excises the physically unnecessary universe somewhat beyond this outer horizon. As an entry level example we study a formulation of the Maxwell equations and causal relations for an outer boundary in that example. We find that an initial central pulse propagates to and through Scri+ in an (hyperboloidal) coordinate time comparable to the pulse width, and that the the numerical evolution remains stable for several times that long. This is a proposal for outer boundary conditions and wave extraction in numerical relativity for which further tests and development would be needed prior to an implementation in full GR.

1. Introduction

The construction of major observatories for gravitational wave astronomy has given a challenge to theoretical and computational astrophysicists to give ever more detailed descriptions of possible sources for detectable gravitational waves. Some indications of these efforts are given in Section 2 below. In the computed models of the coalescence of compact objects (white dwarfs, neutron stars, black holes) using discrete approximations of the Einstein partial differential equations, one of the many problems encountered is the imposition of boundary conditions far from the source of the waves, as discussed in Section 3. In Section 4 we present one approach to this problem which is still in early development. As a test of some aspects of this approach, before any effort to implement it in general relativity, we describe in Section 5 its application to the simpler P.D.E.'s of Maxwell's electromagnetism in relativistic form.

2. Challenge

Gravitational wave astronomy was first proposed by Joseph Weber around 1955. All others before him, such as Einstein, had made preliminary calculations and dismissed the project as impractical. Although Weber had no specific model of a gravitational wave source in mind, he looked at the possibilities for detecting such waves and found that one could, and he did, construct detectors many orders of magnitude more sensitive than had previously been assumed. Dyson, aware of Weber's efforts, made estimates of the frequencies and intensities of gravitational waves that should be emitted by a binary neutron star system near coalescence, thus providing one target number for this nascent field. The next essential contributor to gravitational wave astronomy was Richard Isaacson who began studying gravitational waves at the University of Maryland where Weber was constructing his detectors. Isaacson, after a few years postdoctoral research, became a program director at the U.S. National Science Foundation (NSF) where his office provided NSF funding for gravitation research. After some experience in such bureaucracy (including important mentoring by Marcel Bardon) he played an important role in two essential projects for GW astronomy. The first was the establishment of U.S. supercomputing centers[a]

[a] A significant offshoot of this project came from the National Center for Supercomputing Applications at Illinois where Larry Smarr, an early worker in numerical relativity, was director. It was there that Mosaic, the first graphic browser, was created; it expanded the use of the world wide web by many orders of magnitude.

as he knew that computing on a huge scale would be essential for both the modelling of possible sources of gravitational waves, and for data analysis once detectors were built. The second was the LIGO project (Laser Interferometric Gravitational Observatory). With Isaacson's continual guidance, Kip Thorne and Rainer Weiss were eventually able to produce a successful proposal for this project, which is now in its first long science run at essentially design sensitivity.

The more recent history of the LIGO project, and its parallels and collaborators around the world, is described elsewhere in this Proceedings by Lindblom [1]. But one continuing component of the effort is the production of detailed descriptions of astrophysical processes which should be able to produce detectable quantities of gravitational waves. One possible source is the inspiral and coalescence of compact objects to form black holes. In some parts of this process there are analytic approximations which give the best method for modelling the process, e.g., low velocity or post-Newtonian approximations when the objects are at medium to large separations, and small perturbations of the final black hole ("ring-down") as the coalescence reaches its final stages. But the close encounter stage between the slow inspiral and final ring-down is expected to require solving, for the spacetime metric, the full Einstein equations as discretized partial differential equaions using large scale computing. This is called numerical relativity. These computations are urgently needed by the observers, in part to help in extracting the gravitational wave signals from obscuring noise, but importantly to interpret the signals in terms of the astrophysical events producing them. This task has proved much more difficult than early workers had expected, and a multitude of problems have been met, many of them now solved, with many reported at a recent NASA meeting [2]. The work described below is a proposal for an approach, which has not yet been applied to Einstein equations, to one of these numerical problems — the treatment of the outer boundary of the computational model.

3. Outer Boundary Problem

In the numerical solution of partial differential equations (PDE's), the spacetime manifold is replaced by a finite grid of points at which field values are evolved in a succession of time steps. PDE's typically require not only initial values, but also boundary conditions, in order to define a unique solution. On a finite grid there will correspondingly be some boundary points which require special treatment. Examples include both points which represent the grid's approximation to spatial infinity, and those which mark apparent horizons inside black holes if the computation does not attempt to evolve the dynamics inside the black hole. Among the problems that must be faced when imposing boundary conditions on timelike boundaries are: (a) Constraints may impose relations among fields at the boundary; (b) Gauge modes may not be causal; and (c) Outgoing characteristic modes must not be restricted.

We explore here an approach to the treatment of the boundary modelling spatial infinity which treats it in a way parallel to the methods being developed to discard the invisible physics inside black holes. The physics far from the source of a gravitational wave may not be invisible, but on the scale of computations presently contemplated it is incomprehensible, so we propose to treat it as invisible. That is, for a GW source such as coalescing black holes, the computation of the dynamics of these black holes cannot be expected to also include descriptions of neighboring galaxies, our Galaxy, and the cosmologically distant regions into which most of the waves will propagate without being absorbed. Nor would this be useful, as once the waves are many wavelengths beyond their source, and so spread over large volumes that they are weak enough to be treated in linear approximation, there is nothing to be gained by following their progress with the full Einstein equations rather than as linear waves, usually in a geometrical optics limit. Thus all that is desired from a numerical relativity treatment of GW sources is that they provide a waveform, and an antenna pattern, for a point source in linear gravitation theory. Our proposal below suggest a way to achieve this end which has some different emphases from other schemes with similar objectives.

4. Our Proposal

Our proposed approach to the outer boundary problem emphasizes four tools: (a) Hyperboloidal time slices; (b) Conformal compactification; (c) Artificial cosmology; and (d) Tilting light cones beyond I^+ .

4.1. *Hyperboloidal Slicing*

Begin with Minkowski spacetime

$$d\tilde{s}^2 = -dT^2 + dX^2 + dY^2 + dZ^2 \tag{1}$$

where T, X, Y, Z are the usual rectangular coordinates and $R^2 = X^2+Y^2+Z^2 \equiv X^i X^i$. We change the time coordinate from T to u to obtain a slicing by hyperboloids:

$$T = su - s + \sqrt{s^2 + R^2} \quad . \tag{2}$$

The $u =$ const hypersurfaces are spacelike hyperboloids

$$[T - s(u-1)]^2 - R^2 = s^2 \quad . \tag{3}$$

For large R these constant u hypersurfaces approximate the null cones $R = T - s(u-1)$. The s parameter gives the time lag ΔT between the apex of such lightcones and the $R = 0$ point on the hyperboloid. The parameter s also sets the range in R where the hyperboloids depart only slowly from the $T =$ const slices.

4.2. *Conformal Compactification*

With only the coordinate change from T to u the Minkowski metric becomes

$$d\tilde{s}^2 = -(s^2 + R^2)du^2 + (\frac{sdR}{\sqrt{s^2+R^2}} - Rdu)^2 + R^2 d\Omega^2 \tag{4}$$

using spherical space coordinates. The coordinate speed of light outward is then found to be

$$\begin{aligned} c_{\text{out}} &= s^{-1} dR/du \\ &= \sqrt{1+(R/s)^2}\left[(R/s) + \sqrt{1+(R/s)^2}\right] \end{aligned} \tag{5}$$

which for large R gives $c_{\text{out}} \approx 2(R/s)^2$ which tends to infinity for large R. [The coordinate velocity of light is infinite on a light cone where emission and reception of light occur on the same null slice.] Thus just replacing T by u as the time coordinate would require infinitesimal time steps according to the Courant condition. As illustrated by Misner, Scheel, and Lindblom [3], this problem can be cured using AnMR (Analytic Mesh Refinement), a coordinate transformation that assigns a finite value to $R = \infty$. Thus we set

$$\frac{X^i}{s} = \frac{x^i}{1 - \frac{1}{4}r^2} \quad . \tag{6a}$$

$$\frac{T}{s} = u + \frac{\frac{1}{2}r^2}{1 - \frac{1}{4}r^2} \quad . \tag{6b}$$

where $R^2 = X^2 + Y^2 + Z^2 \equiv X^i X^i$ and $r^2 = x^2 + y^2 + z^2 \equiv x^i x^i$. This brings I^+ in to $r = 2$ and in equation (6b) simply restates equation (2) as a direct transformation from $TXYZ$ to $uxyz$ coordinates.

This transformation leads to a *Minkowski* spacetime metric

$$d\tilde{s}^2 = (s^2/q^2)[-(1 + r^2/4)^2 du^2 + (dx^i - x^i\, du)^2] \tag{7}$$

where

$$q = 1 - r^2/4 \tag{8}$$

This metric is singular at $r = 2$ which corresponds to $R = \infty$; it has an outgoing radial coordinate speed of light

$$c_{\text{out}} = (1 + r/2)^2 \tag{9}$$

which remains bounded for $r \leq 2$ as desired.

Dropping the singular factor s^2/q^2 gives a metric which extends smoothly beyond $r = 2$ which is $R = \infty$.

$$ds^2 = -(1 + r^2/4)^2 du^2 + (dx^i - x^i\, du)^2 \tag{10}$$

This metric is *conformally* equivalent to Minkowski spacetime and has the same light cone structure. It is not flat but has the same coordinate speed of light.

Some physical laws can be restated equivalently in such a comformally modified spacetime. The scalar wave equation has a "conformally covariant" form in which changes in the field, and in the d'Alembertian, allow solutions using the ds^2 metric to generate solutions using the $d\tilde{s}^2$ metric. The Maxwell equations, if written carefully, read identically in both metrics: "conformally invariant". The restructuring needed to get Einstein equation solutions from those of related equations with "conformally regulated" boundary conditions should be studied. For this restructuring a starting point would be the Wald's review [4, Equation 11.1.16]. Many other approaches combining hyperboloidal slicings and conformal compactification have been suggested. Many are now more fully developed than the one we are suggesting here. See Husa et al. [5] and earlier reviews [6] and [7].

4.3. *Artificial Cosmology*

Although conformally regulated Minkowski spacetime (equation 10) allows wave equations to be formulated as finite difference equations all the way out to and including $R = \infty$, boundary conditions there can still be touchy. The ingoing coordinate speed of light there is

$$c_{\text{in}} = -(1 - r/2)^2 \tag{11}$$

which is negative ($dr/du < 0$ on these null directions) except at the horizon I^+ where it is zero. Thus $r =$ const is always a timelike hypersurface except at $r = 2$, i.e., I^+. Thus this outer boundary is very different from the future boundary $u =$ const at each stage of the solution, which is a spacelike hypersurface in the future of the initial conditions. It should be much easier to handle the outer boundary if it also were a spacelike hypersurface in the future of the initial conditions. While preparing [3] Mark Scheel asked CWM in 2003 whether a spacelike outer boundary could be found. Somewhat later a relatively physical way to do this [8] was proposed.

Artificial cosmology replaces the asymptotically flat distant region of spacetime by a de Sitter cosmological horizon at a computationally convenient large distance from the wave source. To modify Minkowski spacetime in this way we take

$$d\tilde{s}^2 = -dT^2 + dX^2 + dY^2 + dZ^2 + (R^2/L^2)(dT - dR)^2 \tag{12}$$

in which the constant R hypersurface $R = L$ is a null surface, the de Sitter horizon. Using the same coordinate transformation and dropping the same conformal factor as before leads to a metric in standard $3+1$ form:

$$ds^2 = -\alpha^2 du^2 + \gamma_{ij}(dx^i + \beta^i du)(dx^j + \beta^j du) \quad . \tag{13}$$

But now the artificial cosmological constant $3/L^2$ appears in the metric components, which we list to help you see that this conformally regulated metric is smooth for all values of $r > 0$.

$$\alpha^2 = \frac{\left(1 + \frac{1}{4}r^2\right)^2}{1 + \left(\frac{s}{L}S\right)^2}, \quad S(r) = \frac{r}{\left(1 + \frac{1}{2}r\right)^2}, \tag{14a}$$

$$\beta^i = -x^i \frac{1 + \left(\frac{s}{L}\right)^2 S}{1 + \left(\frac{s}{L}S\right)^2} \quad , \tag{14b}$$

$$\gamma_{ij} = \delta_{ij} + \left(\frac{sS}{Lr}\right)^2 x^i x^j \quad . \tag{14c}$$

What we gain from this is that the light cones are tilted slightly more outward at $r = 2$ so the $R = \infty$ boundary I^+ becomes not a null hypersurface, but a spacelike surface in the future of the initial conditions. Consequently, could that, or an hypersurface slightly beyond it, be set as the computationaly boundary, *no boundary conditions* would need be set on any causally propagating field components. The inner edge of the light cone $(dr/du)_{\text{null}} = c_{\text{in}}$ points outward for a narrow region around $r = 2$ as seen in Figure 1 here where $L = 10s$. Figure 2 gives a broader picture of

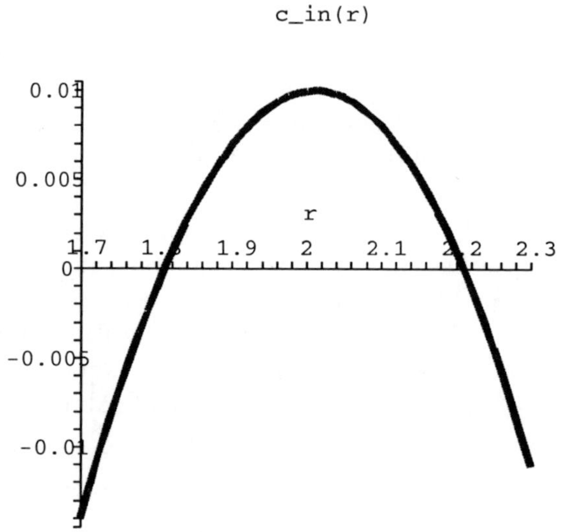

Fig. 1. The inward coordinate speed of light for the conformally regulated de Sitter metric of equations (13) and (14). Where this velocity dr/du is positive, even the inner edge of the lightcone points outward.

the inner and outer edges of the light cone (coordinate velocity of light). It shows the coordinate speeds for light rays on the inward side of the light cone $c_{\rm in}$, on the outward side of the light cone $c_{\rm out}$, and for a timelike center of the light cone $v_{\rm center} = -\beta$ normal to the time slices of constant u.

4.4. *Excising 'das All'*

Were it possible to halt the discrete computation at a spherical boundary (which some do), artificial cosmology might suffice to let the computation be bounded at, say, $r = 2.1$. But the resources available for a first test problem dictated a cubic computational boundary, e.g., at $|x^i| = 2.1$. Then to make $r > 2$ ignorable for the evolutions of the physical domain at $r < 2$ we found it useful to modify the metric beyond I^+ to make the light cones point away from the physical region everywhere there. Then excision boundary conditions such as those used to discard the physically irrelevant regions inside black holes, could here be used to discard any dynamics beyond the domain modelled by the computational grid. In work completed after this ICGA7 meeting we found that the results, although satisfactory for durations several times longer than the sample wave pulse width as reported

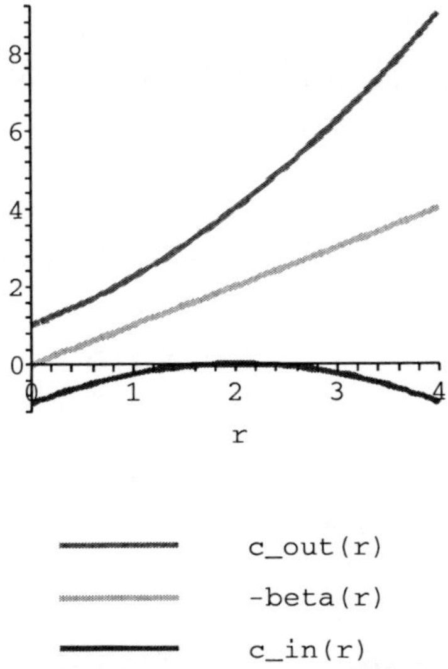

Fig. 2. Coordinate speeds for the de Sitter lightcone: inner edge (c_{in}), normal to the spacelike time slices ($-\beta$), and outer edge (c_{out}). For the Minkowski metric the graph looks very similar, except that c_{in} is there never positive.

here, and moving smoothly out through I^+, were not controlled for very long times. Our conclusion [9] was that further development of this approach would better be done using a spherical outer boundary rather than trying to understand all possibilities for modifying the irrelevant laws of physics in the outer space ('das All') beyond I^+ to make the numerical evolution proceed more smoothly.

5. Maxwell Example

As a test case where some of these ideas could be explored with limited resources, we chose to apply them to the numerical solution of Maxwell's equations in the conformally regulated de Sitter metric. The Maxwell equations have the advantage of being conformally invariant so that no adjustments need be made in going from the original de Sitter metric $d\tilde{s}^2$ of equation (12) to the conformally regulated de Sitter metric ds^2 of equations (13) and (14).

5.1. *Maxwell Equations*

We use various components of the 4-dimensional Maxwell fields $F_{\mu\nu}$ and $\mathfrak{F}^{\mu\nu} = \sqrt{-g}g^{\mu\alpha}g^{\nu\beta}F_{\alpha\beta}$. In particular we define, $\mathfrak{F}^{0i} = \mathfrak{D}^i$, $\mathfrak{F}^{ij} = \mathfrak{H}^{ij} = [ijk]H_k$ and $F_{i0} = E_i$, $F_{ij} = B_{ij} = [ijk]\mathfrak{B}^k$. Here $[ijk]$ is the completely antisymmetric symbol $[ijk] = 0, \pm 1$ with $[123] = +1$. The fundamental fields in the formulation are $\mathfrak{B}^i$ and $\mathfrak{D}^i$. The familiar four dimensional equations

$$\partial_\nu \mathfrak{F}^{\mu\nu} = 0 = \partial_{[\alpha}F_{\beta\gamma]} \tag{15}$$

are then rewritten in $3+1$ form:

Constraint equations

$$\partial_i \mathfrak{B}^i = 0 = \partial_i \mathfrak{D}^i \quad ; \tag{16}$$

Evolution equations

$$\partial_t \mathfrak{B}^i = -[ijk]\partial_j E_k, \quad \partial_t \mathfrak{D}^i = [ijk]\partial_j H_k \quad ; \tag{17}$$

Constituitive relations

$$\begin{aligned} E_i &= (\alpha/\sqrt{\gamma})\gamma_{ij}\mathfrak{D}^j + [ijk]\beta^j \mathfrak{B}^k \\ H_i &= (\alpha/\sqrt{\gamma})\gamma_{ij}\mathfrak{B}^j - [ijk]\beta^j \mathfrak{D}^k \quad . \end{aligned} \tag{18}$$

Note that the metric appears explicitly only in the constituitive relations (18), and there only conformally invariant combinations of metric components appear: $(\alpha/\sqrt{\gamma})\gamma_{ij}$ and β^i.

5.2. *Initial Conditions*

We use the same Minkowski solution as Baumgarte [10] and subsequently Fiske [11].

$$A = A_\mu dX^\mu = f(R,T)\frac{1}{R^2}(X\,dY - Y\,dX) \tag{19}$$

$$f = (\frac{1}{R} - 2\lambda U)\exp(-\lambda U^2) - (\frac{1}{R} + 2\lambda V)\exp(-\lambda V^2) \tag{20}$$

where $U = T - R$ and $V = T + R$. This is then converted to compactifying coordinates $uxyz$. Then we evaluate $\mathfrak{B}^i$ and $\mathfrak{D}^i$ in this conformally regulated Minkowski analytic solution for use as initial data with the regulated de Sitter metric. Using a solutions of Maxwell's equations in the Minkowski metric to supply initial conditions for a solution in the de Sitter metric is possible since the constraint equations (16) do not involve the metric.

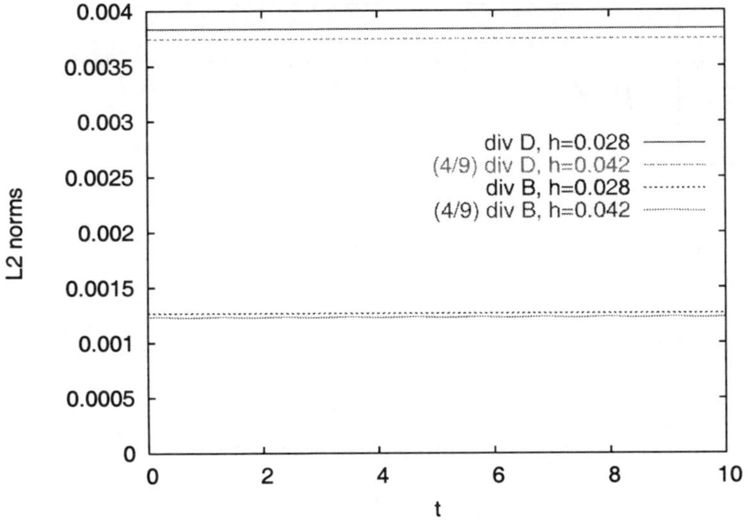

Fig. 3. The L2 norms (over the region $r \leq 2$) of the constraints plotted vs. time. The lower resolution data is multiplied by the factor appropriate to demonstrate second order convergence. The initial values arise from $\mathcal{B}$ and $\mathcal{D}$ fields which are exact analytic solutions of the constraints, evaluated at the grid points and then differenced to form the constraints.

5.3. *Numerical Results*

To show that the numerical computations were approximately solving a wave equation, we present some convergence data. Figure 3 explicitly shows a two-point convergence test on the constraints over the physical region $r \leq 2$, which are seen to be second-order convergent to zero and nearly constant.

Figure 4 shows that the wave pulse exits through I^+ smoothly, and even in close agreement between the numerically evolved field in de Sitter spacetime and the exact analytic solution in Minkowski spacetime. This remarkable agreement is possible because the outgoing speed of light c_{out} is unchanged by the de Sitter modifications while energy conservation conspires to keep the magnitudes consistent.

In Figure 5 one sees a more complete view of the field $\mathfrak{B}^z$ as function of time $t = u$ along a radial line in the equatorial plane. [This field is rotationally symmetric about the z-axis.] To enlarge the scale near the outer radius compared to that at near the origin, the field has been multiplied by r^2 before plotting.

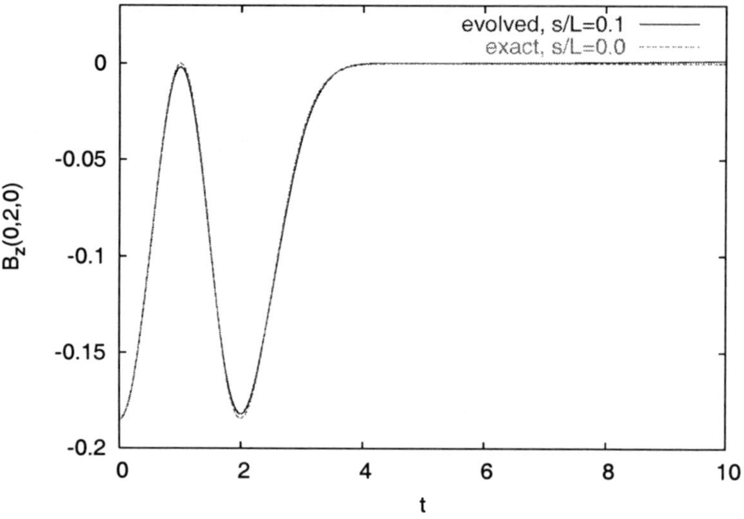

Fig. 4. A field component $\mathfrak{B}^z$ is plotted against (hyperboloidal) time u at a point on I^+. The solid line (red online) gives the numerically evolved solution in de Sitter spacetime, while the dashed line (green online) gives the analytic solution in Minkowski spacetime.

5.4. *Numerical Methods*

A key point in our approach to this problem is that we do not require special numerical techniques to handle the compactified spacetime at or within I^+. Conventional methods of Cauchy evolution for spacelike time slices were used.

- 3D uniform grid
- Second order finite differencing.
- Iterated Crank-Nicholson time steps
- One sided second order differencing used at the computational boundary on cube faces. Equivalent to the methods employed for excision of black holes inside BH apparent horizons.

Acknowledgments

This work was supported in part by NASA Space Sciences Grant ATP02-0043-0056. JvM and DRF were also supported in part by the Research Associateship Programs Office of the National Research Council. CWM gratefully acknowledges the hospitality of the AEI, where much of this work was developed, and the hospitality of National Central University,

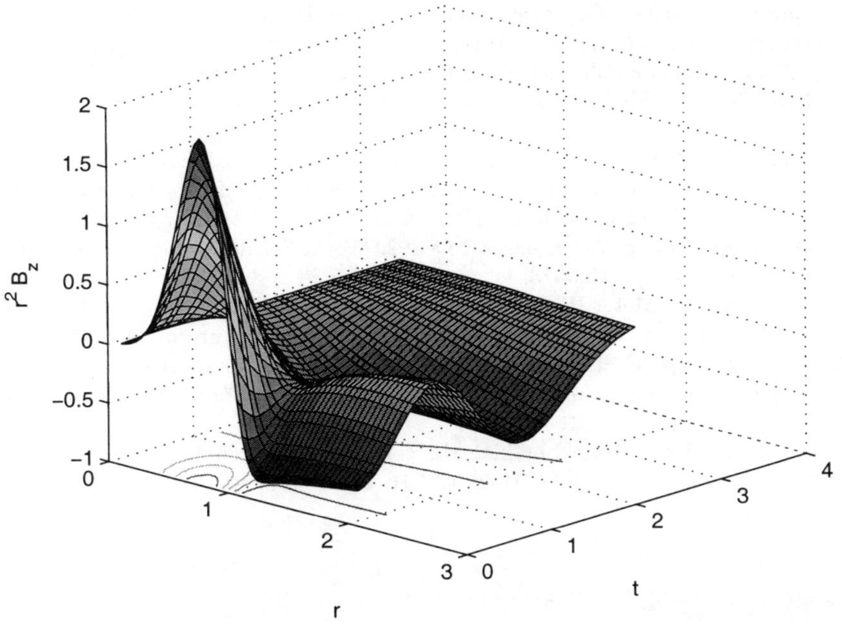

Fig. 5. The numerically evolved field component $\mathfrak{B}^z$ is displayed along a radial line from the origin to I^+ from the initial time $t \equiv u = 0$ until $u = 4$. The values along the line $r = 2$ restate data in Figure 4.

Chungli, Taiwan, in supporting the presentation of this paper at the ICGA7 conference.

References

1. Lee Lindblom. Gravitational wave astronomy. In Chiang-Mei Chen, James M. Nester, and Jong-Ping Hsu, editors, *Proceedings of the VII Asia-Pacific International Conference on Gravitation and Astrophysics*, Singapore, 2006. World Scientific.
2. Numerical relativity 2005: Compact binaries. [Online links] 2005. URL `http://astrogravs.gsfc.nasa.gov/conf/numrel2005/presentations`.
3. Charles W. Misner, Mark Scheel, and Lee Lindblom. Wave propagation with hyperboloidal slicings. [Online report] June 2003. URL `http://online.kitp.ucsb.edu/online/gravity03/misner`.
4. Robert M. Wald. *General Relativity.* University of Chicago Press, 1984.
5. Sascha Husa, Carsten Schneemann, Tilman Vogel, and Anil Zenginoglu. Hyperboloidal data and evolution. 2005. arXiv:gr-qc/0512033.

6. Jörg Frauendiener. Conformal infinity. *Living Rev. Relativity*, 7, 2004. [Online article]. URL `http://www.livingreviews.org/lrr-2004-1`.
7. Jeffrey Winicour. Characteristic evolution and matching. *Living Rev. Relativity*, 8, 2005. URL `http://www.livingreviews.org/lrr-2005-10`. [Online article].
8. Charles W. Misner. Hyperboloidal slices and artificial cosmology for numerical relativity. In James T Liu, Michael J Duff, Kellogg S Stelle, and Richard P Woodard, editors, *DESERFEST: A Celebration of the Life and Works of Stanley Deser*, pages 180–192. World Scientific, Singapore, March 2006. arXiv:gr-qc/0409073.
9. James R. van Meter, David R. Fiske, and Charles W. Misner. Excising das All: Evolving Maxwell waves beyond Scri. 2006. arXiv:gr-qc/0603034.
10. A. M. Knapp, E. J. Walker, and T. W. Baumgarte. Illustrating stability properties of numerical relativity in electrodynamics. *Phys. Rev.*, D65:064031, 2002. arXiv:gr-qc/0201051.
11. David R. Fiske. Toward making the constraint hypersurface an attractor in free evolution. *Phys. Rev.*, D 69:047501, 2004. arXiv:gr-qc/0304024.

Improved numerical stability of rotating black hole evolution calculations

Hwei-Jang Yo

Department of Physics, National Cheng-Kung University, Tainan 701, Taiwan

We experiment with modifications of the BSSN form of the Einstein field equations (a reformulation of the ADM equations) and demonstrate how these modifications affect the stability of numerical black hole evolution calculations. We use excision to evolve rapid-rotating Kerr-Schild black holes, and obtain accurate and stable simulations for specific angular momenta J/M of up to about $0.9M$.

1. Introduction

Binary black holes are among the most promising sources for the gravitational wave laser interferometers currently under development, including LIGO, VIRGO, GEO, TAMA and LISA. The identification and interpretation of possible signals requires theoretically predicted gravitational wave templates. For the late epoch of the binary inspiral, numerical relativity is the most promising tool for the computation of such templates.

There was a huge leap for Binary black hole evolutions in the last year: it includes the successes with the puncture methods[1,2,3] and the excision method[4]. These successes give a firm confidence in the community about modelling of the gravitational source and extracting the information of gravitation radiation from simulations.

Since the gravitational field is strong in the vicinity of black holes, it is still a difficulty to maintain a long term stability for its evolution simulation, especially with a specific angular momentum $\to 1M$. In this paper we continue our previous work[5] to experiment with adding the new constraints that appear in the BSSN formulation to the evolution equation of the new auxiliary functions. These modifications can fairly enhance the stability of the evolution of black hole.

2. Adjusting the BSSN equations

The BSSN formulation has been described in detail in previous papers[5]. We will discuss here only about the newly improvements. For a solution of the BSSN equations to be equivalent with a solution of the ADM equations, the new auxiliary variables have to satisfy new constraint equations. In particular, $\tilde{A}_{ij}$ has to be traceless

$$\mathcal{A} \equiv \tilde{\gamma}^{ij}\tilde{A}_{ij} = 0, \tag{1}$$

the determinant of the conformally related metric $\tilde{\gamma}_{ij}$ has to be unity

$$\mathcal{D} \equiv \det(\tilde{\gamma}_{ij}) - 1 = 0, \tag{2}$$

and the conformal connection functions $\tilde{\Gamma}^i$ have to satisfy the identity

$$\mathcal{G}^i \equiv \tilde{\Gamma}^i - \tilde{\Gamma}^i_g = \tilde{\Gamma}^i - \tilde{\gamma}^{jk}\tilde{\Gamma}^i{}_{jk} = 0. \tag{3}$$

These conditions can be viewed as new constraints, in addition to the Hamiltonian and momentum constraints.

In an unconstrained evolution calculation, the constraints are monitored only as a code check. It has been proved to be advantageous, however, either to enforce at least some of the constraints during the evolution, or to add evolution constraint equations to the evolution equations.

Of crucial importance for the stability of our code are our constraint additions to the $\tilde{\Gamma}^i$ evolution equation. The evolution equation for $\partial_t\tilde{\Gamma}^i$ has the terms

$$\partial_t\tilde{\Gamma}^i = \frac{2}{3}\tilde{\Gamma}^i\beta^j{}_{,j} - \tilde{\Gamma}^j\beta^i{}_{,j} + \cdots \tag{4}$$

Looking, for example, at the x-component of this equation,

$$\partial_t\tilde{\Gamma}^x = \left(\frac{2}{3}\beta^j{}_{,j} - \beta^x{}_{,x}\right)\tilde{\Gamma}^x + \cdots \tag{5}$$

we see that if $\frac{2}{3}\beta^j{}_{,j} - \beta^x{}_{,x} > 0$, then $\partial_t\tilde{\Gamma}^x$ contains a term tending to produce exponential growth. We lessen the possibility of an instability caused by these terms by using (3) to replace (5) with

$$\begin{aligned}
\partial_t\tilde{\Gamma}^x &= \left(\frac{2}{3}\beta^j{}_{,j} - \beta^x{}_{,x}\right)\tilde{\Gamma}^x_g - \left(\frac{2}{3}\lambda_A\left|\beta^j{}_{,j}\right| + \lambda_B\left|\beta^x{}_{,x}\right|\right)\mathcal{G}^x + \cdots \\
&= \frac{2}{3}\left(\beta^j{}_{,j} + \lambda_A\left|\beta^j{}_{,j}\right|\right)\tilde{\Gamma}^x_g - \frac{2}{3}\lambda_A\left|\beta^j{}_{,j}\right|\tilde{\Gamma}^x \\
&\quad - \left(\beta^x{}_{,x} - \lambda_B\left|\beta^x{}_{,x}\right|\right)\tilde{\Gamma}^x_g - \lambda_B\left|\beta^x{}_{,x}\right|\tilde{\Gamma}^x + \cdots,
\end{aligned} \tag{6}$$

where $\lambda_A > 0$, $\lambda_B > 0$, and we makes the similar modifications for the evolution equations of $\tilde{\Gamma}^y$ and $\tilde{\Gamma}^z$. Note that the "exponential" terms in the above equation (i.e. the terms proportional to $\tilde{\Gamma}^x$) are now guaranteed to be exponential *decay* terms.

Alcubierre *et al.* [6] found improved behavior when they enforce the constraint $\mathcal{T} = 0$. Yo *et al.* [5] found it useful to enforce $\mathcal{T} = \mathcal{D} = 0$. Duez *et al.* [7] instead applied the reasoning above to modify the evolution equations for the diagonal terms of $\tilde{\gamma}_{ij}$ and $\tilde{A}_{ij}$. For example, in the equation for $\tilde{\gamma}_{xx}$, they found that the terms

$$\partial_t \tilde{\gamma}_{xx} = \left(2\beta^x{}_{,x} - \frac{2}{3}\beta^j{}_{,j}\right)\tilde{\gamma}_{xx} + \cdots \tag{7}$$

can be replaced by

$$\begin{aligned}\partial_t \tilde{\gamma}_{xx} = &\frac{2}{3}\left[\lambda_C \left|\beta^j{}_{,j}\right| - \beta^j{}_{,j}\right] G_{xx} - \frac{2}{3}\lambda_C \left|\beta^j{}_{,j}\right| \tilde{\gamma}_{xx} \\ &+ 2\left[\beta^x{}_{,x} + \lambda_D \left|\beta^x{}_{,x}\right|\right] G_{xx} - 2\lambda_D \left|\beta^x{}_{,x}\right| \tilde{\gamma}_{xx} + \cdots ,\end{aligned} \tag{8}$$

where $\lambda_C > 0$, $\lambda_D > 0$, and G_{xx} is the value of $\tilde{\gamma}_{xx}$ as computed from the five other independent components of $\tilde{\gamma}_{ij}$, assuming $\mathcal{D} = 0$. They performed the same substitution for $\tilde{\gamma}_{yy}$ and $\tilde{\gamma}_{zz}$. In a similar fashion, they modified the evolution of $\tilde{A}_{xx}$ from

$$\partial_t \tilde{A}_{xx} = \cdots + \left(2\beta^x{}_{,x} - \frac{2}{3}\beta^j{}_{,j} + \alpha K\right)\tilde{A}_{xx} \tag{9}$$

to

$$\begin{aligned}\partial_t \tilde{A}_{xx} = &\frac{2}{3}\left[\lambda_E \left|\beta^j{}_{,j}\right| - \beta^j{}_{,j}\right] H_{xx} - \frac{2}{3}\lambda_E \left|\beta^j{}_{,j}\right| \tilde{A}_{xx} \\ &+ 2\left[\beta^x{}_{,x} + \lambda_F \left|\beta^x{}_{,x}\right|\right] H_{xx} - 2\lambda_F \left|\beta^x{}_{,x}\right| \tilde{A}_{xx} \\ &+ \left[\alpha K + \lambda_G \left|\alpha K\right|\right] H_{xx} - 2\lambda_G \left|\alpha K\right| \tilde{A}_{xx} + \cdots ,\end{aligned} \tag{10}$$

where $\lambda_E > 0$, $\lambda_F > 0$, and $\lambda_G > 0$, and H_{xx} is the value of $\tilde{A}_{xx}$ computed from the five other independent components of $\tilde{A}_{ij}$ assuming $\mathcal{T} = 0$. The similar modifications were done for $\tilde{A}_{yy}$ and $\tilde{A}_{zz}$.

In this work we extend the modifications for the 3-metric variables to the off-diagonal terms. For example, for $\tilde{\gamma}_{xy}$, we modify its evolution equation from

$$\partial_t \tilde{\gamma}_{xy} = \left(\beta^x{}_{,x} + \beta^y{}_{,y} - \frac{2}{3}\beta^j{}_{,j}\right)\tilde{\gamma}_{xy} + \cdots \tag{11}$$

to

$$\begin{aligned}\partial_t \tilde{\gamma}_{xy} = &\frac{2}{3}\left[\lambda_C \left|\beta^j{}_{,j}\right| - \beta^j{}_{,j}\right] G_{xy} - \frac{2}{3}\lambda_C \left|\beta^j{}_{,j}\right| \tilde{\gamma}_{xy} \\ &+ \left[\beta^x{}_{,x} + \lambda_D \left|\beta^x{}_{,x}\right|\right] G_{xy} - \lambda_D \left|\beta^x{}_{,x}\right| \tilde{\gamma}_{xy} \\ &+ \left[\beta^y{}_{,y} + \lambda_D \left|\beta^y{}_{,y}\right|\right] G_{xy} - \lambda_D \left|\beta^y{}_{,y}\right| \tilde{\gamma}_{xy} + \cdots ,\end{aligned} \tag{12}$$

where G_{xy} is the value of $\tilde{\gamma}_{xy}$ as computed from the five other independent components of $\tilde{\gamma}_{ij}$, assuming $\mathcal{D} = 0$. The similar modifications were done for $\tilde{\gamma}_{xz}$, and $\tilde{\gamma}_{yz}$. We find that the modifications is helpful in stabilizing the evolution of a fast-rotating black hole.

It is not suitable to apply similar substitutions to the off-diagonal terms of the traceless extrinsic curvature from the traceless constraint, $\mathcal{T} = 0$, due to the possibly vanishing coefficients attached to these off-diagonal terms in the constraint.

We also experimented with the following adjustment inspired from the work of Yoneda and Shinkai [8], namely,

$$\partial_t \tilde{\gamma}_{ij}^{\text{new}} = \partial_t \tilde{\gamma}_{ij} - \lambda_H \alpha \left(\tilde{\gamma}_{k(i} \partial_{j)} \mathcal{G}^k - \frac{1}{3} \tilde{\gamma}_{ij} \partial_k \mathcal{G}^k \right). \tag{13}$$

One subtlety needs to be noted to make this modification work, ie, the substitution of $\tilde{\gamma}_{ij}$ with G_{ij} should be activated when one term is likely to be the exponentially-growing term. This modification has an inherent noise about the order of magnitude 10^{-12} (compared with the usual one 10^{-16}) although it is helpful in stabilization.

3. Numerical Result

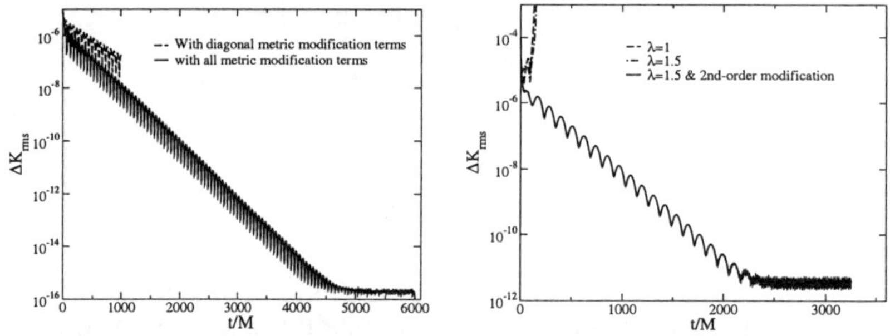

Fig. 1. In the left panel, the the r.m.s. of the change in the trace of extrinsic curvature between consecutive time steps as functions of time in the rotating cases with $a = 0.7M$. The right panel shows the changes in the rotating cases with $a = 09M$.

Our numerical implementation follows very closely the recipe of our previous work[5]. We use centered differencing everywhere except for the advection terms on the shift. For these terms, a second-order upwind scheme is used along the shift direction[6]. We adopt "1+log" slicing to specify the lapse α. The shift β^i is determined either from the "Gamma-driver" condition. On the outer boundaries of the numerical grid we impose a radiative boundary condition. For the excision regions, we adopt the copying method[6].

In the left panel of Fig. 1 we compare the r.m.s. of changes in K for different cases with $a = 0.7M$. We show that the modifications for the off-diagonal metric components do enhance fairly the stability of the evolution.

Instabilities become even harder to control for $a = 0.9M$. The larger angular momentum leads to larger numerical error, which by itself makes the simulations more demanding. In the right panel of Fig. 1 we compare the r.m.s. of the change in K for different parameter settings and modifications. We find that this instability can be controlled by adding the modification (13) although the modification offers an error with the magnitude of $\sim 10^{-12}$.

We modify the BSSN evolution equation for the new auxiliary conformal connection functions by adding their constraint equation as well as the conformal metric and the traceless extrinsic curvature, and also experiment with adjustments of the other evolution equations suggested by [8]. And the modifications shows their effects on the stabilization in the black hole evolution simulations.

Acknowledgments

It is a pleasure to thank Stuart Shapiro and Thomas Baumgarte for helpful discussions. This work was supported in part by the National Science Council of the R.O.C. (Taiwan) under grant No. NSC94-2112-M-006-014. Some of the calculations were performed at the National Center for High-performance Computing in Taiwan.

References

1. B. Brügmann, W. Tichy, N. Jansen, *Phys. Rev. Lett.* **92**, 211101 (2004).
2. J.G. Baker, J. Centrella, D. Choi, M. Koppitz, J. van Meter, gr-qc/0511103 (2006).
3. M. Campanelli, C.O. Lousto, P. Marronetti, Y. Zlochower, *Phys. Rev. Lett.* **96**, 111101 (2006).
4. F. Pretorius, *Phys. Rev. Lett.* **95**, 121101 (2005).
5. H.J. Yo, T.W. Baumgarte, S.L. Shapiro, *Phys.Rev.* **D66**, 084026 (2002).
6. M. Alcubierre and B. Barügmann, *Phys.Rev.* **D63**, 104006 (2001).
7. M.D. Duez, S.L. Shapiro, H.J. Yo, *Phys.Rev.* **D69**, 104016 (2004).
8. G. Yoneda, H. Shinkai *Phys.Rev.* **D66**, 124003 (2002).

Cosmology

Integrable Cosmological Models in DD and Variations of Fundamental Constants

V. N. Melnikov

Center for Gravitation and Fundamental Metrology, VNIIMS, and Institute of Gravitation and Cosmology, Peoples' Friendship University of Russia, 46 Ozernaya Str., Moscow, 119361, Russia

E-mail: melnikov@phys.msu.su, melnikov@vniims.ru

Discovery of present acceleration of the Universe, dark matter and dark energy problems are great challenges to modern physics, which may bring to a new revolution. Integrable multidimensional models of gravitation and cosmology make up one of the proper approaches to study basic issues and, in particular, strong field objects, the Early and present Universe and black hole physics [1,2]. Problems of the absolute G measurements and its possible time and range variations, which are reflections of the unification problem are discussed. A need for further measurements of G and its possible variations (also in space) is pointed out.

1. Introduction

The necessity of studying multidimensional models of gravitation and cosmology [1,2] is motivated by several reasons. First, the main trend of modern physics is the unification of all known fundamental physical interactions: electromagnetic, weak, strong and gravitational ones. During the recent decades there has been a significant progress in unifying weak and electromagnetic interactions, some more modest achievements in GUT, supersymmetric, string and superstring theories.

Now, theories with membranes, p-branes and more vague M-theory are being created and studied. Having no definite successful theory of unification now, it is desirable to study the common features of these theories and their applications to solving basic problems of modern gravity and cosmology. Moreover, if we really believe in unified theories, the early stages of the Universe evolution and black hole physics, as unique superhigh energy regions and possibly even low energy stage, where we observe the present acceleration, are the most proper and natural arena for them.

Second, multidimensional gravitational models, as well as scalar-tensor theories of gravity, are theoretical frameworks for describing possible temporal and range variations of fundamental physical constants [3,4,5,6]. These ideas have originated from the earlier papers of E. Milne (1935) and P. Dirac (1937) on relations between the phenomena of micro- and macro-worlds, and up till now they are under thorough study both theoretically [7,8] and experimentally [9,10]. The possible discovery of the fine structure constant variations is now at a critical further investigation.

Lastly, applying multidimensional gravitational models to basic problems of modern cosmology and black hole physics, we hope to find answers to such long-standing problems as singular or nonsingular initial states, creation of the Universe, creation of matter and its entropy, cosmological constant, coincidence problem, origin of inflation and specific scalar fields which may be necessary for its realization, isotropization and graceful exit problems, stability and nature of fundamental constants [4,11,12], possible number of extra dimensions, their stable compactification, new revolutionary data on present acceleration of the Universe, dark matter and dark energy etc.

Bearing in mind that multidimensional gravitational models are certain generalizations of general relativity which is tested reliably for weak fields up to 0.0001 and partially in strong fields (binary pulsars), it is quite natural to inquire about their possible observational or experimental windows. From what we already know, among these windows are [11]:

– possible deviations from the Newton and Coulomb laws, or new interactions,

– possible variations of the effective gravitational constant with a time rate smaller than the Hubble one,

– possible existence of monopole modes in gravitational waves,

– different behavior of strong field objects, such as multidimensional black holes, wormholes and p-branes,

– standard cosmological tests,

– possible non-conservation of energy in strong field objects and accelerators, if braneworld ideas about gravity in the bulk turn out to be true etc.

Since modern cosmology has already become a unique laboratory for testing standard unified models of physical interactions at energies that are far beyond the level of existing and future man-made accelerators and other installations on Earth, there exists a possibility of using cosmological and astrophysical data for discriminating between future unified schemes.

As no accepted unified model exists, in our approach [1,2,13,14] we adopted simple (but general from the point of view of number of dimensions) models, based on multidimensional Einstein equations with or without sources of different nature:

– cosmological constant,
– perfect and viscous fluids,
– scalar and electromagnetic fields,
– their possible interactions,
– dilaton and moduli fields,
– fields of antisymmetric forms (related to p-branes) etc.

Our program's main objective was and is to obtain exact self-consistent solutions (integrable models) for these models and then to analyze them in cosmological, spherically and axially symmetric cases. In our view this is a natural and most reliable way to study highly nonlinear systems.

As our model [1,2] we use n Einstein spaces of constant curvature with sources as (m+1)-component perfect fluid, (or fields or form-fields,), cosmological or spherically symmetric metric, manifold as a direct product of factor-spaces of arbitrary dimensions. Then, in harmonic time gauge we show that Einstein multidimensional equations are equivalent to Lagrange equations with non-diagonal in general minisuperspace metric and some exponential potential. After diagonalization of this metric we perform reduction to sigma-model [16] and Toda-like systems [17], further to Liouville, Abel, generalized Emden-Fowler Eqs. etc. and try to find exact solutions. We suppose that behavior of extra spaces is the following: they are constant, or dynamically compactified, or like torus, or large, but with barriers, walls etc.

So, we realized the program in arbitrary dimensions (from 1988), see [1,2,13,14]:

In cosmology:

obtained exact general solutions of multidimensional Einstein equations with sources:

– Λ, Λ + scalar field [18] (e.g. nonsingular, dynamically compactified, inflationary);

– perfect fluid, PF + scalar field (e.g. nonsingular, inflationary solutions);

– viscous fluid (e.g. nonsingular, generation of mass and entropy, quintessence and coincidence in the 2-component model);

– stochastic behavior near the singularity, billiards in Lobachevsky space, $D = 11$ is critical, φ destroys billiards [19].

For all above cases Ricci-flat solutions above are obtained for any n, also with curvature in one factor-space; with curvatures in 2 factor-spaces only for total $N = 10, 11$;

– fields: scalar, dilatons, forms of arbitrary rank (98) - inflationary, with acceleration, nonsingular, Λ generation by forms (p-branes) [20];

– first billiards for dilaton-forms (p-branes) interaction [21];

– quantum variants (solutions of WDW-equation [18,22]) for nearly all above cases where classical solutions were obtained;

– dilatonic fields with potentials, billiard behavior for them also [24];

– using same methods also 4D models of dust or perfect fluid with scalar fields and potentials (acceleration, coincidence etc.)

Solutions depending on r in any dimensions:

– generalized Schwarzchild, generalized Tangerlini (BHs are singled out), also with minimal scalar field (no BHs);

– generalized Reissner-Nordstrom (BHs also are singled out), the same plus φ (no BHs);

– multi-temporal;

– for dilaton-like interaction of φ and electromagnetic fields (BHs exist only for a special case) [25];

– stability studies (stable solutions only for BH case above) [25];

– the same for dilaton-forms (p-branes) interaction, stability found only in some cases, e.g. for one form in particular [26,27].

2. Multidimensional Models

The earlier papers on multidimensional gravity and cosmology dealt with multidimensional Einstein equations, perfect fluid as a source [28] and with a block-diagonal cosmological or spherically symmetric metric defined on the manifold $M = R \times M_0 \times \cdots \times M_n$ of the form

$$g = -dt \otimes dt + \sum_{r=0}^{n} a_r^2(t) g^r \tag{1}$$

where (M_r, g^r) are Einstein spaces, $r = 0, \ldots, n$. In some of them a cosmological constant and simple scalar fields were also used [18].

Such models are usually reduced to pseudo-Euclidean Toda-like systems with the Lagrangian [17]:

$$L = \frac{1}{2} G_{ij} \dot{x}^i \dot{x}^j - \sum_{k=1}^{m} A_k e^{u_i^k x^i} \tag{2}$$

and the zero-energy constraint $E = 0$.

It should be noted that pseudo-Euclidean Toda-like systems are not well-studied yet. There exists a special class of equations of state that gives rise to Euclidean Toda models [17].

Later exact solutions with fields of forms ("branes") as sources became more actual [29,30,31]. In our other papers several classes of the exact solutions for the multidimensional gravitational model governed by the Lagrangian

$$\mathcal{L} = R[g] - 2\Lambda - h_{\alpha\beta} g^{MN} \partial_M \varphi^\alpha \partial_N \varphi^\beta - \sum_a \frac{1}{n_a!} \exp(2\lambda_{a\alpha}\varphi^\alpha)(F^a)^2, \quad (3)$$

were considered [32,33]. Here g is metric, $F^a = dA^a$ are forms of ranks n_a and φ^α are scalar fields and Λ is a cosmological constant (the matrix $h_{\alpha\beta}$ is invertible).

For certain field contents with distinguished values of total dimension D, ranks n_a, dilatonic couplings λ_a and $\Lambda = 0$ such Lagrangians appear as "truncated" bosonic sectors (i.e. without Chern-Simons terms) of certain supergravitational theories or low-energy limit of superstring models [34,35].

In our review [14] certain classes of p-brane solutions to field equations corresponding to the Lagrangian (4), obtained by us earlier, were presented.

These solutions have a block-diagonal metrics defined on D-dimensional product manifold, i.e.

$$g = e^{2\gamma} g^0 + \sum_{i=1}^{n} e^{2\phi^i} g^i, \qquad M_0 \times M_1 \times \ldots \times M_n, \quad (4)$$

where g^0 is a metric on M_0 (our space) and g^i are fixed Ricci-flat (or Einstein) metrics on M_i (internal space, $i > 0$). The moduli γ, ϕ^i and scalar fields φ^α are functions on M_0 and fields of forms are also governed by several scalar functions on M_0. Any F^a is supposed to be a sum of monoms, corresponding to electric or magnetic p-branes (p-dimensional analogues of membranes), i.e. the so-called composite p-brane ansatz is considered [29,30]. In non-composite case we have no more than one monom for each F^a.)

$p = 0$ corresponds to a particle, $p = 1$ to a string, $p = 2$ to a membrane etc. The p-brane world-volume (world-line for $p = 0$, world-surface for $p = 1$ etc) is isomorphic to some product of internal manifolds: $M_I = M_{i_1} \times \ldots \times M_{i_k}$ where $1 \le i_1 < \ldots < i_k \le n$ and has dimension $p + 1 = d_{i_1} + \ldots + d_{i_k} = d(I)$, where $I = \{i_1, \ldots, i_k\}$ is a multi-index describing the location of p-brane and $d_i = \dim M_i$. Any p-brane is described by the triplet (p-brane index) $s = (a, v, I)$, where a is the color index labelling the form F^a, $v = e(lectric), m(agnetic)$ and I is the multi-index defined above. For the electric and magnetic branes corresponding to form F^a the world-volume dimensions are $d(I) = n_a - 1$ and $d(I) = D - n_a - 1$, respectively.

The sum of this dimensions is $D-2$. For $D = 11$ supergravity we get $d(I) = 3$ and $d(I) = 6$, corresponding to electric $M2$-brane and magnetic $M5$-brane.

In [16] the model under consideration was reduced to gravitating self-interacting sigma-model with certain constraints imposed. The sigma-model representation for non-composite electric case was obtained earlier in [29,30], for electric composite case see also [36].

The σ-model Lagrangian, obtained from (4), has the form [16]

$$\mathcal{L}_\sigma = R[g^0] - \hat{G}_{AB} g^{0\mu\nu} \partial_\mu \sigma^A \partial_\nu \sigma^B - \sum_s \varepsilon_s \exp(-2U^s) g^{0\mu\nu} \partial_\mu \Phi^s \partial_\nu \Phi^s - 2V, \qquad (5)$$

where $(\sigma^A) = (\phi^i, \varphi^\alpha)$, V is a potential, $(\hat{G}_{AB})$ are components of (truncated) target space metric, $\varepsilon_s = \pm 1$,

$$U^s = U^s_A \sigma^A = \sum_{i \in I_s} d_i \phi^i - \chi_s \lambda_{a_s \alpha} \varphi^\alpha$$

are linear functions, Φ^s are scalar functions on M_0 (corresponding to forms), and $s = (a_s, v_s, I_s)$. Here parameter $\chi_s = +1$ for the electric brane ($v_s = e$) and $\chi_s = -1$ for the magnetic one ($v_s = m$).

A pure gravitational sector of the sigma-model was considered earlier in our paper [37].

A family of general cosmological type p-brane solutions with n Ricci-flat internal spaces was considered in [38], where also a generalization to the case of $(n-1)$ Ricci-flat spaces and one Einstein space of non-zero curvature (say M_1) was obtained. These solutions are defined up to solutions to Toda-type equations and may be obtained using the Lagrange dynamics following from our sigma-model approach [22]. The solutions from [38] contain a subclass of spherically symmetric solutions (for $M_1 = S^{d_1}$). Special solutions with orthogonal and block-orthogonal sets of U-vectors were considered earlier in [22] and [13,38], respectively. For non-composite case see [32,33] and references therein.

In [22] the reduction of p-brane cosmological type solutions to Toda-like systems was first performed. General classes of p-brane solutions (cosmological and spherically symmetric ones) related to Euclidean Toda lattices associated with Lie algebras (mainly $\mathbf{A_m}$, $\mathbf{C_m}$ ones) were obtained in [39,40,41,42,43].

A class of space-like brane (S-brane) solutions (related to Toda-type systems) with product of Ricci-flat internal spaces and S-brane solutions with special orthogonal intersection rules were considered in [44,45] and solutions with accelerated expansion (e.g. with power-law and exponential behavior of scale factors) were singled out.

Scalar fields play an essential role in modern cosmology. They are attributed to inflationary models of the early universe and the models describing the present stage of the accelerated expansion as well. A lot of them appear due to reduction from many to our 4 dimensions. There is no unique candidate for the potential of the minimally coupled scalar field. Typically a potential is a sum of exponents. Such potentials appear quite generically in a large class of theories: multidimensional, Kaluza-Klein models, supergravity and string/M - theories.

Single exponential potential was extensively studied within Friedmann-Robertson-Walker (FRW) model containing both a minimally coupled scalar field and a perfect fluid with the linear barotropic equation of state. The attention was mainly focussed on the qualitative behavior of solutions, stability of the exceptional solutions to curvature and shear perturbations and their possible applications within the known cosmological scenario such as inflation and scaling ("tracking") . The energy-density of the scalar field scales with that of the perfect fluid. Using our methods for multidimensional cosmology the problem of integrability by quadratures of the model in 4-dimensions was studied. Four classes of general solutions, when the parameter characterizing the steepness of the potential and the barotropic parameter obey some relations, were found [46]. For the case of multiple exponential potential of the scalar field and dust integrable model in 4D was obtained in [47] (see also [48]).

As to scalar fields with multiple exponential potential in any dimensions, a wide class of exact solutions was obtained in our papers [49,50]. In [24] a behavior of this system near the singularity was studied using a billiard approach suggested earlier in our papers [19,51], some systems with maximal number of branes, where oscillating behavior is absent, see [21]. A number of S-brane solutions were found in [44,45,52], where solutions with acceleration were also singled out.

Details for 2-component D-dimensional integrable models for general and some spacial EOS see in [53,54,55]. Quite different model with dilaton, branes and cosmological constant and static internal spaces was investigated in [20], where possible generation of the effective cosmological constant by branes was demonstrated. Model with variable equations of state see in [56], where general solutions and in particular with acceleration of our space and compactification of internal spaces were found. Non-singular (bouncing) solutions were obtained in [50].

3. Fundamental physical constants

1. In any physical theory we meet constants which characterize the stability properties of different types of matter: of objects, processes, classes of processes and so on. These constants are important because they arise independently in different situations and have the same value, at any rate within accuracies we have gained nowadays. That is why they are called fundamental physical constants (FPC) [3,11]. It is impossible to define strictly this notion. It is because the constants, mainly dimensional, are present in definite physical theories. In the process of scientific progress some theories are replaced by more general ones with their own constants, some relations between old and new constants arise. So, we may talk not about an absolute choice of FPC, but only about a choice corresponding to the present state of physical sciences. In some unified models, i.e. in multidimensional ones, they may be related in some manner. From the point of view of these unified models the above mentioned ones are the low energy ones.

All these constants are known with different *accuracies*. See this data and also their classification, number and possible role of Planck parameters (units) in [11,12].

2. The problem of the gravitational constant G measurement and its stability is a part of a rapidly developing field, called gravitational-relativistic metrology (GRM). It has appeared due to the growth of measurement technology precision, spread of measurements over large scales and a tendency to the unification of fundamental physical interaction [6], where main problems arise and are concentrated on the gravitational interaction.

There are three problems related to G, which origin lies mainly in unified models predictions [57]:

1) absolute G measurements, 2) possible time variations of G, 3) possible range variations of G – non-Newtonian, or new interactions.

Absolute measurements of G. There are many laboratory determinations of G with errors of the order 10^{-3} and only several [58] on the level of 10^{-4}. The most recent and precise G measurements do not agree between themselves and some differ from the CODATA values of 1986-2004 (in $10^{-11} \cdot m^3 \cdot kg^{-1} \cdot s^{-2}$):

1. Gundlach and Merkowitz, 2000 (USA) [59]: $G = (6,674215 \pm 0,000092)$
2. Armstrong, 2002 (New Zealand, MSL): $G = (6,6742 \pm 0,0007)$.

3. O.Karagioz (Moscow, Russia, 2003): $G = (6.6729 \pm 0.0005)$.
4. Luo Zhun (China, Wuhan): (1998): $G = (6,6699 \pm 0,0007)$.
5. T. Quinn et al., 2001 (BIPM): $G = (6,6693 \pm 0,0009)$ (1),
$G = (6,6689 \pm 0,0014)$ (2).
6. Schlamminger et.al., 2002(CH): $G = (6.7404 \pm 33ppm)$.

But, from 2004 CODATA gives:

$$G = 6.6742(10) \cdot 10^{-11} \cdot m^3 \cdot kg^{-1} \cdot s^{-2}. \qquad (6)$$

So, we see that we are not too far (a little more one order) from Cavendish, who obtained value of G two centuries ago at the level 10^{-2}. The situation with the measurement of the absolute value of G is really different from atomic constants. This means that either the limit of terrestrial accuracies of defining G has been reached or we have some new physics entering the measurement procedure [6]. The first means that, maybe we should turn to space experiments to measure G [11,10], and and second means that a more thorough study of theories generalizing Einstein's general relativity or unified theories is necessary. For more details see [11,12].

3. *Time Variations of G.* The problem of variations of FPC arose with the attempts to explain the relations between micro- and macro-world phenomena. Dirac was the first to introduce (1937) the so-called "Large Numbers Hypothesis". After the original *Dirac hypothesis* some new ones appeared (Gamov, Teller, Landau, Terazawa, Staniukovich etc., see [3,11]) and also some generalized theories of gravitation and multidimensional ones admitting the variations of an effective gravitational coupling.

In [60,61,4] the conception was worked out that variations of constants are not absolute but depend on the system of measurements (choice of standards, units and devices using this or that fundamental interaction). Each fundamental interaction through dynamics, described by the corresponding theory, defines the system of units and the corresponding system of basic standards, e.g. atomic and gravitational (ephemeris) seconds.

Earlier reviews of some hypotheses on variations of FPC and experimental tests can be found in [3,4,6].

But the most strict data were obtained by A. Schlyachter in 1976 (Russia) from an analysis of the ancient natural nuclear reactor data in Gabon, Oklo, because the event took place $2 \cdot 10^9$ years ago. The recent review see in [62].

There appeared some data on a possible variation of α on the level of 10^{-16} at some z [63]. Other groups do not support these results. Also appeared data on possible violation of m_e/m_p (Varshalovich et al.) The problem may be that even if they are correct, all these results are mean values of variations at some epoch of the evolution of the Universe (certain z interval). In essence variations may be different at different epochs (if they exist at all) and at the next stage observational data should be analyzed with the account of evolution of corresponding ("true"?)cosmological models.

We know that scalar-tensor and multidimensional theories are frameworks for these variations. So, one of the ways to describe variable gravitational coupling is the introduction of a *scalar field* as an additional variable of the gravitational interaction. It may be done by different means (e.g. Jordan, Brans-Dicke, Canuto and others). We have suggested a variant of gravitational theory with a conformal scalar field (Higgs-type field [64,3]) where Einstein's general relativity may be considered as a result of spontaneous symmetry breaking of conformal symmetry and obtained the estimation on the level of R. Hellings result [65]:

$$|\dot{G}/G| < (2 \pm 4) \cdot 10^{-12}\ \text{year}^{-1}. \tag{7}$$

More recent E.V.Pitjeva's result, Russia [66], based on satellites and planets motion data is:

$$|\dot{G}/G| < (0 \pm 2) \cdot 10^{-12} year^{-1} \tag{8}$$

Some new results on pulsars and BBN are on the level of 10^{-12} per year. But, the most reliable ones are based on lunar laser ranging (Muller et al, 1993, Williams et al, 1996, Nordtvedt, 2003) and spacecraft tracking (Hellings). They are not better than 10^{-12} per year.

Here, once more we see that there is a need for corresponding theoretical and experinmental studies. Probably, future space missions like Earth SEE-satellite [9,10,11,12] or missions to other planets and lunar laser ranging will be a decisive step in solving the problem of temporal variations of G and determining the fate of different theories which predict them.

Different theoretical schemes lead to temporal variations of the effective gravitational constant:

(1) Models and theories, where G is replaced by $G(t)$.
(2) Numerous scalar-tensor theories of Jordan-Brans-Dicke type where G depending on the scalar field $\sigma(t)$ appears [67].
(3) Gravitational theories with a conformal scalar field [60,61,3,64,68].
(4) Multidimensional theories in which there are dilaton fields and effective scalar fields appearing in our spacetime from additional dimensions

[1,72,69,70,71]. They may help also in solving the problem of a variable cosmological constant from Planckian to present values.

As was shown in [4,72,1] temporal variations of FPC are connected with each other in *multidimensional models* of unification of interactions. So, experimental tests on $\dot{\alpha}/\alpha$ may at the same time be used for estimation of $\dot{G}/G$ and vice versa. Moreover, variations of G are related also to the cosmological parameters ρ, Ω and q which gives opportunities of raising the precision of their determination.

As variations of FPC are closely connected with the behaviour of internal scale factors, it is a direct probe of properties of extra dimensions and the corresponding unified theories [7,8,1].

6. *Non-Newtonian interactions, or range variations of G*. Nearly all modified theories of gravity and unified theories predict also some deviations from the Newton law (inverse square law, ISL) or composition-dependent violations of the Equivalence Principle (EP) due to appearance of new possible massive particles (partners) [4]. Experimental data exclude the existence of these particles at nearly all ranges except less than *millimeter* and also at *meters and hundreds of meters* ranges. Recent analysis of experimental bounds and new possible limits on ISL violation from planets, pulsar and LLR data were done in [73].

In the Einstein theory G is a true constant. But, if we think that G may vary with time, then, from a relativistic point of view, it may vary with distance as well. In GR massless gravitons are mediators of the gravitational interaction, they obey second-order differential equations and interact with matter with a constant strength G. If any of these requirements is violated, we come in general to deviations from the Newton law with range (or to generalization of GR).

In [5] we analyzed several classes of such theories.In all these theories some effective or real masses appear leading to Yukawa-type (or power-law) deviations from the Newton law, characterized by strength α and range λ.

Recently some new attempts appeared: modifications of the Newton law at large ranges (MOND etc.), models to explain small acceleration at $a > a_0$ of Pioneer 10 and 11 spacecrafts (Pioneer anomaly, etc.)

There exist some model-dependant estimations of these forces. The most well-known one belongs to Scherk (1979) from supergravity where the graviton is accompanied by a spin-1 partner (graviphoton) leading to an additional repulsion.

Some p-brane models (ADD, branewolds) also predict non-Newtonian additional interactions of both Yukawa or power-law, in particular in the less than mm range, what is intensively discussed nowadays [12,74]. About PPN parameters for multidimensional models with p-branes see [13,14].

We saw that there are three problems connected with G. There is a promising new multi-purpose space experiment SEE - Satellite Energy Exchange [9,10], which addresses all these problems and may be more effective in solving them than other laboratory or space experiments.

We studied many aspects of the SEE-project [10,11] and the general conclusion is that realization of the SEE-project may improve our knowledge of G, G-dot and $G(r)$ by 3-4 orders.

Acknowledgments

The author is grateful to Prof. J. Nester and Dr. C.-M. Chen for their hospitality during the conference at NCU, Taiwan. The work was supported in part by DFG, Univ. Paris-7 and RFBR.

References

1. V.N. Melnikov, "Multidimensional Classical and Quantum Cosmology and Gravitation. Exact Solutions and Variations of Constants." CBPF-NF-051/93, Rio de Janeiro, 1993;
V.N. Melnikov, in: "Cosmology and Gravitation", ed. M. Novello, Editions Frontieres, Singapore, 1994, p. 147.
2. V.N. Melnikov, "Multidimensional Cosmology and Gravitation", CBPF-MO-002/95, Rio de Janeiro, 1995, 210 p.;
V.N. Melnikov. In: *Cosmology and Gravitation. II*, ed. M. Novello, Editions Frontieres, Singapore, 1996, p. 465.
V.N.Melnikov. "Exact Solutions in Multidimensional Gravity and Cosmology III." CBPF-MO-03/02, Rio de Janeiro, 2002, 297 pp.
3. K.P. Staniukovich and V.N. Melnikov, "Hydrodynamics, Fields and Constants in the Theory of Gravitation", Energoatomizdat, Moscow, 1983, 256 pp. (in Russian). See English translation of first 5 sections in:
V.N. Melnikov, Fields and Constraints in the Theory of Gravitation, CBPF MO-02/02, Rio de Janeiro, 2002, 145 pp.
4. V.N. Melnikov, *Int. J. Theor. Phys.* **33**, 1569 (1994).
5. V. de Sabbata, V.N. Melnikov and P.I. Pronin, *Prog. Theor. Phys.* **88**, 623 (1992).
6. V.N. Melnikov. General-Relativistic Metrology. In: "Gravitational Measurements, Fundamental Metrology and Constants", eds. V. de Sabbata and V.N. Melnikov, Kluwer Academic Publ., Dordtrecht, 1988, p. 283.
7. V.D. Ivashchuk and V.N. Melnikov, *Nuovo Cimento* **B 102**, 131 (1988).
8. K.A. Bronnikov, V.D. Ivashchuk and V.N. Melnikov, *Nuovo Cimento* **B 102**, 209 (1988).

9. A. Sanders and W. Deeds. *Phys. Rev.* **D 46**, 480 (1992).
10. A.D.Alexeev, K.A.Bronnikov, N.I.Kolosnitsyn, M.Yu.Konstantinov, V.N.Melnikov and A.G.Radynov. *Izmeritelnaya tekhnika,* 1993, N 8, p.6; N 9, p.3; N 10, p.6; 1994, N1, p.3. *Int.J.Mod.Phys.*, 1994, **D 3**, N 4, p.773.
P.N. Antonyuk, K.A.Bronnikov and V.N.Melnikov. *Astron.Lett.*, 1994, **20**, N 1, p.59. *Grav. Cosm.*, 1997, **3**, N 4(12).
K.A.Bronnikov, M.Yu.Konstantinov and V.N.Melnikov, *Grav. Cosm.*, 1996, **2**, N 4(8), p.361. *Izmerit. Tekhnika*, 1996, N 5, p.3.
A.D. Alexeev, V.N.Melnikov et al. Ibid., 1997, N 10, p.3. *Grav.Cosm.*, 1999, **5**, N1(17), p.67.
A.J.Sanders, V.N.Melnikov et al. Ibid, 1997, **3**, p.287. *Meas.Sci.Technol.*, 1999, **10**, p.514. *Class.Quant.Grav.*, 2000, **17**, N 1.
V.N.Melnikov. *Science in Russia*, 2000, N 6, p.3.
11. V.N.Melnikov. Gravity as a Key Problem of the Millennium. Proc.2000 NASA/JPL Conference on Fundamental Physics in Microgravity, NASA Document D-21522, 2001, p.4.1-4.17, Solvang, CA, USA.
12. S.A. Kononogov and V.N. Melnikov, The Fundamental Physical Constants, the Gravitational Constant, and the SEE Space Experiment Project. Izm. Tekhnika **6**, 1 (2005) - in Russian; English translation: Measurement Techniques **48**, 6, 521 (2005).
13. V.D.Ivashchuk and V.N.Melnikov. Multidimensional cosmological and spherically symmetric solutions with intersecting p-branes. In Lecture Notes in Physics, Vol. 537, "Mathematical and Quantum Aspects of Relativity and Cosmology Proceedings of the Second Samos Meeting on Cosmology, Geometry and Relativity held at Pythagoreon, Samos, Greece, 1998, eds: S. Cotsakis, G.W. Gibbons., Berlin, Springer, 2000; gr-qc/9901001.
14. V.D. Ivashchuk and V.N. Melnikov, Exact solutions in multidimensional gravity with antisymmetric forms, topical review, *Class. Quantum Grav.* **18**, R82-R157 (2001); hep-th/0110274.
15. V.N. Melnikov. Multidimensional Cosmology with Different Matter Sources. Proc. RESCEU-99, Tokyo, 1999. In :Birth and Evolution of the Universe. Eds. K.Sato and M.Kawasaki, Univ. Acad. Press, Tokyo, 2001, pp.45-54.
16. V.D. Ivashchuk and V.N. Melnikov, Sigma-model for the Generalized Composite p-branes, hep-th/9705036; *Class. Quant. Grav.* **14**, 3001-3029 (1997).
17. V.R. Gavrilov, V.D. Ivashchuk and V.N. Melnikov, *J. Math. Phys.* **36**, 5829 (1995).
18. U. Bleyer, V.D. Ivashchuk, V.N. Melnikov and A.I. Zhuk, "Multidimensional Classical and Quantum Wormholes in Models with Cosmological Constant", gr-qc/9405020; *Nucl. Phys.* **B 429**, 117 (1994).
19. V.D. Ivashchuk and V.N. Melnikov Billiard Representation for Multidimensional Cosmology with Multicomponent Perfect Fluid near the Singularity. Clas.Quant.Grav., 1995, v.12, N3, p.809.
20. M.A. Grebeniuk, V.D. Ivashchuk and V.N. Melnikov, "Multidimensional Cosmology for Intersecting *p*-branes with Static Internal Spaces", *Grav. & Cosmol.*, **4**, No 2(14), 145 (1998).
21. V. D. Ivashchuk, V. N. Melnikov, D. Singleton. On avoiding cosmological

oscillating behavior for S-brane solutions with diagonal metrics. Phys. Rev. D72, 103511 (2005).
22. V.D. Ivashchuk and V.N. Melnikov, Multidimensional classical and quantum cosmology with intersecting p-branes, *J. Math. Phys.*, **39**, 2866-2889 (1998).
23. V.N. Melnikov. Fundamental Physical Constants and Multidimensional Black Holes. Proc. Nara Int. Conf., Nara, Japan, Grav. Cosm, 2000, v 6, N 2(22), p.81.
24. H. Dehnen, V.D. Ivashchuk and V.N. Melnikov, Billiard representation for multidimensional multi-scalar cosmological model with exponential potentials, *GRG,* v.36, issue 7, 1563-1578, 2004; hep-th/0312317.
25. K.A. Bronnikov and V.N. Melnikov, *Ann. Phys. (N.Y.)* **239**, 40 (1995).
26. K.A. Bronnikov and V.N. Melnikov. *Grav. & Cosmol.* **5**, 306 (1999).
27. K.A. Bronnikov and V.N. Melnikov, p-Brane Black Holes as Stability Islands, *Nucl. Phys.* **B 584**, 436-458 (2000).
28. V.D. Ivashchuk and V.N. Melnikov, Multidimensional cosmology with m-component perfect fluid. *Int. J. Mod. Phys.* **D 3** (1994), 795; gr-qc/9403063.
29. V.D. Ivashchuk and V.N. Melnikov, Intersecting p-Brane Solutions in Multidimensional Gravity and M-Theory, hep-th/9612089; *Grav. and Cosmol.* **2**, No 4, 297-305 (1996).
30. V.D. Ivashchuk and V.N. Melnikov, *Phys. Lett. B* **403**, 23-30 (1997).
31. V.N. Melnikov. P-branes, Extra Dimensions and their Observational Windows. In: "Cosmology and Gravitation". Eds. M.Novello and S.E. Perez Bergliaffa. Proc. X-th Brasilian School of Cosmology and Gravitation. AIP Conf. Proc., 2003, v. 668, pp. 187-197.
32. M.A. Grebeniuk, V.D. Ivashchuk and V.N. Melnikov, Integrable Multidimensional Quantum Cosmology for Intersecting p-Branes, *Grav. and Cosmol.* **3**, No 3 (11), 243-249 (1997), gr-qc/9708031.
33. K.A. Bronnikov, M.A. Grebeniuk, V.D. Ivashchuk and V.N. Melnikov, Integrable Multidimensional Cosmology for Intersecting p-branes, *Grav. and Cosmol.* **3**, No 2(10), 105-112 (1997).
34. E. Cremmer, B. Julia, and J. Scherk, *Phys. Lett.* **B76** 409 (1978).
35. M.B. Green, J.H. Schwarz and E. Witten, Superstring Theory (Cambridge University Press., Cambridge, 1987).
36. V.D. Ivashchuk, V.N. Melnikov and M. Rainer, Multidimensional Sigma-Models with Composite Electric p-branes, gr-qc/9705005; *Gravit. Cosm.* **4**, No 1(13), 73-82 (1998).
37. V.D. Ivashchuk and V.N. Melnikov, "Multidimensional Gravity with Einstein Internal Spaces", hep-th/9612054; *Grav. & Cosmol.* **2**, 177 (1996).
38. V.D.Ivashchuk and V.N.Melnikov. Cosmological and Spherically Symmetric Solutions with Intersecting p-branes. *J. Math. Phys.*, 1999, **40** (12), 6558.
39. V.R. Gavrilov and V.N. Melnikov, Toda Chains with Type A_m Lie Algebra for Multidimensional Classical Cosmology with Intersecting p-branes, In : Proceedings of the International seminar "Curent topics in mathematical cosmology", (Potsdam, Germany , 30 March - 4 April 1998), Eds. M. Rainer and H.-J. Schmidt, World Scientific, 1998, p. 310; hep-th/9807004.
40. V.R. Gavrilov and V.N. Melnikov, Toda Chains Associated with Lie Algebras

A_m in Multidimensional Gravitation and Cosmology with Intersecting *p*-branes, *Theor. Math. Phys.* **123**, No 3, 374-394 (2000) (in Russian).

41. S. Cotsakis, V.R. Gavrilov and V.N. Melnikov, Spherically Symmetric Solutions for p-Brane Models Associated with Lie Algebras, *Grav. and Cosmol.* **6**, No 1 (21), 66-75 (2000).
42. V.D.Ivashchuk and V.N.Melnikov. Black hole p-brane solutions for general intersection rules. *Grav. and Cosmol.* **6**, No 1 (21), 27-40 (2000); hep-th/9910041.
43. V.D.Ivashchuk and V.N.Melnikov. Toda p-brane black holes and polynomials related to Lie algebras. *Class. Quant. Grav.* **17** 2073-2092 (2000).
44. V.D. Ivashchuk, V.N. Melnikov and A.B. Selivanov, Composite S-brane solutions on product of Ricci-flat spaces, *GRG*, v.36, issue 7, 1593, 2004.
45. V.D. Ivashchuk, V.N. Melnikov and S.-W. Kim, S-brane solutions with acceleration, in models with forms and multiple exponential potential, *Grav.& Cosm.* 2004, v. 10, N 1-2, pp. 141-148; hep-th/0405009.
46. H. Dehnen, V.R. Gavrilov and V.N. Melnikov, General solutions for a flat Friedmann universe filled with a perfect fluid and a scalar field with an exponential potential, *Grav. Cosmol.*, **8**, 4 (32), (2003), p. 189-195.
47. V.R. Gavrilov, V.N. Melnikov and S.T. Abdyrakhmanov, Flat Friedmann Universe filled by dust and scalar field with multiple exponential potential, *Gen. Rel. Gravit.*, (2004), v.36, issue 7, 1579-1592.
48. V.N. Melnikov. Gravity in Diverse Dimensions. In: Inquiring the Universe, Essays to celebrate Prof. M. Novello's Jublee. Eds. J. M. Salim, S. E. Perez Bergliaffa, L. A. Oliveira and V. A. De Lorenci, Frontier Group, 2003, pp.187-206.
49. V.D. Ivashchuk, V.N. Melnikov and A.B. Selivanov, Cosmological solutions in multidimensional model with multiple exponential potential, *JHEP* 0309 (2003) 059; hep-th/0309027.
50. J.-M. Alimi, V.D. Ivashchuk and V.N. Melnikov, Non-singular solutions in multidimensional model with scalar fields and exponential potential, *Grav. Cosm.* 2005, v.11, N1-2, p. 111.
51. V.D. Ivashchuk and V.N. Melnikov, Billiard representation for multidimensional cosmology with intersecting p-branes near the singularity. *J. Math. Phys.*, **41**, No 8, 6341-6363 (2000); hep-th/9904077.
52. H. Dehnen, V.D. Ivashchuk and V.N. Melnikov. Composite electric S-brane solutions with a maximum number of branes and several scalar fields, Grav.& Cosm., v.11, N4, p. 345.
53. V.N. Melnikov and V.R. Gavrilov, 2-component cosmological models with perfect fluid: exact solutions. In: The Gravitational Constant: Generalized Gravitational Theories and Experiments, V. de Sabbata, G.Gillies and V.N. Melnikov eds., 2004, Kluwer Academic Publishers, p. 247-268.
54. V.R. Gavrilov and V.N. Melnikov, D-dimensional integrable 2-component viscous cosmology, *Grav. Cosmol.*,**7**, No 4, (2001), 301-307.
55. V.R. Gavrilov and V.N. Melnikov, Modern trends in multidimensional gravity and cosmology, In: Advances in the Interplay Between Quantum and Gravity Physics, P.G. Bergmann and V. de Sabbata (eds.), Kluwer Academic Pub-

lishers, 2002, p. 285-315.
56. J.-M. Alimi, V. R. Gavrilov and V. N. Melnikov, Multicomponent perfect fluid with variable parameters in n Ricci-flat spaces, Proceedings of ICGA6, 6-9 October 2003, Seoul, Korea *J. Korean Phys. Soc.* 2004, v.44, p. S148.
57. V.N. Melnikov. Variations of G and SEE Project. Proc. Rencontre de Moriond-99: Gravitational Waves and Experimental Gravity. Editions Frontieres, 1999.
58. V.N. Melnikov. Gravity and Cosmology as Key Problems of the Millennium. In: Proc. Int. Conf. "Einstein Siecle", Herman Publ., Paris, 2006.
59. J.H.Gundlach and S.M.Merkowitz, Phys.Rev.Lett. 85 (2000) 2869-2872.
60. V.N. Melnikov and K.P. Staniukovich. In: "Problems of Gravitation and Elementary Particle Theory (PGEPT)", Atomizdat, Moscow, 1978, v.9, p.3 (in Russian).
61. N.A. Zaitsev and V.N. Melnikov. In: PGEPT, 1979, v.10, p.131.
62. K.A. Bronnikov and S.A. Kononogov. Possible variations of the fine structure constant α and their metrological significance. Submitted to Metrologia.
63. J. Webb et al., Phys. Rev. Lett. 87, 091301 (2001).
64. K.A. Bronnikov, V.N. Melnikov and K.P. Staniukovich. ITP-68-69, Kiev, 1968.
65. R. Hellings, *Phys. Rev. Lett.* **51**, 1609 (1983).
66. E.V.Pitjeva. In: Dynamics and Astrometry of Natural and Artificial Celestial Bodies. Kluwer Acad. Publ., Netherlands, 1997, p. 251.
67. K.A. Bronnikov, V.N. Melnikov and M. Novello. Possible time variations of G in scalar-tensor theories of gravity. Gravit. & Cosm., 2002, v.8, Suppl.II, p.18-21.
68. V.N. Melnikov et al., *GRG* **17**, 63 (1985).
69. V.D. Ivashchuk and V.N. Melnikov. Problems of G and Multidimensional Models. In Proc. JGRG11, Eds. J. Koga et al., Waseda Univ., Tokyo, 2002, pp. 405-409.
70. V.N. Melnikov. Time Variations of G in Different Models. Proc. 5 Int. Friedmann Seminar, Joao Pessoa, Brazil. Int.J.Mod.Phys.A, 2002, 17, 4325-4334.
71. H. Dehnen, V.D. Ivashchuk, S.A.Kononogov and V.N. Melnikov). On time variation of G in multidimensional models with two curvatures. Grav.& Cosm., v.11, N4, p. 340. =
72. V.N. Melnikov. In: "Results of Science and Technology. Gravitation and Cosmology". Ed.: V.N. Melnikov, 1991, v.1, p.49 (in Russian).
73. N.I. Kolosnitsyn and V.N.Melnikov. Test of Inverse Square Law Through Prec)ession of Orbits. GRG, v.36, issue 7, 1619-1624, 2004.
74. K.A. Bronnikov, S.A. Kononogov, V.N. Melnikov. Brane world corrections to Newton's law. 2006, to appear in GRG; gr-qc/0601114.

Coupled Quintessence and CMB

Seokcheon Lee
Institute of Physics, Academia Sinica, Taipei, Taiwan 11529, R.O.C.
E-mail: skylee@phys.sinica.edu.tw

We revise the stability of the tracking solutions and briefly review the potentials of quintessence models. We discuss the evolution of linear perturbations for $V(\phi) = V_0 \exp(\lambda\phi^2/2)$ potential in which the scalar field is non-minimally coupled to cold dark matter. We consider the effects of this coupling on both cosmic microwave background temperature anisotropies and matter perturbations. We find that the phenomenology of this model is consistent with current observations up to the coupling power $n_c \leq 0.01$ while adopting the current parameters measured by WMAP, $\Omega_\phi^{(0)} = 0.76$, $\Omega_{cdm}^{(0)} = 0.191$, $\Omega_b^{(0)} = 0.049$, and $h = 0.70$. Upcoming cosmic microwave background observations continuing to focus on resolving the higher peaks may put strong constraints on the strength of the coupling.

1. Introduction

If we treat Type Ia supernovae (SNe Ia) as standardized candles, then the Hubble diagram of them shows that the expansion of the Universe is currently accelerating [1]. Combining measurements of the acoustic peaks in the angular power spectrum of the cosmic microwave background (CMB) anisotropy [2] and the matter power spectrum of large scale structure (LSS) which is inferred from galaxy redshift surveys like the Sloan Digital Sky Survey (SDSS) [3] and the 2-degree Field Galaxy Redshift Survey (2dFGRS) [4] has also confirmed that a component with negative pressure (dark energy) should be added to the matter component to make up the critical density today.

A quintessence field is a dynamical scalar field leading to a time dependent equation of state (EOS), ω_ϕ. The possibility that a scalar field at early cosmological times follows an attractor-type solution and tracks the evolution of the visible matter-energy density has been explored [5]. This may help alleviate the severe fine-tuning associated with the cosmological constant problem. However we need to investigate the tracking condition and its stability at the matter dominated epoch carefully [6].

Are there experimental ways of checking for the existence or absence of dark energy in the form of quintessence? There are several different observational effects of matter coupling to the scalar field on CMB spectra and matter power spectrum compared to the minimally coupled models [7]. And this can be used to check the existence of quintessence.

This paper is organized as follows. We briefly investigate the condition and the stability of tracker solution and review the potentials of quintessence models in the next section. In Sec. III, we show the coupling effects on CMB and matter power spectrum. We conclude in the last section.

2. Quintessence Models and Tacker Solutions

Many models of quintessence have a tracker behavior, which solves the "coincidence problem" (*i.e.* initial condition). In these models, the quintessence field has a density which closely tracks (but is less than) the radiation density until matter-radiation equality, which triggers quintessence to start behaving as dark energy, eventually dominating the Universe. This naturally sets the low scale of the dark energy. However the present small value of dark energy density still cannot be solved with quintessence ("fine-tuning problem").

Since the energy density of the scalar field generally decreases more slowly than the matter energy density, it appears that the ratio of the two densities must be set to a special, infinitesimal value in the early Universe in order to have the two densities nearly coincide today. To avoid this initial conditions problem we focus on tracker fields.

2.1. *Tracking Condition*

We rely details of this section on the reference [6] due to the shortage of space. We introduce new quantity θ which is related to the ratio of the kinetic energy and the potential energy of the scalar field :

$$2\theta = \ln \frac{KE}{PE} = \ln \frac{1+\omega_\phi}{1-\omega_\phi} \tag{1}$$

θ can have any value and especially positive θ means the kinetic energy dominated era and negative θ indicates the potential energy dominated one. Now we can define EOS as the function of this new quantity θ :

$$\omega_\phi = \tanh\theta \tag{2}$$

The "tracker equation" ($\Gamma \equiv V''V/(V')^2$) can be expressed by θ ;

$$\Gamma = 1 + \frac{3}{2}\frac{(A\omega_r - \omega_\phi)}{(1+\omega_\phi)}\frac{(1-\Omega_\phi)}{(3+\tilde{\theta})} - \frac{1-\omega_\phi}{2(1+\omega_\phi)}\frac{\tilde{\theta}}{(3+\tilde{\theta})} - \frac{1}{(1+\omega_\phi)}\frac{\tilde{\tilde{\theta}}}{(3+\tilde{\theta})^2} \tag{3}$$

where ω_r is EOS of the radiation, $A = a_{eq}/(a + a_{eq})$, a_{eq} is the scale factor when the energy density of the radiation and that of the matter become equal, and tilde means the derivative with respect to $x = \ln a = -\ln(1+z)$ [a]. This equation looks like quite different from the well known tracker equation. But when we choose the early Universe constraints ($A \simeq 1$ and $\Omega_\phi \simeq 0$) we can get the well known tracker equation.

$$\begin{aligned}\Gamma &\simeq 1 + \frac{3}{2}\frac{(\omega_r - \omega_\phi)}{(1+\omega_\phi)}\frac{1}{(3+\tilde{\theta})} - \frac{1-\omega_\phi}{2(1+\omega_\phi)}\frac{\tilde{\theta}}{(3+\tilde{\theta})} - \frac{1}{(1+\omega_\phi)}\frac{\tilde{\tilde{\theta}}}{(3+\tilde{\theta})^2} \\ &= 1 + \frac{(\omega_r - \omega_\phi)}{2(1+\omega_\phi)} - \frac{1+\omega_r - 2\omega_\phi}{2(1+\omega_\phi)}\frac{\tilde{\theta}}{(3+\tilde{\theta})} - \frac{1}{(1+\omega_\phi)}\frac{\tilde{\tilde{\theta}}}{(3+\tilde{\theta})^2}\end{aligned} \tag{4}$$

When we can ignore the change of θ (*i.e.* when ω_ϕ is almost constant), we can get the tracking solution. From the above tracking equation we can check this condition.

$$\Gamma \simeq 1 + \frac{1}{2}\frac{(A\omega_r - \omega_\phi)}{1+\omega_\phi}(1-\Omega_\phi) \tag{5}$$

This equation can be rearranged to see the behavior of the EOS as following.

$$\omega_\phi \simeq \frac{\omega_r A(1-\Omega_\phi) - 2(\Gamma-1)}{(1-\Omega_\phi) + 2(\Gamma-1)} \tag{6}$$

To investigate this equation more carefully, we define the new quantities.

$$Q = (1-\Omega_\phi) \tag{7}$$
$$F = (\Gamma - 1) \tag{8}$$

where Q shows the energy information of the Universe and F depends on the form of the given potential. With these we can rewrite the equation (6).

$$\omega_\phi \simeq \omega_r \frac{Q}{Q+2F}A - \frac{2F}{Q+2F} \tag{9}$$

[a] Where we put the present value of scale factor, $a^{(0)}$ as one.

In the reference [5], this equation is expressed as :

$$\omega_\phi \simeq \frac{\omega_r - 2F}{1 + 2F} \tag{10}$$

This equation (10) can be true only when $Q, A \simeq 1$, which can be satisfied at the early Universe and not at the late one. So with this equation, it is not proper to check the evolution of tracking solutions at late Universe. Instead we should use the equation (9) to check the evolution of the tracking solutions. Before checking the properties of this equation, we should notice that Q is always positive and has the interval as $0 \leq Q \leq 1$. F can be positive or negative based on the given shape of the potential.

2.2. *Stability Of Tracker Solution*

We need to check that solutions with ω_ϕ,which is not equal to the tracker solution value (ω_0) can converge to the tracker ones (*i.e.* Are tracker solutions stable?). To check this we need to check the small deviation ($\delta\omega$) of the tracker solution of EOS.

$$\omega_\phi = \omega_0 + \delta\omega \tag{11}$$

If we insert this into the tracker equation (3), then we have following.

$$\ddot{\delta\omega} + \frac{3}{2}\Big[(A\omega_r - \omega_0)(1-\Omega_\phi) + (1-\omega_0)\Big]\dot{\delta\omega} + \frac{9}{2}(1-\omega_0)\Big[(1+A\omega_r)(1-\Omega_\phi)\Big]\delta\omega = 0 \tag{12}$$

where we use the tracking condition (5). The general solution to this nonlinear differential equation cannot be obtained analytically. But this equation can be simplified as follow in the early Universe constraints.

$$\ddot{\delta\omega} + \frac{3}{2}\Big[(1+\omega_r) - 2\omega_0\Big]\dot{\delta\omega} + \frac{9}{2}(1-\omega_0)(1+\omega_r)\delta\omega \simeq 0 \tag{13}$$

The solution of this equation is

$$\delta\omega \propto a^{\gamma_1} \tag{14}$$

where

$$\gamma_1 = -\frac{3}{2}\Big[\frac{1}{2}(1+\omega_r) - \omega_0\Big] \pm \frac{i}{2}\sqrt{18(1+\omega_r)(1-\omega_0) - 9\Big[\frac{1}{2}(1+\omega_r) - \omega_0\Big]^2} \tag{15}$$

The real part of this is negative for ω_0 less than 2/3. So $\delta\omega$ will decays exponentially and solution reaches to the tracking one. In addition to this it also oscillates due to the second term. For the late Universe case, we can

Table 1. Quintessence models.

Potential	Reference	Properties
$V_0 \exp(-\lambda\phi)$	Ratra & Peebles (1988), Wetterich (1988) [5]	$\omega = \lambda^2/3 - 1$
	Ferreira & Joyce (1998) [5]	$\lambda > 5.5 - 4.5, \Omega < 0.1 - 0.15$
$V_0/\phi^\alpha, \alpha > 0$	Ratra & Peebles (1988) [5]	$\omega > -0.7$
$m^2\phi^2, \lambda\phi^4$	Frieman et al (1995) [8]	PNGB, $M^4[\cos(\phi/f) + 1]$
$V_0(\exp M_p/\phi - 1)$	Zlatev, Wang & Steinhardt (1999) [9]	$\Omega_m \geq 0.2, \omega < -0.8$
$V_0 \exp(\lambda\phi^2)/\phi^\alpha$	Brax & Martin (1999,2000) [10]	$\alpha \geq 11, \omega \simeq -0.82$
$V_0(\cosh\lambda\phi - 1)^p$	Sahni & Wang (2000) [11]	$p < 1/2, \omega < -1/3$
$V_0 \sinh^{-\alpha}(\lambda\phi)$	Sahni & Starobinsky (2000) [12],	early time : inverse power
	Ureña-López & Matos (2000) [13]	late time : exponential
$V_0(e^{\alpha\kappa\phi} + e^{\beta\kappa\phi})$	Barreiro, Copeland & Nunes (2000) [14]	$\alpha > 5.5, \beta < 0.8, \omega < -0.8$
$V_0[(\phi - B)^\alpha + A]e^{-\lambda\phi}$	Albrecht & Skordis (2000) [15]	$\omega \sim -1$
$V_0 \exp[\lambda(\phi/M_p)^2]$	Lee, Olive, & Pospelov (2004) [16]	$\omega \sim -1$
$V_0 \cosh[\lambda\phi/M_p]$	Lee, Olive, & Pospelov (2004) [16]	$\omega \sim -1$

change this equation. In the late Universe A goes to zero and Ω_ϕ is not zero. With these we can modify the general equation (12).

$$\ddot{\delta\omega} + \frac{3}{2}\Big[(1 + \Omega_\phi\omega_0) - 2\omega_0\Big]\tilde{\delta\omega} + \frac{9}{2}(1 - \omega_0)(1 - \Omega_\phi)\delta\omega \simeq 0 \qquad (16)$$

We can repeat the similar step to find the solution of this equation if we assume that Ω_ϕ is almost constant.

$$\delta\omega \propto a^{\gamma_2} \qquad (17)$$

where

$$\gamma_2 = -\frac{3}{2}\Big[\frac{1}{2}(1+\Omega_\phi\omega_0) - \omega_0\Big] \pm \frac{i}{2}\sqrt{18(1 + \Omega_\phi)(1 - \omega_0) - 9\Big[\frac{1}{2}(1 + \Omega_\phi\omega_0) - \omega_0\Big]^2} \qquad (18)$$

The real part of this solution can be negative if ω_0 satisfies following.

$$\omega_0 < \frac{1}{(2 - \Omega_\phi)} \qquad (19)$$

where $1 \leq (2 - \Omega_\phi) \leq 2$ for the entire history of Universe.

2.3. *Quintessence Potentials*

We display the potentials of the quintessence models in Table 1. Any detail of each model can be found in each reference.

3. Coupled Quintessence

The general equation for the interaction of a light scalar field φ with matter is,

$$S_\phi = \int d^4x\sqrt{-g}\left\{\frac{\bar{M}^2}{2}[\partial^\mu\phi\partial_\mu\phi - R] - V(\phi) - \frac{B_{Fi}(\phi)}{4}F^{(i)}_{\mu\nu}F^{(i)\mu\nu}\right.$$
$$\left. + \sum_j[\bar{\psi}_j i\not{D}\psi_j - B_j(\phi)m_j\bar{\psi}_j\psi_j]\right\}. \tag{20}$$

The coupling gives rise to the additional mass and source terms of the evolution equations for CDM and scalar field perturbations. This also affects the perturbation of radiation indirectly through the background bulk and the metric perturbations [7,17]. The value of the energy density contrast of the CDM (Ω_c) is increased in the past when the coupling is increased. We specify the potential and the coupling as in the reference [7,16].

$$V(\phi) = V_0 \exp\left(\frac{\lambda\phi^2}{2}\right), \qquad \exp[B_c(\phi)] = \left(\frac{b_c + V(\phi)/V_0}{1+b_c}\right)^{n_c} \tag{21}$$

3.1. *CMB*

Now, we investigate the effects of non-minimal coupling of a scalar field to the CDM on the CMB power spectrum. Firstly, the Newtonian potential at late times changes more rapidly as the coupling increases. This leads to an enhanced ISW effect. Thus we have a relatively larger C_ℓ at large scales (*i.e.* small ℓ). Thus, if the CMB power spectrum normalized by COBE, then we will have smaller quadrupole [18]. This is shown in the first panel of Figure 1. One thing that should be emphasized is that we use different parameters for the ΛCDM and the coupled quintessence models to match the amplitude of the first CMB anisotropy peak. The parameter used for the quintessence model is indicated in Figure 1 (*i.e.* $\Omega_\phi^{(0)} = 0.76$, $\Omega_m^{(0)} = 0.191$, $\Omega_b^{(0)} = 0.049$, and $h = 0.7$, where h is the present Hubble parameter in the unit of $100\text{kms}^{-1}\text{Mpc}^{-1}$). However, these parameters are well inside the $1\,\sigma$ region given by the WMAP data. We use the WMAP parameters for the ΛCDM model (*i.e.* $\Omega_\phi^{(0)} = 0.73$, $\Omega_m^{(0)} = 0.23$, $\Omega_b^{(0)} = 0.04$, and $h = 0.72$) [b]. In both models we use the same spectral index $n_s = 1$. The heights of the acoustic peaks at small scales (*i.e* large ℓ) can be affected by the

[b] Our data prefers WMAP 3 year data to WMAP 1 year one.

following two factors. One is the fact that the scaling of the CDM energy density deviates from that of the baryon energy density. Therefore for the given CDM and baryon energy densities today, the energy density contrast of baryons at decoupling ($\Omega_b^{(ls)}$) is getting lower as the coupling is being increased. This suppresses the amplitude of compressional (odd number) peaks while enhancing rarefaction (even number) peaks. The other is that for models normalized by COBE, which approximately fixes the spectrum at $\ell \simeq 10$, the angular amplitude at small scales is suppressed in the coupled quintessence. This is shown in the second panel of Figure 1. The third peak in this model is smaller than that in the ΛCDM model.

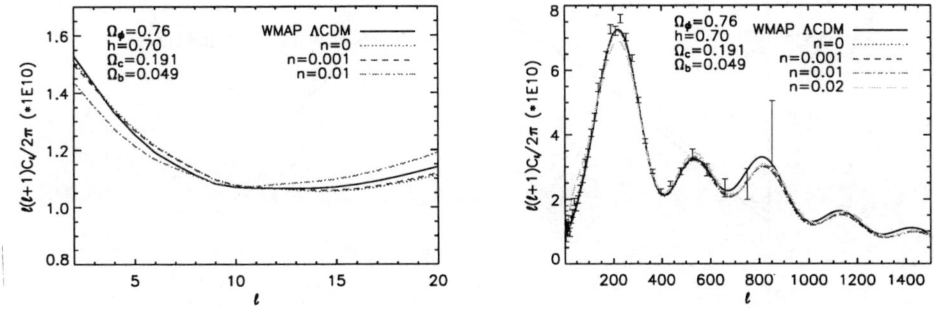

Fig. 1. (a) CMB large-scale anisotropy power spectra of ΛCDM (solid line), minimally coupled $n_c = 0$ (dotted line), and non-minimally coupled $n_c = 10^{-3}, 10^{-2}$ (dashed, dash-dotted line respectively) quintessence models. (b) Same spectra for the entire scales.

3.2. *Matter Power Spectrum*

The coupling of quintessence to the CDM can change the shape of matter power spectrum because the location of the turnover corresponds to the scale that entered the Hubble radius when the Universe became matter-dominated. This shift on the scale of matter and radiation equality is indicated

$$a_{eq} \simeq \frac{\rho_r^{(0)}}{\rho_c^{(0)}} \exp[B_c(\phi_0) - B_c(\phi_{eq})], \tag{22}$$

where $\rho_r^{(0)}$ and $\rho_c^{(0)}$ are the present values of the energy densities of radiation and CDM respectively, and the approximation comes from the fact that the present energy density of CDM is bigger than that of baryons

($\rho_c^{(0)} > \rho_b^{(0)}$). This is indicated in Figure 2. Increasing the coupling shifts the epoch of matter-radiation equality further from the present, thereby moving the turnover in the power spectrum to smaller scale. If we define k_{eq} as the wavenumber of the mode which enters the horizon at radiation-matter equality, then we will obtain

$$k_{eq} = \frac{2\pi}{\eta_{eq}}. \tag{23}$$

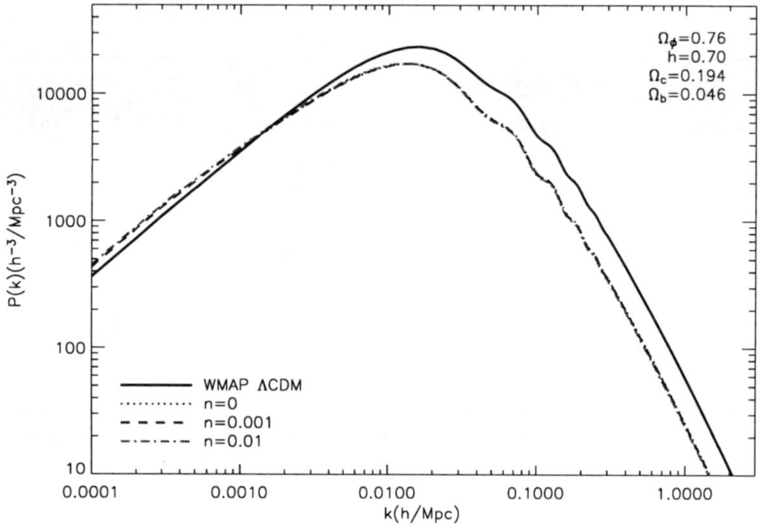

Fig. 2. Matter power spectra for the models using the same parameters in Figure 1.

4. Conclusions

We have investigated the tracking condition of the quintessence models and their stability. We have shown that it is necessary to distinguish the tracking condition at the matter dominated epoch and at the radiation dominated one.

We have considered the CMB anisotropy spectrum and the matter power spectrum for the non-minimally coupled models. Additional mass and source terms in the Boltzmann equations induced by the coupling give the rapid changes of the Newtonian potential Φ and enhance the ISW effect in the CMB power spectrum. The modification of the evolution of the CDM, $\rho_c = \rho_c^{(0)} a^{-3+\xi}$, changes the energy density contrast of the CDM at early epoch. We have adopted the current cosmological parameters measured by WMAP within 1σ level. With the COBE normalization and the WMAP data we have found the constraint of the coupling $n_c \leq 0.01$. The locations and the heights of the CMB anisotropy peaks have been changed due to the coupling. Especially, there is a significant difference for the heights of the second and the third peaks among the models. Thus upcoming observations continuing to focus on resolving the higher peaks may constrain the strength of the coupling. The suppression of the amplitudes of the matter power spectra could be lifted by a bias factor. However, a detailed fitting is beyond the scope of this paper. The turnover scale of the matter power spectrum may be also used to constrain the strength of the coupling n_c.

Acknowledgments

We thank ICGA7 organizing committee, J. M. Nester, C-M. Chen, and J-P. Hsu for their hospitality and for organizing such a nice meeting.

References

1. A. G. Riess *et al.* [Supernova Search Team Collaboration], Astron. J. **116**, 1009 (1998) [astro-ph/9805201]; Astrophys. J. **607**, 665 (2004) [astro-ph/0402512]; S. Perlmutter *et al.* [Supernova Cosmology Project Collaboration], Astrophys. J. **517**, 565 (1999) [astro-ph/9812133]; N. A. Bahcall, J. P. Ostriker, S. Perlmutter, and P. J. Steinhardt, Science **284**, 1481 (1999) [astro-ph/9906463]; T. Padmanabhan and T. R. Choudhury, Mon. Not. Roy. Astron. Soc. **344**, 823 (2003) [astro-ph/0212573]; Astron. Astrophys. **429**, 807 (2005) [astro-ph/0311622]; P. Astier *et al.*, [astro-ph/0510447].
2. C. L. Bennett *et al.*, Astrophys. J. Suppl . Ser **148**, 1 (2003) [astro-ph/0302207]; D. N. Spergel *et al.*, Astrophys. J. Suppl. Ser. **148**, 175 (2003) [astro-ph/0302209].
3. M. Tegmark *et al.*, Phys. Rev. D **69**, 103501 (2004) [astro-ph/0310723].
4. M. Colless *et al.*, Mon. Not. Roy. Astron. Soc. **328**, 1039 (2001) [astro-ph/0106498]; [astro-ph/0306581].
5. B. Ratra and P. J. E. Peebles, Phys. Rev. D **37**, 3406 (1988); Astrophys. J. **325**, L117 (1988); C. Wetterich, Nucl. Phys. B **302**, 668 (1988); P. G. Ferreira and M. Joyce, Phys. Rev. D **58**, 023503 (1998) [astro-ph/9711102].
6. S. Lee, PhD Thesis, University of Minnesota, (2005).
7. S. Lee, G-C. Liu, and K-W. Ng, Phys. Rev. D **73**, 083516 (2006) [astro-ph/0601333].
8. J. Frieman, C. Hill, A. Stebbins, and I. Waga, Phys. Rev. Lett. **75**, 2077 (1995) [astro-ph/9505060].

9. I. Zlatev, L. Wang, and P. J. Steinhardt, Phys. Rev. Lett. **82**, 896 (1999) [astro-ph/9807002].
10. P. Brax and J. Martin, Phys. Lett. B **468**, 40 (1999) [astro-ph/9905040] ; Phys. Rev. D **61**, 103502 (2000) [astro-ph/9912046].
11. V. Sahni and L. Wang, Phys. Rev. D **62**, 103517 (2000) [astro-ph/9910097].
12. V. Sahni and A. A. Starobinsky, Int. J. Mod. Phys. D **9**, 373 (2000) [astro-ph/9904398].
13. L. A. Urena-Lopez and T. Matos, Phys. Rev. D **62**, 081302 (2000) [astro-ph/0003364].
14. T. Barreiro, E. J. Copeland, and N. J. Nunes Phys. Rev. D **61**, 127301 (2000) [astro-ph/9910214].
15. A. Albrecht and C. Skordis Phys. Rev. Lett. **84**, 2076 (2000) [astro-ph/9908085].
16. S. Lee, K. A. Olive, and M. Pospelov Phys. Rev. D **70**, 083503 (2004) [astro-ph/0406039].
17. R. Bean, Phys. Rev. D **64**, 123516 (2001) [astro-ph /0104464].
18. M. White and E. F. Bunn, Astrophys. J **450**, 477 (1995); Erratum-ibid. **477**, 460 (1995) [astro-ph/9503054].

Cosmic Lee-Yang Force, Dark Energy and Accelerated Wu-Doppler Effect

Jong-Ping Hsu and Zhenhua Ning*
Department of Physics,
University of Massachusetts Dartmouth
North Dartmouth, MA 02747, USA
e-mail: jhsu@umassd.edu

In 1955, Lee and Yang discussed a new massless gauge field based on the established conservation of baryon number. They predicted the existence of a very weak repulsive force between baryons. Such a repulsive long-range force may be the physical origin of the dark-energy-induced acceleration of the expansion of the universe. The accelerated Wu transformation of spacetime based on limiting Lorentz and Poincaré invariance is employed to investigate the Wu-Doppler effects in which the source and observer have linearly accelerated motion. The results are applied to discuss experimental detections of the accelerated expansion of the universe.

1. Introduction

After the discovery of accelerated expansion of the universe, there are many discussions and speculations regarding its physical origin and cause. [1] It is natural to assume that the cosmic force involves an additional repulsive force between two ordinary objects, beside the Newtonian attractive gravitational force. Suppose we consider a simple two-body system. In a static approximation, the total 'cosmic force' F_C between two objects can be written phenomenologically as a combination of the gravitational force and a new force $Bf(r)$.

If the cosmological constant is assumed for Einstein's field equation, it could lead to a long-range repulsive force between matter. In a suitable

*Present address: Department of Physics, University of Illinois at Urbana-Champaign, IL 61801, USA.

approximation, the cosmological constant will lead to a quadratic potential $\propto r^2$ and modify the Newtonian law of motion as follows: [1]

$$\frac{d^2\mathbf{r}}{dt^2} = \mathbf{g} + \Omega_{\lambda o} H_o^2 \mathbf{r}, \quad or \quad F_C = -\frac{GM_1M_2}{r^2} + Bf(r), \tag{1}$$

where $\mathbf{g}$ is the relative gravitational acceleration produced by the distribution of ordinary matter, $\Omega_{\lambda o}$ is a constant 'dark-energy density,' and H_o is the Hubble constant. The cosmic force F_C* is for two bodies.

In 1955 Lee and Yang predicted an extremely weak and long-range repulsive force between baryons, which was based on a U(1) gauge symmetry associated with the established conservation of baryon number. [2] Using the results of Eötvös experiments, [3,4] its strength was estimated to be much smaller than that of the gravitational force. Since the accelerated expansion of the universe has been detected, it is interesting to see whether the new cosmological force $Bf(r)$ between two distant galaxies could be related to the gauge symmetry and Lee-Yang force associated with the conservations of the baryon number. [5] Moreover, the established conservation laws of the electron-lepton number also implies the existence of a long-range repulsive force between electrons.

Furthermore, if the U(1) gauge invariant Lagrangian is modified to involve a spacetime derivative of the field strength, the resultant potential has properties similar to that of the 'dark energy' implied by the cosmological constant in Einstein's equation. Both of these conservation laws for baryon and electron-lepton quantum numbers are experimentally well established in particle physics, [6] in contrast to the cosmological constant.

2. Cosmic Lee-Yang Repulsive Force

The new cosmic Lagrangian of baryons should be expressed in terms of up- and down-quarks with baryon number 1/3. It also involves two massless U(1) gauge fields, $B^\mu(x)$ and $L^\mu(x)$, associated with the conserved baryon and electron-lepton numbers respectively. The new gauge invariant cosmic Lagrangian L_C can be written as [2,7]

$$L_C = L_{q\psi} + L_{BL}, \tag{2}$$

where $L_{q\psi}$ is the usual quark-electron Lagrangian [6] and L_{BL} is given by

$$L_{BL} = -\frac{1}{4} B_{\mu\nu} B^{\mu\nu} - \frac{1}{4} L_{\mu\nu} L^{\mu\nu}$$

$$B_{\mu\nu} = \partial_\mu B_\nu - \partial_\nu B_\mu, \quad L_{\mu\nu} = \partial_\mu L_\nu - \partial_\nu L_\mu.$$

In analogy with electrodynamics, the leptonic gauge field $L_\mu(x)$ is coupled to the current $j_\mu(x) = \overline{\psi}(x)\gamma_\mu\psi(x)$ with the coupling constant g_e; and the gauge field $B^\mu(x)$ is coupled to the up- and down-quark currents $J^q_\mu = \overline{u}_n\gamma_\mu u_n + \overline{d}_n\gamma_\mu d_n$ with the coupling constant $g_b/3$, where the index n denotes summation over color quantum numbers. The structure of the quark-electron Lagrangian $L_{q\psi}$ is formally the same as that in quantum electrodynamics.

The force between two ordinary objects with masses M_1 and M_2 should involve the repulsive Lee-Yang force between quarks and between electrons in all galaxies in the universe, provided galaxies are not made of anti-matter. The Lee-Yang forces originated from baryon and lepton number conservation imply that $f(r) = 1/r^2$ in the total cosmic force $F_C{}^*$ in (1). Thus, the total force F_C between M_1 and M_2 is [7]

$$F_C = -\frac{GM_1M_2}{r^2}(1-Q), \qquad Q = \frac{g_b^2 + g_e^2 Z_1 Z_2/(A_1A_2)}{Gm_p^2}, \tag{3}$$

where A_1 (A_2) and Z_1 (Z_2) are respectively the mass number and the atomic number of the body with mass M_1 (M_2). We can assume that the magnetic dipole-like interaction is negligible because the net spin of atoms in the bodies is usually averaged to nearly zero. [2]

Apart from the simplest Lagrangian which is quadratic in the fields strength $B_{\mu\nu}$ in (2), there is another simple gauge invariant Lagrangian L_D which is quadratic in $\partial_\lambda B_{\mu\nu}$. It is interesting that this simple gauge invariant Lagrangian can lead to a linear potential, $\propto r$, which resembles to the quadratic potential associated with the cosmological constant. Let us consider this simple gauge invariant Lagrangian: [7]

$$L_D = -\frac{L_s^2}{4}\partial_\lambda B_{\mu\nu}\partial^\lambda B^{\mu\nu} + L_{q\psi} + \frac{g_b}{3}B^\mu J^q_\mu + ..., \tag{4}$$

where $J^q_\mu = \overline{u}_n\gamma_\mu u_n + \overline{d}_n\gamma_\mu d_n$. The generalized gauge field equation takes the form

$$-\partial^2\partial_\alpha B^{\alpha\beta} + g'_b J^\beta = 0, \qquad g'_b = g_b/(3L_s^2), \tag{5}$$

where the field strength $B_{\mu\nu}$ satisfies the Bianchi identity,

$$\partial^\lambda B^{\mu\nu} + \partial^\mu B^{\nu\lambda} + \partial^\nu B^{\lambda\mu} = 0. \tag{6}$$

In the 'Coulomb-like gauge' $\partial_k B^k = 0$, the static exterior potential satisfies the equation $\nabla^2\nabla^2 B^0 = 0$. The static gauge potential B^0 produced by, say, some point-like sources can be written in the form

$$B_0 = \frac{A'}{r} + B'r + C'r^2. \tag{7}$$

This result may be considered as a generalization of the original Lee-Yang static potential $\propto r^{-1}$ which can be derived from the Lagrangian (2). Different r-dependent terms in (7) corresponds to sources with different types of singularity at $r = 0$ or different boundary conditions at infinity, $r \to \infty$. For large distances, the last term $C'r^2$ dominates.

However, from the viewpoint of quantum field theory, it appears natural to have the usual source in (5) represented by a delta-function, then the solution for the potential B_0 is proportional to r rather than r^2. For example, in quantum electrodynamics, the static Coulomb potential produced by a delta-function source can be related to the Fourier transform of a special Feynman propagator of the photon: $1/(4\pi r) = \int [(1/\mathbf{k}^2) exp(-i\mathbf{k}\cdot\mathbf{r})] d^3k/(2\pi)^3$. This is basic and significant because the electric force is originated from the exchange of virtual photons between two charges, according to quantum electrodynamics in which the photon satisfies a second order differential equation. However, the present theory based on the gauge invariant Lagrangian (4) is different. Since the field equation of B_μ in (5) is of the fourth order, the corresponding Fourier transform of $(1/\mathbf{k}^2)^2$ leads to the static potential of the form,

$$\int_{-\infty}^{\infty} \frac{1}{(\mathbf{k}^2)^2} e^{i\mathbf{k}\cdot\mathbf{r}} d^3k = -\pi^2 r, \tag{8}$$

in the sense of generalized functions. [8] The potential (8) leads to a constant acceleration caused by the repulsive force and modifies Newtonian law,

$$\frac{d^2\mathbf{r}}{dt^2} = \mathbf{g} - \mathbf{g}_u, \tag{9}$$

where $|\mathbf{g}_u| = B' \approx g_b^2/(72\pi L_s^2)$ is the universal constant acceleration associated with the fourth-order gauge invariant field equation. It implies the existence of a universal constant acceleration between any two galaxies, provided other forces (e.g., the 'magnetic-type' force) are negligible. This is an interesting prediction of U(1) gauge symmetry together with the fourth-order gauge invariant field equation (5). [9] Such a constant cosmic repulsive force can be stronger than the usual gravitational force at large distances and can lead to the accelerated expansion of the universe.

3. The Wu Transformation of Spacetime Coordinates and Wave 4-Vector

A method of studing the constant cosmic repulsive force and the related accelerated expansion of the universe is using the Wu-Doppler effect which involves constant-linear-accelerations of observers and sources.

Suppose a distant galaxy is idealized as an atom which is located at the origin of a CLA frame $F(w', x', y', z')$ which moves with a velocity $\beta' = \beta'_o + \alpha'_o w'$ and a constant acceleration α_o along the +x-direction. Suppose the observer is at rest in another CLA frame $F(w, x, y, z)$ which moves with a velocity $\beta = \beta_o + \alpha_o w$ and an acceleration α_o. From the viewpoint of an inertial frame $F_I(w_I, x_I, y_I, z_I)$, both CLA frames F and F' move with constantlinear-acceleration. We assume as usual that the origins of all three frames coincide at time $w_I = w = w' = 0$.

The accelerated Wu transformations for these frames are given by [10]

$$
\begin{aligned}
w_I = \gamma\beta(x + \frac{1}{\alpha_o\gamma_o^2}) - \frac{\beta_o}{\alpha_o\gamma_o} &= \gamma'\beta'(x' + \frac{1}{\alpha'_o\gamma'^2_o}) - \frac{\beta'_o}{\alpha'_o\gamma'_o}, \qquad (10)\\
x_I = \gamma(x + \frac{1}{\alpha_o\gamma_o^2}) - \frac{1}{\alpha_o\gamma_o} &= \gamma(x + \frac{1}{\alpha'_o\gamma'^2_o}) - \frac{1}{\alpha'_o\gamma'_o},\\
y_I = y = y', \qquad & z_I = z = z'.
\end{aligned}
$$

By differentation of these transformations, we obtain the transformations of the differentials dx_I^μ, dx^μ and dx'^μ,

$$
\begin{aligned}
dw_I &= \gamma(Wdw + \beta dx) = \gamma'(W'dw' + \beta' dx'), \qquad (11)\\
dx_I &= \gamma(dx + \beta W dw) = \gamma'(dx' + \beta' W' dw'),\\
dy_I &= dy = dy', \qquad dz_I = dz = dz';\\
W &= \gamma^2(\gamma_o^{-2} + \alpha_o x), \qquad W' = \gamma'^2(\gamma'^{-2}_o + \alpha'_o x'),
\end{aligned}
$$

It can be shown that the contravariant momentum 4-vector p^μ (or the contravariant wave 4-vector k^μ) transforms like the coordinate differential dx^μ,based on the invariant action $S = -\int m ds$, $ds^2 = W^2 dw^2 - dx^2 - dy^2 - dz^2$. Thus we have

$$
\begin{aligned}
k_I^0 &= \gamma(Wk^0 + \beta k_x) = \gamma'(W'k'^0 + \beta' k'_x), \qquad (12)\\
k_{Ix}^0 &= \gamma(k_x + \beta W k^0) = \gamma'(k'_x + \beta' W' k'^0),\\
k_{Iy} &= k_y = k'_y, \qquad k_{Iz} = k_z = k'_z,
\end{aligned}
$$

where $k^\mu = (k^0, k^1, k^2, k^3) = (k^0, k_x, k_y, k_z)$. The wave 4-vector satisfies the invariant relation $k_\mu k^\mu = W^2 k^{02} - \mathbf{k}^2 = 0$, $k_\mu = g_{\mu\nu}k^\nu$, $g_{\mu\nu} = (W^2, -1, -1, -1)$.

Let us consider the Wu-Doppler effect of the wavelength $\lambda = 2\pi/|\mathbf{k}|$, where $\mathbf{k} = (k_x, 0, 0)$. Eq. (12) leads to the Wu-Doppler effect,

$$\frac{1}{\lambda_I} = \frac{1}{\lambda}\gamma(1+\beta) = \frac{1}{\lambda'}\gamma'(1+\beta'), \tag{13}$$

$$\gamma = \frac{1}{\sqrt{1-\beta^2}}, \qquad \gamma' = \frac{1}{\sqrt{1-\beta'^2}}$$

because $W'k'^0 = |k'_x|$ and $Wk^0 = |k_x|$. Since the atom is at rest in the CLA frame F', let us denote $\lambda' = \lambda'_o$. The observed wavelength in F_I is λ_I,

$$\lambda_I = \lambda'_o \frac{1}{\gamma'(1+\beta')} = \lambda'_o \sqrt{\frac{1-\beta'}{1+\beta'}} \tag{14}$$

The accelerated frame F is not equivalent to the inertial frame F_I. The wavelength λ'_o emitted by an atom at rest in the CLA frame F' is, strictly speaking, different from the wavelength λ_{Io} emitted from the same kind of atom at rest in the inertial frame F_I. However, if the acceleration α'_o is sufficiently small, one has the approximate relation $\lambda_{Io} = \lambda'_o$. In this approximation one has

$$\lambda_I = \lambda_{Io}\sqrt{\frac{1-\beta'}{1+\beta'}}, \qquad \beta' = \beta'_o + \alpha'_o w' \tag{15}$$

In such an approximation, the Wu-Doppler shift of the wavelength depends only on the velocity $\beta' = \beta'_o + \alpha'_o w'$, similar to that in special relativity. The only difference is that the velocity β' in (15) is time dependent. The acceleration α'_o has a very small numerical value because it is related to the usual acceleration a'_o (measured in cm/s^2) by the relation $\alpha'_o = a'_o/c^2$, where c is the speed of light in an inertial frame. The numerical value of λ'_o can be obtained either by experiment or by theoretical calculations of atomic energy level in a CLA frame. [11]

The Wu-Doppler effect in the accelerated expansion of the universe is more complicated because the observed wavelength was emitted from the source in F' frame billions years ago and both the observer and the source have accelerations. Thus, the result (15) cannot be applied to the expansion of the universe.

First, suppose the atom at rest at the origin of F' emitted a light wave with the wavelength λ'_o at time $w' = w'_{em}$. The corresponding time $w_{I,em}$ in the inertial frame F_I is given by

$$w_{I,em} = [\gamma'\beta'(x' + \frac{1}{\alpha'_0 {\gamma'}_0^2}) - \frac{\beta'_0}{\alpha'_0\gamma'_0}]|_{x'=0,w'=w'_{em}} \tag{16}$$

At that instant of time, the wavelength measured in F_I (near the moving source) is

$$\frac{1}{\lambda_{I,em}} = \frac{1}{\lambda'_o}\gamma'(1+\beta') \tag{17}$$

This light wave with $\lambda_{I,em}$ propagates for billions of years in vacuum without changing its wavelength, as observed in the inertial frame F_I. Suppose this wavelength $\lambda_{I,ob} = \lambda_{I,em}$ is measured by observers in the CLA frame F after billions of years at time $w_I = w_{I,ob} \approx w_{I,em} + billions \quad of \quad years$. According to the the Wu-Doppler effect (13), we have

$$\frac{1}{\lambda_{I,ob}} = \frac{1}{\lambda_{I,em}} = \frac{1}{\lambda_{ob}}\gamma(1+\beta)|_{w=w_{ob}}, \tag{18}$$

where w_{ob} is related to $w_{I,ob}$ by the relation

$$\begin{aligned} w_{I,ob} &= [\gamma\beta(x + \frac{1}{\alpha_0\gamma_0^2}) - \frac{\beta_0}{\alpha_0\gamma_0}]|_{w=w_{ob}} \\ &= [\gamma'\beta'(x' + \frac{1}{\alpha'_0\gamma'^2_0}) - \frac{\beta'_0}{\alpha'_0\gamma'_0}]|_{w'=w'_{ob},x'=0} \end{aligned} \tag{19}$$

Thus, the Wu-Doppler effect for the light wave emitted at time $w' = w'_{em}$ and observed at time $w' = w'_{ob} \approx w'_{em} + billion \quad years$ is given by (17) and (18):

$$\begin{aligned} \frac{1}{\lambda_{I,ob}} &= \frac{1}{\lambda_{ob}}\gamma(1+\beta)|_{w=w_{ob}}, \\ &= \frac{1}{\lambda'_o}\gamma'(1+\beta')|_{w'=w'_{ob}} \end{aligned} \tag{20}$$

This equation by itself is not very useful because it involves three unknown quantities, $\lambda'_o, \beta, \beta'$. To get more information, the observer in F can make a second observation, say, 50 years later, $w_I = w_{I,ob} + 50years \equiv w_{I,ob2}$, we have

$$\frac{1}{\lambda_{ob2}} = \frac{1}{\lambda'_0}\frac{\gamma'(1+\beta')|_{w'=w'_{em2}}}{\gamma(1+\beta)|_{w=w_{ob2}}} \tag{21}$$

where $w'_{em2} = w'_{em} + 50years$. It follows from (20) and (21) that the Wu-Doppler effect as measured in the CLA frame F is

$$\Delta\lambda = \lambda_{ob2} - \lambda_{ob} = \lambda'\{\frac{\gamma(1+\beta)|_{w=w_{ob2}}}{\gamma'(1+\beta')|_{w'=w'_{em2}}} - \frac{\gamma(1+\beta)|_{w=w_{ob}}}{\gamma'(1+\beta')|_{w'=w'_{em}}}\} \tag{22}$$

as measured in the CLA frame F. Let us assume the universal cosmic acceleration, $\alpha'_o = \alpha_o$, as suggested by (9). This implies that the two CLA frames F and F' are "equivalent" in the sense that if they start from the same initial velocity, then they would be the same CLA frame, so that the wavelength λ_o of the radiation emitted from the atom at rest in F should be the same as λ'_o in (14) $\lambda_0 = \lambda'_0$.

Therefore, the result (22) enables us to obtain $\Delta\lambda = \lambda_{ob2} - \lambda_{ob}$

$$\Delta\lambda = \lambda_0 \frac{\gamma_0(1+\beta_0)}{\gamma'_0(1+\beta'_0)}\{\alpha_0\gamma_0^2(w_{ob2} - w_{ob}) - \alpha_0{\gamma'}_0^2(w'_{em2} - w'_{em})\}. \quad (23)$$

$$\approx \lambda_0 \frac{\gamma_0(1+\beta_0)}{\gamma'_0(1+\beta'_0)}\{(\alpha_0\gamma_0^2 - \alpha_0{\gamma'}_0^2)(w_{ob2} - w_{ob})\}.$$

It is highly unlikely that we will ever know for sure the initial velocities of our galaxy and the distant galaxy which emitted light. Therefore, one can only say that there is an accelerated expansion of the universe because the observed Wu-Doppler effect is time dependent. However, the value of acceleration cannot be determined unambiguously by the accelerated Wu-Doppler effect, unless the initial velocities β_o and β'_o are known.

The work was supported by the J.S. Research Fund of UMD Foundation.

References

1. P. J. E. Peebles and B. Ratra, Rev. Mod. Phys. **75**, 559 (2003).
2. T. D. Lee and C. N. Yang, Phys. Rev. **98**, 1501 (1955).
3. P. G. Roll, R. Krotkov, and R. H. Dicke, Ann. Phys. (N. Y.) **26**, 442 (1964); V. B. Braginsky and V. I. Panov, Zh. Eksp. Teor. Fiz. **61**, 875 (1971) [Sov. Phys. JEPT **34** 463 (1971).
4. J. P. Hsu, Phys. Rev. **D17** 3164 (1978). The test of weak- interaction effects by Eötvös experiment was discussed.
5. Jong-Ping Hsu and Leonardo Hsu, UMassD preprint (2005); see also J. P. Hsu and D. Fine, *100 Years of Gravity and Accelerated Frames–The Deepoest Insights of Einstein and Yang-Mills* (World Scientific, 2005) p. xxiii.
6. T. D. Lee, *Particle Physics and Introduction to Field Theory* (Harwood, Chur, 1981) pp. 210-213.
7. Jong-Ping Hsu, Mod. Phys. Lett. **20** 2855 (2005).
8. I. M. Gel'fand and G. E. Shilov, *Generalized Functions* vol. 1 (Academic Press, New York, 1964), p. 363.
9. For a discussion of quark confinement based on the foiurth-order field equations, see Jong-Ping Hsu, Nuovo Cimento **91B**, 100 (1986).
10. Jong-Ping Hsu and Leonardo Hsu, Nuovo Cimento **112**, 575 (1997) and Chin. J. Phys. **35**, 407 (1997). This is called the Wu transformation to honor Ta-You Wu's idea of a kinematic approach to finding an accelerated transformation in a spacetime with a vanishing Riemann curvature tensor.
11. Jong-Ping Hsu, *Einstein's Relativity and Beyond* (World Scientific, 2000) p. 310; A. Ernst and J. P. Hsu, Chin. J. Phys. **39**, 122 (2001).

Accelerating Expansion from Inhomogeneities?*

Je-An Gu
Department of Physics, National Taiwan University,
Taipei 10617, Taiwan, R.O.C.
E-mail: jagu@phys.ntu.edu.tw

It is the common consensus that the expansion of a universe always slows down if the gravity provided by the energy sources therein is attractive. To examine this point we find counter-examples for a spherically symmetric dust fluid described by the Lemaitre-Tolman-Bondi solution. As suggested by these counter-intuitive examples, the effects of inhomogeneities on the evolution of the space-time geometry (such as the cosmic evolution) should be restudied, and the intuition about general relativity is yet to be built.

1. Introduction

Since the end of the last century (i.e. 1990s), as benefiting from the advanced high-resolution astronomical observations, there has been a huge progress in cosmology. One of the most amazing and mysterious discoveries is the accelerating expansion of the present universe. The existence of the cosmic acceleration at the present epoch was first suggested in 1998 by type Ia supernova (SN Ia) data[1,2], and was further reinforced recently by updated SN Ia data[3–6] and WMAP measurement[7] of cosmic microwave background (CMB). Most of the explanations for this mysterious phenomenon invoke some exotic energy source, as generally called "dark energy", which provides significant negative pressure and repulsive gravity to drive the expansion to accelerate.

Both the existence of the present cosmic acceleration and the need to invoke dark energy to describe it are based on the Friedmann-Robertson-Walker (FRW) cosmology where the universe is assumed to be homogeneous and isotropic and is accordingly described by the Robertson-Walker (RW)

*This work is in collaboration with Chia-Hsun Chuang and W-Y. P. Hwang, and is supported by the National Science Council, Taiwan, R.O.C. (NSC 94-2811-M-002-048, NSC 94-2112-M-002-036, NSC 94-2752-M-002-007-PAE and NSC 94-2112-M-002-029).

metric. But, apparently, our universe is NOT homogeneous and isotropic. (Galaxies, the solar system, and we are definitely far from being homogeneous.) Nevertheless, at large scales (say, hundreds Mpc), after averaging in space, the universe is roughly homogeneous and isotropic. Many people therefore believe that for describing the large-scale or cosmological-scale phenomena (such as the evolution our universe) homogeneity and isotropy is a good assumption and the RW metric is a good approximation. This simple model invoking RW metric for a homogeneous and isotropic universe has been widely used in cosmology, even though so far there is no rigorous proof supporting the validity of this simplification.

The above is a concern about the interpretation of SN Ia data in the FRW cosmology, i.e., including the existence of the present cosmic acceleration and the existence of dark energy. More precisely, this concern corresponds to the following two questions:

(1) Do SN Ia data really suggest the existence of the cosmic acceleration?
(2) Do we really need dark energy to generate accelerating expansion?

In the work to be presented in this article, we are particularly interested in the second question.

It is the common consensus that normal matter (such as protons, neutrons, electrons, etc.) can only provide attractive gravity and therefore should always slow down the cosmic expansion, i.e.,

$$\text{Normal Matter} \Longrightarrow \text{Attractive Gravity} \Longrightarrow \text{Deceleration}\,. \tag{1}$$

Thus, to explain the surprising, mysterious phenomenon of the accelerating expansion, it seems inevitable to invoke dark energy. In our work we try to challenge this common consensus, in particular, through studying carefully the effect of inhomogeneity.

It was suggested that the inhomogeneities of the universe might be able to induce the cosmic acceleration.[8–27] Is this proposal of the inhomogeneity-induced accelerating expansion possible? There are two directions for studying this issue. One is to prove the no-go theorem forbidding it. The other is to find counter-examples to the common consensus in Eq. (1), which is the approach we took in our study.

For countering the common intuition about Eq. (1) we found examples[28] of accelerating expansion in the case of the spherically symmetric dust fluid described by the Lemaitre-Tolman-Bondi (LTB) solution[29]. Note that these examples, although probably having nothing to do with the reality, are to reveal the mathematical possibility of inducing accelerating expansion from inhomogeneities, thereby giving us a hint about how our understanding of the interplay of gravity and the space-time (geometry) evolution may go wrong. They open a new perspective for understanding the cosmic evolution. In the following sections I will introduce the LTB solution and study the examples of two kinds of acceleration based on this metric.

2. Lemaitre-Tolman-Bondi (LTB) Solution

The LTB solution[29] is an exact solution of the Einstein equations for a spherically symmetric dust fluid. The metric can be written in a form similar to that of the RW metric, as follows:

$$ds^2 = -dt^2 + a^2(t,r)\left[\left(1+\frac{a_{,r}\,r}{a}\right)^2 \frac{dr^2}{1-k(r)r^2} + r^2 d\Omega_2^2\right]. \qquad (2)$$

With this metric the Einstein equations can be reduced to two equations:

$$\left(\frac{\dot{a}}{a}\right)^2 = -\frac{k(r)}{a^2} + \frac{\rho_0(r)}{3a^3}, \qquad (3)$$

$$\rho(t,r) = \frac{(\rho_0 r^3)'}{6a^2 r^2 (ar)_{,r}}. \qquad (4)$$

The solution of Eq. (3) can be written parametrically by using the variable, $\eta = \int dt/a$, as follows:

$$a(\eta,r) = \frac{\rho_0(r)}{6k(r)}\left[1-\cos\left(\sqrt{k(r)}\,\eta\right)\right], \qquad (5)$$

$$t(\eta,r) = \frac{\rho_0(r)}{6k(r)}\left[\eta - \frac{1}{\sqrt{k(r)}}\sin\left(\sqrt{k(r)}\,\eta\right)\right] + t_b(r). \qquad (6)$$

In the above equations there are three arbitrary functions: $k(r)$, $\rho_0(r)$ and $t_b(r)$. With different choices of these three functions, we get different LTB solutions.[a] In search of examples of accelerating expansion for a dust fluid described by the LTB solution, we tried a variety of LTB solutions corresponding to different choices of these three functions and eventually found examples among these tedious trials.[28]

3. Acceleration Induced by Inhomogeneity

For a given length quantity L, the expansion rate and the deceleration parameter of this length are defined as follows: $H \equiv \dot{L}/L$; $q \equiv -\ddot{L}/(LH^2) =$

[a]When $a(t,r)$ and these three functions have no dependence on the radial coordinate r, we retrieve the RW metric from Eq. (2) and the Friedmann equation from Eq. (3).

$-\ddot{L}L/\dot{L}^2$. In our study we consider two different length quantities corresponding to two definitions of acceleration. One is the proper distance of two points in space, corresponding to the acceleration to be called "Line Acceleration". The other is the cubic root of the volume of a domain in space, corresponding to the "Domain Acceleration", which has been used in Refs. 9, 11, 23 and 30. In the following I will study the examples of acceleration we found in Ref. 28 for both definitions.

3.1. *Line Acceleration*

For studying the inhomogeneity-induced acceleration based on the LTB solution, due to the spherical symmetry, there can be inhomogeneity and the resultant acceleration only in the radial direction. Thus, for the line acceleration we focus on the radial proper distance between the origin ($r = 0$) and the point at $r = r_L$: $L_r(t) \equiv \int_0^{r_L} \sqrt{g_{rr}}dr$.

Regarding the choice of the three arbitrary functions, $k(r)$, $\rho_0(r)$ and $t_b(r)$, for simplicity, here we introduce inhomogeneity only through $k(r)$:

$$k(r) = -\frac{(h_k + 1)(r/r_k)^{n_k}}{1 + (r/r_k)^{n_k}} + 1\,, \tag{7}$$

while employing constant $\rho_0(r)$ and vanishing $t_b(r)$. With this choice we have six free parameters to tune: $(t, r_L, \rho_0, r_k, n_k, h_k)$. In search of examples of the line acceleration in the radial direction, we surveyed this six-dimensional parameter space and eventually found them. One of the examples with significant acceleration, $q_r \sim -1$, is as follows: $(t, r_L, \rho_0, r_k, n_k, h_k) = (1, 1, 1, 0.7, 20, 1)$.[b] For detailed investigations of this example and for more examples, in particular, showing how q_r changes with the six parameters, see Ref. 28.

3.2. *Domain Acceleration*

In the following we study the domain acceleration for a spherical domain, $0 < r < r_D$, with the volume V_D.

We first made the same choice of $k(r)$, $\rho_0(r)$ and $t_b(r)$ as that for the line acceleration, but found no domain acceleration. Contradicting our result, in Ref. 30 it was claimed that the examples of the domain acceleration were found with the choice of the function $k(r)$ as a step function, i.e.,

[b]Note that in the FRW cosmology the deceleration parameter of the present universe, as suggested by observational data, is on the order of unity and is negative in sign.

with infinitely large n_k in Eq. (7). There is a mistake in the calculations of the volume V_D in Ref. 30, where the authors ignored the volume at the transition point $r = r_k$ that is actually nonzero (even though $r = r_k$ is a "point" in the coordinate space) and cannot be ignored. After taking the volume at $r = r_k$ back into consideration, we found no domain acceleration.

We then consider a non-trivial function for $t_b(r)$:

$$t_b(r) = -\frac{h_{tb}(r/r_t)^{n_t}}{1 + (r/r_t)^{n_t}}, \tag{8}$$

while invoking constant $\rho_0(r)$ and the same $k(r)$ in Eq. (7). With this choice we have nine free parameters to tune: $(t, r_D, \rho_0, r_k, n_k, h_k, r_t, n_t, h_{tb})$. In search of examples of the domain acceleration, we surveyed this nine-dimensional parameter space and eventually found them. One of these examples with significant acceleration, $q_D \sim -1$, is as follows: $(t, r_D, \rho_0, r_k, n_k, h_k, r_t, n_t, h_{tb}) = (0.1, 1.1, 10^5, 0.9, 40, 40, 0.9, 40, 10)$. For detailed investigations of this example and for more examples, in particular, showing how q_D changes with the nine parameters, see Ref. 28.

4. Discussions

Regarding the energy density distribution, in every example of acceleration we found, the spherically symmetric dust fluid consists of three regions: two roughly homogeneous regions — the inner over-density region with positive $k(r)$ and smaller $a(t,r)$ and the outer under-density region with negative $k(r)$ and larger $a(t,r)$ — and one transition or junction region, where the inhomogeneity locates, of these two homogeneous regions.

For further understanding the acceleration, the quantity $\partial_t^2 \sqrt{g_{rr}}$, which indicates the acceleration/deceleration status of an infinitesimal radial line element, is one of the good quantities to study. The regions with positive/negative $\partial_t^2 \sqrt{g_{rr}}$ make positive/negative contribution to acceleration. In every example of acceleration we found, there exists a region with positive $\partial_t^2 \sqrt{g_{rr}}$ that coincides with the inhomogeneous transition/junction region quite well. This result reveals the strong correlation between acceleration and inhomogeneity, thereby giving a strong support for the suggestion that inhomogeneity can induce accelerating expansion.

For the quantity $\partial_t^2 \sqrt{g_{rr}}$ to be positive (accordingly making a positive contribution to acceleration), we found a necessary and sufficient condition: $(a^2/r)_{,r} > 0$. This condition tells us that a positive contribution to acceleration is made in the place where $a(t,r)$ increases sufficiently fast with the radial coordinate r.

5. Summary and Perspectives

Against the common intuition and consensus in Eq. (1), we found counter-examples of the line and the domain acceleration, which are based on the LTB solution for a dust fluid of spherical symmetry. These examples strongly support the suggestion that inhomogeneity can induce acceleration.

These examples raise two issues worthy of further investigations:

(1) Can inhomogeneities explain "cosmic acceleration"?
(2) How to understand these counter-intuitive examples?

Regarding the first issue, the most important is whether the inhomogeneities of our universe can explain SN Ia data. If yes, the next step is to see, according to observational data, how the universe evolves, in particular, whether the cosmic acceleration exists or not, in this cosmological model taking inhomogeneities into consideration. Even if the inhomogeneities can not solely explain SN Ia data, the effects of inhomogeneities on the cosmic evolution should be restudied.

Regarding the second issue, the common intuition about Eq. (1) may actually stem from the Newtonian gravity, and is valid only for describing the gravitational interaction between particles or energy density fluctuations/clumps in a (fixed) background space-time geometry, but has nothing to do with the evolution the background space-time that is described by general relativity. Here involve two different entities: (i) particle motion and (ii) background space-time evolution. Both should play a role in the acceleration/deceleration of a length. However, the common intuition about Eq. (1) may touch only the particle motion, and is therefore invalid in general, especially when the general-relativity effect is important or dominates.

How about the intuition about general relativity? More precisely, for an arbitrarily given energy-momentum tensor, can we make a proper guess at the behavior of the space-time geometry in an intuitive way? In my personal opinion (if excluding the intuition from the Newtonian gravity) we have no intuition (purely) about general relativity. To further understand the cosmic evolution and other topics involving general-relativity effects, the intuition about general relativity should be rebuilt.

After 90 years, general relativity is still not fully understood!

References

1. S. Perlmutter *et al.* [Supernova Cosmology Project Collaboration], *Astrophys. J.* **517**, 565 (1999) [arXiv:astro-ph/9812133].

2. A. G. Riess *et al.* [Supernova Search Team Collaboration], *Astron. J.* **116**, 1009 (1998) [arXiv:astro-ph/9805201].
3. J. L. Tonry *et al.* [Supernova Search Team Collaboration], *Astrophys. J.* **594**, 1 (2003) [arXiv:astro-ph/0305008].
4. R. A. Knop *et al.* [The Supernova Cosmology Project Collaboration], *Astrophys. J.* **598**, 102 (2003) [arXiv:astro-ph/0309368].
5. B. J. Barris *et al.*, *Astrophys. J.* **602**, 571 (2004) [arXiv:astro-ph/0310843].
6. A. G. Riess *et al.* [Supernova Search Team Collaboration], *Astrophys. J.* **607**, 665 (2004) [arXiv:astro-ph/0402512].
7. C. L. Bennett *et al.*, *Astrophys. J. Suppl.* **148**, 1 (2003) [arXiv:astro-ph/0302207]; D. N. Spergel *et al.* [WMAP Collaboration], *Astrophys. J. Suppl.* **148**, 175 (2003) [arXiv:astro-ph/0302209].
8. M. N. Celerier, *Astron. Astrophys.* **353**, 63 (2000) [arXiv:astro-ph/9907206].
9. T. Buchert, *Gen. Rel. Grav.* **9**, 306 (2000) [arXiv:gr-qc/0001056].
10. H. Iguchi, T. Nakamura and K. i. Nakao, *Prog. Theor. Phys.* **108**, 809 (2002) [arXiv:astro-ph/0112419].
11. D. Palle, *Nuovo Cim.* **117B**, 687 (2002) [arXiv:astro-ph/0205462].
12. D. J. Schwarz, arXiv:astro-ph/0209584.
13. R. H. Brandenberger, arXiv:hep-th/0210165.
14. G. Bene, V. Czinner and M. Vasuth, arXiv:astro-ph/0308161.
15. S. Rasanen, *JCAP* **0402**, 003 (2004) [arXiv:astro-ph/0311257].
16. S. Rasanen, *JCAP* **0411**, 010 (2004) [arXiv:gr-qc/0408097].
17. E. W. Kolb, S. Matarrese, A. Notari and A. Riotto, *Phys. Rev.* **D71**, 023524 (2005) [arXiv:hep-ph/0409038].
18. E. Barausse, S. Matarrese and A. Riotto, *Phys. Rev.* **D71**, 063537 (2005) [arXiv:astro-ph/0501152].
19. E. W. Kolb, S. Matarrese, A. Notari and A. Riotto, arXiv:hep-th/0503117.
20. A. Notari, arXiv:astro-ph/0503715.
21. S. Rasanen, arXiv:astro-ph/0504005.
22. J. W. Moffat, arXiv:astro-ph/0505326.
23. E. W. Kolb, S. Matarrese and A. Riotto, arXiv:astro-ph/0506534.
24. T. Buchert, *Class. Quant. Grav.* **22**, L113 (2005) [arXiv:gr-qc/0507028].
25. H. Alnes, M. Amarzguioui and O. Gron, arXiv:astro-ph/0512006.
26. K. Bolejko, arXiv:astro-ph/0512103.
27. R. Mansouri, arXiv:astro-ph/0512605.
28. C.-H. Chuang, Je-An Gu and W-Y. P. Hwang, arXiv:astro-ph/0512651.
29. G. Lemaitre, *Annales Soc. Sci. Brux. Ser. I Sci. Math. Astron. Phys.* **A53**, 51 (1933); R. C. Tolman, *Proc. Nat. Acad. Sci.* **20**, 169 (1934); H. Bondi, *Mon. Not. Roy. Astron. Soc.* **107**, 410 (1947).
30. Y. Nambu and M. Tanimoto, arXiv:gr-qc/0507057.

Thermalization in the Inflationary Universe*

Sang Pyo Kim
Department of Physics
Kunsan National University
Kunsan 573-701, Korea
E-mail: sangkim@kunsan.ac.kr

The inflationary Big Bang model based on general relativity, inflation and cosmological principle is consistent with the current observational data from WMAP. In this talk I review some issues in the inflationary Big Bang model and discuss hot thermal inflation in detail.

1. Introduction

A century ago Einstein proposed general relativity, a theory of curved spacetime and matter distribution, which is aesthetically formulated by the Einstein equation $G_{\mu\nu} = 8\pi T_{\mu\nu}$. A matter distribution warps the spacetime, which in turn governs the motion of the matter. General relativity since then has been confirmed in many tests and observations. The most challenging arena for general relativity is black holes in small scale and cosmology in large scale. General relativity has been the founding framework for modern cosmology. Since Hubble's observation of expansion of the universe in nineteen twenties, cosmology has advanced to a precise science at the turn of twenty century and the new millennium, culminating in observations by Wilkinson Microwave Anisotropy Probe (WMAP).[1,2]

The universe, if dated backward in time, would become smaller, hotter and denser in the past. This implies that the universe would have had some origin, Big Bang model being the champion for such a history of the universe. General relativity, the dynamical law of the universe, and the cosmological principle, the uniform distribution of matters on all scales, are two key ingredients of Big Bang model (for review and references, see Ref. 3). The expansion of the universe and the abundances of light elements

*This work is supported by the Korea Research Foundation Grant funded by the Korea Government (MOEHRD) under Grant No. KRF-2004-041-C00840

such as H, He, Li, etc. are evidences for Big Bang model. Big Bang model, however, does not explain in a natural way the homogeneity, isotropy of the universe and the large and small structures such as galaxies, clusters, superclusters, great wall and so on. Moreover, Big Bang model is plagued with monopole and fine tuning problems from the underlying particle physics.

Inflation models not only overcome most of problems of Big Bang model but also fit well with observations (for review and references, see Ref. 4). In the early evolution the universe would have undergone a period of inflation to stretch the spacetime fluctuations to provide scale invariant density perturbations and to dilute monopoles and other heavy particles contradicting with current observations. And immediately after and/or long after inflation, the inflationary universe should fit with Big Bang model to explain matter contents of the present universe. Therefore, a complete and consistent theory of cosmology requires both inflation and Big Bang, inflationary Big Bang, a variant of which is thermal inflation.

2. Inflation Models

Inflation models introduced to overcome the problems of Big Bang model have a period of (quasi-) exponential or power-law expansion of the universe at the Planck or GUT scale. There are several scenarios of inflation: old inflation by Guth,[5] new inflation by Linde[6] and by Albrecht and Steinhardt,[7] chaotic inflation by Linde.[8] There are hybrid inflation and many others. Inflation models predict that the density of energy plus all kinds of matters is close to the critical value and the geometry of the universe is almost flat. They also predict that the power spectrum of the primordial density perturbations are scale-invariant (Harrison-Zeldovich) and the cosmic microwave background radiations are homogeneous and isotropic with a tiny anisotropy.

The universe is described remarkably simply by the homogeneous and isotropic Friedmann-Robertson-Walker universe with a scale factor $a(t)$ and an inflaton (homogeneous scalar field). The corresponding Friedmann equation (time-time component of the Einstein equation) takes the form

$$\left(H = \frac{\dot{a}}{a}\right)^2 + \frac{k}{a^2} = \frac{8\pi}{3}\left(\frac{\dot{\phi}^2}{2} + V(\phi)\right), \tag{1}$$

where H is the Hubble parameter and k takes 1, 0 and -1 for a spatially closed, flat and open universe. The inflaton satisfies the field equation

$$\ddot{\phi} + 3H\dot{\phi} + V'(\phi) = 0. \tag{2}$$

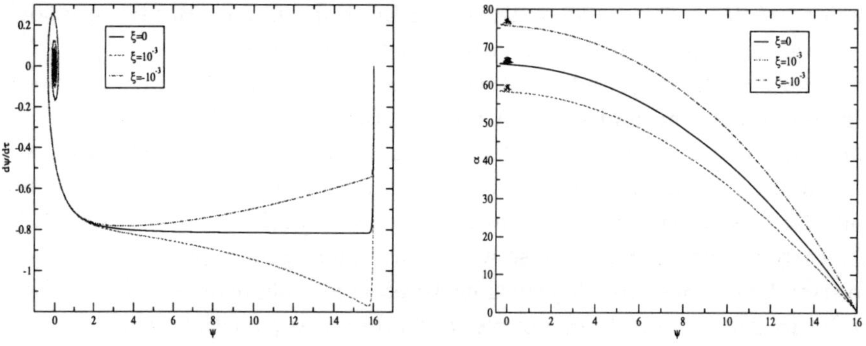

Fig. 1. (Left panel) The phase diagram of $\psi = \sqrt{8\pi G}\phi$ and $d\psi/d\tau$, the inflaton scaled in Planck units, in the nonminimal scalar field theory $V = m^2\phi^2/2 + \xi R\phi^2/2$ for $\xi = -10^{-3}$ (top curve), $\xi = 0$ (middle curve) and $\xi = 10^{-3}$ (bottom curve). (Right panel) The number of e-foldings. All adopted from Ref. 9.

Various inflation models are classified by potentials V and initial states of the inflation.

Chaotic inflation, in which large quantum fluctuations away from the vacuum drive inflation, can be realized by a wide class of potentials. The damping term $3H\dot{\phi}$ drags the inflaton and the potential energy then drives the expansion of the universe. Chaotic inflation is shown in the generalized gravity motivated by many fundamental theories with the potential

$$V(\phi) = \frac{m^2}{2}\phi^2 + \frac{\xi R}{2}\phi^2. \tag{3}$$

Figure 1 shows the phase diagram of the scaled inflaton and the number of e-foldings for different values of nonminimal coupling ξ.[9] In the left panel of Fig. 1, the inflaton deviated from the vacuum by large quantum fluctuations is attracted toward the vacuum and instantaneously gains the kinetic energy (the vertical segment of the curve). As the velocity increases, the damping also increases and drags the inflaton to result in slow-rolling toward the vacuum (almost horizontal segment). During slow-rolling over the potential wall the potential energy dominates the kinetic energy in the Friedmann equation (1) and the universe expands (quasi-) exponentially (right panel of Fig. 1). After inflation, the inflaton oscillates around the vacuum and finally settles down to the vacuum (the in-spiral curve in the left panel).

3. Semiclassical Gravity

The right context to study the evolution of the universe is quantum gravity, by which even the origin of the universe may be explained. We do not know the consistent theory of quantum gravity yet, though string theory or canonical quantum gravity may be a viable theory. Our present universe, however, is precisely described by classical general relativity. Thus, there should be a transition from quantum gravity to classical gravity:

$$\hat{G}_{\mu\nu} = 8\pi\hat{T}_{\mu\nu} \Longrightarrow G_{\mu\nu} = 8\pi\langle\hat{T}_{\mu\nu}\rangle \Longrightarrow G_{\mu\nu} = 8\pi T_{\mu\nu} \tag{4}$$

The quantum-to-classical transition can be understood from the point of view of the wave function of the universe.[10,11,12,13] Gravity at Planck scale should have first decohered and become classical but matters maintained quantum nature, the so-called semiclassical gravity. Quantum matters finally gain classicality through decoherence mechanism.[14]

The inflation period belongs to the semiclassical regime, where a classical spacetime equated to the expectation value of stress-energy tensor and matter fields obey the time-dependent functional Schrodinger equation

$$G_{\mu\nu} = 8\pi\langle\hat{T}_{\mu\nu}\rangle, \quad i\frac{\delta}{\delta t}\Psi = \hat{H}_m(t)\Psi. \tag{5}$$

Due to the expansion of the spacetime, matter fields evolve out of equilibrium. The nonequilibrium quantum fields can be described by the time-dependent functional equation[15,16] or by the Schwinger-Keldysh closed-time path integral or Hartree-Fock method or 1/N expansion (for references, see Ref. 15). Here, an issue is what quantum state the inflaton does have just before inflation. Most inflation models assume non-thermal states to drive the inflation. In new inflation, the inflaton slowly rolls down the potential from the initial false vacuum toward the true vacuum. In chaotic inflation, the quantum fluctuations of the inflaton collectively behave as a coherent state and the potential energy of the coherent state drives the inflation.

4. Hot Thermal Inflation

In chaotic inflation, large quantum fluctuations provide the energy for the slow-roll motion of the inflaton and a sufficient energy density necessary for inflation. Another possibility is that the inflaton might have started from a hot thermal state in the early universe as in Big Bang model but undergone a period of inflation, thus hot thermal inflation. The hot thermal inflation model in non-SUSY theory was proposed by Page and myself.[17] This should not be confused with thermal inflation in SUSY theory[18] and warm inflation with dissipation.[19]

In hot thermal inflation it is thermal fluctuations that drives inflation. The initial thermal equilibrium with a temperature $T_0 = 1/(k\beta_0)$

at the moment t_0 before inflation is described by the density operator $\hat{\rho} = e^{-\beta_0 \hat{H}_m(t_0)}/Z$. For instance, in the minimal massive model ($\xi = 0$ in Eq. (3)),[17] the semiclassical Friedmann equation for a flat universe ($k = 0$) takes the form

$$H^2(t) = \frac{8\pi n(t_0)}{3}\Big(\dot{\varphi}^*(t)\dot{\varphi}(t) + m^2\varphi^*(t)\varphi(t)\Big), \tag{6}$$

where

$$n(m, T_0) = \frac{1}{e^{m\beta_0} - 1} + \frac{1}{2}. \tag{7}$$

The first term in n is the Bose-Einstein distribution and the second term is the vacuum fluctuation. The auxiliary function φ satisfies the same classical field equation

$$\ddot{\varphi}(t) + 3H\dot{\varphi}(t) + m^2\varphi(t) = 0, \tag{8}$$

together with the Wronskian condition from quantization rule

$$a^3(t)[\dot{\varphi}^*(t)\varphi(t) - \dot{\varphi}(t)\varphi^*(t)] = i. \tag{9}$$

The modification from classical theory (1) is the overall factor n, which may correspond to the classical field as $\phi = \sqrt{n}\varphi$. Thus hot thermal inflation shares most of the useful properties of chaotic inflation. As shown in Fig. 1, 60-foldings require $\psi = \sqrt{8\pi}\phi/m_p = 16$, so $\varphi = (8/\sqrt{2\pi}) \times (m_p/\sqrt{n(t_0)})$. For a thermal energy at Planck scale, where the universe was really hot, even a small fluctuation of φ of order one may trigger a sufficient inflation. Thus the chance of inflation is more probable in thermal inflation than in chaotic inflation.

With redefining the field as $\zeta = \sqrt{n}\varphi/a^{3/2}$, the semiclassical Friedmann equation can be written as

$$H^2(t) = \frac{8\pi}{3}\Big[\Big(\dot{\zeta}^* - \frac{3}{2}H\zeta^*\Big)\Big(\dot{\zeta} - \frac{3}{2}H\zeta\Big) + m^2\zeta^*\zeta\Big]e^{3\int H}, \tag{10}$$

where the field ζ satisfies

$$\ddot{\zeta}(t) + (m^2 - U(t))\zeta(t) = 0, \quad U(t) = \frac{9}{4}H^2(t) + \frac{3}{2}\dot{H}(t). \tag{11}$$

Equation (11) is a one-dimensional wave equation in the canonical form with energy $E = m^2$. There are oscillatory or exponential solutions depending on whether $m^2 > U$ or $m^2 < U$. The inflation occurs in the latter case, where ζ undergoes a tunneling motion under the potential barrier U. The critical condition is obtained from $m^2 = U$, which leads to the de Sitter phase

$$a(t) = a_{in}(\cosh mt)^{2/3}. \tag{12}$$

This condition is determined by m and T_0. In the very early evolution before inflation (stiff regime or ultra-relativistic regime), the universe expands according to the power-law $a(t) = a_{st}t^{1/3}$ and U becomes negative. On the other hand, in the later evolution the universe enters the matter-dominated era with the power-law $a(t) = a_{ma}t^{2/3}$ and U approaches zero.

During the interim period when U is positive, the universe inflates and the field slowly varies as $\zeta = (e^{\int \Omega} + ie^{-\int \Omega})/\sqrt{2\Omega}$. As the exponential expansion leads to the scale factor (12), the mass and temperature are constrained by the condition $m = 3\pi n(m, T_0)$. Assuming $3\gamma/2 \gg m$, the Hubble parameter is approximated given by

$$H(t) \approx \left[\frac{8\pi m^2 n(t_0)}{9\gamma}\left(e^{-\frac{2\mu^2}{3\gamma}t} + \frac{9\gamma^2}{m^2}e^{-6\gamma t}\right)\right]^{1/2}. \tag{13}$$

On the other hand, in the matter-dominated era the field oscillates as

$$\zeta(t) = \left(\frac{n(t_0)}{a^3(t)}\right)^{1/2}\varphi(t) = \frac{e^{-i\int\Omega}}{\sqrt{2\Omega}}, \tag{14}$$

and the universe expands as

$$a(t) \approx \left[6\pi m n_0\left(1 + \frac{1}{2m^2t^2}\right)\right]^{1/3} t^{2/3}. \tag{15}$$

Here, the oscillation frequency is

$$\Omega(t) \approx m\left[1 - \frac{1}{(2m^2t^2+1)^2}\left(1 - \frac{1}{4m^2t^2}\right)\right]^{1/2}. \tag{16}$$

The power spectrum for density perturbations has the same scale-invariant form as in chaotic inflation.

5. Preheating

At the end of inflation, the inflaton must transfer its energy to other types of matters except a possibly small vacuum energy for the dark energy. Now a question is how the inflaton decays and settles down the vacuum. Thermalization processes proposed so far are preheating and reheating. Preheating is thermalization at the early stage after inflation. While reheating is thermalization during oscillations after inflation.

Preheating is a thermalization process where the inflaton transfers its energy to coupled fields of light particles.[20,21] The simplest model is the inflaton coupled to the field χ through a quartic interaction

$$L_{int} = -\frac{g}{2}\phi^2(t)\chi^2(t,x) \tag{17}$$

While the inflaton oscillates around the vacuum, χ has an effective coupling $m_{eff}^2 = m_\chi^2 + g^2\phi^2(t)$. Then the inflaton decays explosively due to the oscillating coupling and leads to large fluctuations at low momenta for all coupled scalar fields. Thus the occupation numbers of particles grow exponentially in time. This process is a non-thermal phase transition and a stochastic background of gravitational waves may be generated and possibly provide a new mechanism for baryogenesis.[22]

Turning to hot thermal inflation again,[17] the inflaton also oscillates around the vacuum after inflation. During oscillations after the end of inflation, t_{en}, the position dispersion of the inflaton narrows as

$$(\Delta\phi)^2 = 2n(t_{en})\varphi^*(t)\varphi(t) \approx \frac{n(t_{en})}{\Omega} \times \frac{1}{a^3(t)}, \tag{18}$$

while the momentum dispersion spreads as

$$(\Delta\pi)^2 = 2n(t_{en})a^6(t)\dot{\varphi}^*(t)\dot{\varphi}(t) \approx \frac{n(t_0)}{\Omega}\left(\Omega^2 + \frac{9}{4}H^2\right) \times a^3(t). \tag{19}$$

Though the uncertainty decreases toward the minimum value

$$(\Delta\phi)(\Delta\pi) \approx n(t_{en})\left(1 + \frac{9}{4}\frac{H^2}{\Omega^2}\right)^{1/2}, \tag{20}$$

the inflaton is getting classically correlated in the position space (the wave function is sharply peaked) and the quantum state is squeezed.

In the matter-dominated era, the particle number, $\mathrm{Tr}(\hat{\rho}(t_{en})\hat{N}(t))$, from a thermal state at t_{en} is given by

$$N(t, t_{en}) = 2n(t_{en})a^6(t)|\varphi(t)\dot{\varphi}(t_{en}) - \dot{\varphi}(t)\varphi(t_{en})|^2. \tag{21}$$

In the limit $t_{en} \gg 1/m$, one approximately has

$$N(t, t_{en}) \approx \frac{n(t_{en})(t - t_{en})^2}{4m^2 t_{en}^4}. \tag{22}$$

Thus the number of created particles increases as the oscillation goes on and the inflaton transfers its energy, implying that the inflaton cannot continue the coherent oscillation due to the decay. The duration of stable coherent oscillation is limited by $N(t, t_{en}) \leq q$ for some number q not large enough for catastrophic production. That is, the duration is restricted by[23]

$$\Delta t < \left(\frac{q}{n}\right)^{1/2} \frac{T}{\sqrt{2\pi}}, \tag{23}$$

where $T = 2\pi/m$ is the period of oscillation. In hot thermal inflation, preheating is not an efficient mechanism to thermalize the universe after inflation to fit with Big Bang but the intrinsic thermal energy provides the necessary energy.

6. Conclusion

Inflationary Big Bang model is a theory so far consistent with all data from WMAP and other observations.[1,2] The inflation period is relatively well-understood in various inflation models. But the thermalization process during and/or after inflation, where the inflaton decays to other light particles and reheats the universe, has not been completely understood yet. Preheating and reheating are such thermalization processes. Hot thermal inflation is another variant, in which the initial thermal energy not only provides the inflation but also warms the universe after inflation.[17] To achieve thermalization requires not only the correct Bose-Einstein or Fermi-Dirac distributions but also decoherence of created particles throughout inflation and oscillations. Thermalization may be the last issue for future study.

References

1. D. N. Spergel et al., *Astrophys. J. (Suppl.)* **148**, 175 (2003).
2. D. N. Spergel et al., "Wilkinson Microwave Anisotropy (WMAP) Three Year Results: Implications for Cosmology," astro-ph/0603449 (2006).
3. E. W. Kolb and M. S. Turner, *The Eraly Universe* (Addison-Wesley Pub. Redwood, California, 1989).
4. A. Linde, *Particle Physics and Inflationary Cosmology* (Harwood, Chur, Switzerland, 1990) [hep-th/0503203].
5. A. Guth, *Phys. Rev.* **D23**, 1729 (1981).
6. A. Linde, *Phys. Lett.* **B108**, 1729 (1982).
7. Albrecht and Steinhardt, *Phys. Rev. Lett.* **48**, 1729 (1982).
8. A. Linde, *Phys. Lett.* **B129**, 1729 (1983).
9. S. Koh, S. P. Kim and D. J. Song, *Phys. Rev.* **D71**, 2072 (2005).
10. T. Banks, *Nucl. Phys.* **B429**, 332 (1985).
11. C. Kiefer, *Class. Quantum Grav.* **4**, 1369 (1987).
12. S. P. Kim, *Class. Qauntum Grav.* **13**, 1377 (1996).
13. S. P. Kim, *Phys. Rev.* **D55**, 7511 (1997).
14. D. Giulini, E. Joos, C. Kiefer, J. Kupsch, I.-O. Stamatescu and H. D. Zeh, *Decoherence and the Appearance of a Classical World in Quantum Theory* (Springer, Berlin, 1996).
15. S. P. Kim and C. H. Lee, *Phys. Rev.* **D62**, 125020 (2000).
16. S. Sengupta, F. C. Khanna and S. P. Kim, *Phys. Rev.* **D68**, 105014 (2003).
17. S. P. Kim and D. N. Page, *J. Korean Phys. Soc.* **35**, S660 (1999).
18. D. H. Lyth and E. D. Stewart, *Phys. Rev.* **D53**, 1784 (1996).
19. A. Berera, *Phys. Rev. Lett.* **75**, 3218 (1995).
20. L. Kofman, A. Linde and A. A. Starobinsky, *Phys. Rev. Lett.* **73**, 3195 (1994).
21. J. Traschen and R. Brandenberg, *Phys. Rev.* **D42**, 2491 (1990).
22. D. I. Podolsky, G. N. Felder, L. Kofman and M. Peloso, *Phys. Rev.* **D73**, 023501 (2006).
23. J. K. Kim and S. P. Kim, *J. Korean Phys. Soc.* **33**, 119 (1998).

Dark energy and Wormhole

Sung-Won KIM*

Department of Science Education,
Ewha Women's University, Seoul 120-750, Korea,
**E-mail: sungwon@ewha.ac.kr*

We study the accelerating cosmological model with a static traversable wormhole to see the relation of wormhole to the dark energy. In this model, the phantom energy is considered as the engine of the acceleration of the universe. It is shown that the time to the 'Big Rip', that is derived by using the phantom energy, will be delayed for a sufficient wormhole distribution.

1. Introduction

The wormhole has a structure which is given by two asymptotically flat regions and a bridge connecting these two regions [1]. For the Lorentzian wormhole to be traversable, it requires exotic matter which violates known energy conditions, such as strong, weak, and null enegy conditions. To find reasonable models, there have been studies on generalized models of the wormhole with other matter and/or in various geometries. Among the models, the matter or waves in the wormhole geometry and their effects, such as radiation, are very interesting to us. The scalar field could be considered in wormhole geometry as primary and auxiliary effects [2]. The gravitational perturbation of the wormhole was also studied [3].

Recent data on supernova showed that our universe is being accelerated nowadays by so-called dark energy, and whose nature has not been identified yet, though it constitutes two thirds of our universe [4]. Like the exotic matter, the dark energy violates the 'strong energy condition' so that it can have various exotic properties. Thus, it can be naturally considered and analyzed by using a combined model of the wormhole and dark energy. The general form of the accelerating cosmological model with a wormhole is needed to see the effect of the wormhole and the dynamical behavior of the cosmological model. Similar works on a wormhole and phantom energy, a special form of dark energy, were done [5]. Their solutions' idea came from the model of phantom energy accretion onto a black hole by Babichev *et al.* [6]. There are also other models of a wormhole made by phantom energy [7].

In this paper, we study the accelerating cosmological model with a wormhole and consider the role of wormhole in the acceleration and the other physical properties. Here, we adopt geometrical units, *i.e.*, $G = c = \hbar = 1$.

2. Cosmological model with a wormhole

The spacetime metric for a cosmological model with a static wormhole is given as [8]

$$ds^2 = -e^{2\Phi(r)}dt^2 + R^2(t)\left[\frac{dr^2}{1 - \kappa r^2 - b(r)/r} + r^2(d\theta^2 + \sin^2\theta d\phi^2)\right], \quad (1)$$

where $\Phi(r)$ is the lapse function and $b(r)$ is the wormhole shape function. They are assumed to depend on r only for the static wormhole case while the time dependency is in the scale factor of the universe $R(t)$ only. This is constructed by combining simply two spacetime metrics: the static wormhole spacetime and the Friedmann-Robertson-Walker (FRW) spacetime. This combination is analogous to the case of the Schwarzschild-de Sitter spacetime, which is constructed by using the Schwarzschild and the de Sitter metrics. Here, κ is the sign of the curvature of FRW spacetime. In this case, the spacetime is a rather asymptotically FRW cosmology, and that seems not to be wrong, even though the wormhole is assumed to be asymptotically flat. The reason is that the cosmological model with a traversable wormhole is considered here.

The Einstein equations for $\Phi = 0$ can be written as

$$\rho(r,t) = -\frac{1}{8\pi}\left[\frac{3(\dot{R}^2 + \kappa)}{R^2} + \frac{1}{R^2}\frac{b'}{r^2}\right], \quad (2)$$

$$\tau(r,t) = \frac{1}{8\pi}\left[2\frac{\ddot{R}}{R} + \frac{(\dot{R}^2 + \kappa)}{R^2} + \frac{1}{R^2}\frac{b}{r^3}\right], \quad (3)$$

$$P(r,t) = \frac{1}{8\pi}\left[-2\frac{\ddot{R}}{R} - \frac{(\dot{R}^2 + \kappa)}{R^2} + \frac{1}{2R^2}\left(\frac{b}{r^3} - \frac{b'}{r^2}\right)\right], \quad (4)$$

where $\rho(r,t) = T_{\hat{t}\hat{t}}$ is the energy density, $\tau(r,t) = -T_{\hat{r}\hat{r}}$ is the surface tension, and $P(r,t) = T_{\hat{\theta}\hat{\theta}}$ is the pressure in the orthonormal frame $\hat{\imath}\hat{\jmath}$. A prime denotes differentiation with respect to r, and an overdot denotes differentiation with respect to t. In the isotropic case, as in the FRW cosmological model, it is certain that $\tau = -P$. However, in general, the relation is not satisfied for the case of a spacetime including a wormhole, because of the wormhole breaks isotropy. Here, there are only diagonal terms in Einstein's equations while there can be off-diagonal terms, such as $T_{\hat{t}\hat{r}}$, when $\Phi(r,t)$ and $b(r,t)$ depend on both t and r. The conservation laws, $T^{\mu}{}_{\nu;\mu} = 0$, become

$$\dot{\rho} + (3\rho + 2P - \tau)\frac{\dot{R}}{R} = 0, \tag{5}$$

$$\tau' + (P + \tau)\frac{2}{r} = 0. \tag{6}$$

Since the matter terms ρ, τ, and P depend on both t and r, the following ansatz for the matter parts readily helps us to solve Einstein's equations:

$$R^2(t)\rho(r,t) = R^2(t)\rho_{\rm c}(t) + \rho_{\rm w}(r), \tag{7}$$
$$R^2(t)\tau(r,t) = R^2(t)\tau_{\rm c}(t) + \tau_{\rm w}(r), \tag{8}$$
$$R^2(t)P(r,t) = R^2(t)P_{\rm c}(t) + P_{\rm w}(r). \tag{9}$$

This ansatz allows the separation of the cosmological part and the wormhole part without any coupling. The subscript 'c' indicates the cosmological part depends on time only and 'w' indicates the wormhole part depends on space only. That is, $\rho_{\rm c}, \tau_{\rm c}$, and $P_{\rm c}$ describe the matter distribution of the universe without any wormhole. Therefore, because of the isotropy of the cosmological part, $\tau_{\rm c}(t) = -P_{\rm c}(t)$.

Since our universe is nearly flat, the curvature term is assumed to vanish. Thus the Einstein's equations with the zero curvature, $\kappa = 0$, are

$$R^2\rho_{\rm c} - \frac{3}{8\pi}\dot{R}^2 = \frac{b'(r)}{8\pi r^2} - \rho_{\rm w}(r) = l, \tag{10}$$
$$R^2\tau_{\rm c} - \frac{1}{8\pi}(2R\ddot{R} + \dot{R}^2) = \frac{b(r)}{8\pi r^3} - \tau_{\rm w}(r) = m, \tag{11}$$
$$R^2P_{\rm c} + \frac{1}{8\pi}(2R\ddot{R} + \dot{R}^2) = \frac{1}{8\pi}\frac{1}{2}\left(\frac{b(r)}{r^3} - \frac{b'(r)}{r^2}\right) - P_{\rm w}(r) = n, \tag{12}$$

where l, m, and n are the separation constants and are independent of both t and r. Here, $m = -n$ by the relationship $\tau_{\rm c} = -P_{\rm c}$. The separation constants q, l, m, and n are also determined by using the wormhole matter distribution.

The first conservation law becomes

$$\frac{R^3}{\dot{R}}\left[\dot{\rho}_{\rm c} + (3\rho_{\rm c} + 2P_{\rm c} - \tau_{\rm c})\frac{\dot{R}}{R}\right] = s. \tag{13}$$

From this law, we can get the scale factor dependence of the cosmic density ρ_c as

$$\rho_c = aR^{-3(1+\omega)} + \frac{s}{1+3\omega}R^{-2}. \tag{14}$$

Here, ω is the ratio of the pressure to the energy density, i.e., $P = \omega\rho$, and a is a proper integration constant. The second term of Eq. (14) says that the cosmic matter density is affected by the wormhole with the proper constant s. The constant s can be determined by wormhole distribution. It is natural for the positive definite $s/(1+3\omega)$ to be larger than a when the wormhole effect is dominant. If the separation constant s vanishes, the R dependency of ρ_c becomes just like it is in the FRW case. Fig. 1 shows the R dependence of the density ρ_c when $\omega = -2/3, -1$, and $-4/3$ under the assumption of the positiveness of both a and $s/(1+3\omega)$. The choice of ω in Fig. 1 provides the three cases: $\omega < -1/3, \omega = -1$, and $\omega < -1$. For the accelerating model, $\omega < -\frac{1}{3}$, and when $\omega = -1$, it is the cosmological constant. If $\omega < -1$ it is the so-called, 'phantom energy.' For the case of the phantom energy, the density diverges when R grows, unlike the cases in other models of the dark energy.

From Eq. (10),

$$\begin{aligned} H^2 &= \frac{8\pi}{3}\left(\rho_c - \frac{l}{R^2}\right) \\ &= \frac{8\pi}{3}\left(aR^{-3(1+\omega)} + s^* R^{-2}\right), \end{aligned} \tag{15}$$

where $H = \dot{R}/R$ and $s^* = q/(1+3\omega) - l$. When we compare this with the case of the multi-component model with curvature, s^* plays the role of the curvature $-\kappa$. The additional term that comes from the ansatz Eqs. (7)-(9) shows the R^{-2} dependency, which means that the curvature is affected by the wormhole even though the universe is flat, $\kappa = 0$. This indicates that the wormhole has a non-flat effect on the multiple component universe, which shows the non-trivial topology.

If we add three times of Eq. (12) to Eq. (10)

$$\begin{aligned} \frac{3}{4\pi}\frac{\ddot{R}}{R} &= -(1+3\omega)\rho_c + (l+3n)R^{-2} \\ &= -(1+3\omega)aR^{-3(1+\omega)}. \end{aligned} \tag{16}$$

This shows that apparently the acceleration is not affected by wormhole, and has the same form as in the FRW case. However, when we inspect the deceleration parameter defined by $q = -\ddot{R}R/\dot{R}^2$, the result will be clearer

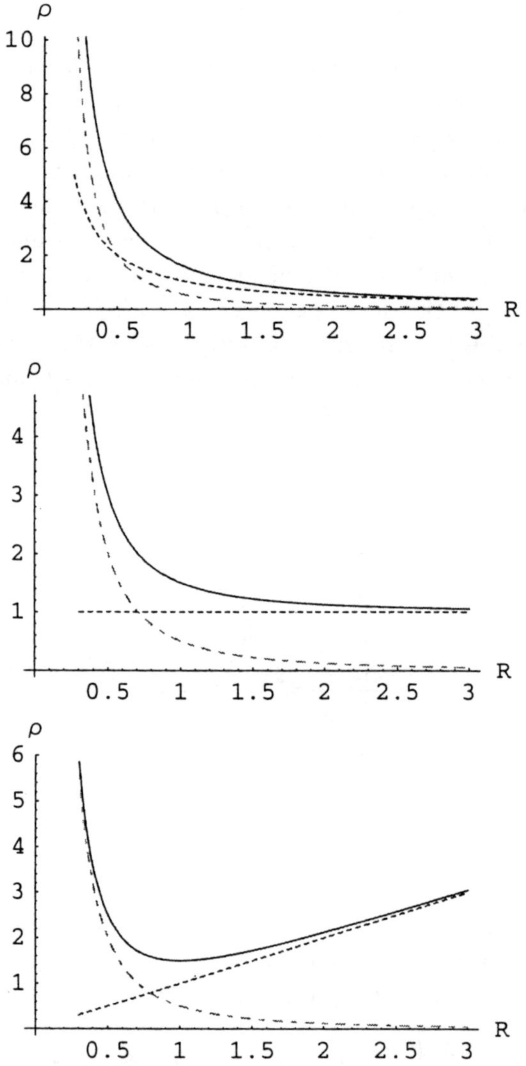

Fig. 1. Plot of ρ in terms of R for $\omega = -\frac{2}{3}, -1$ and $-\frac{4}{3}$ from the top. The dotted lines denote the first term in Eq. (2.14), and the dashed line the second R^{-2} term.

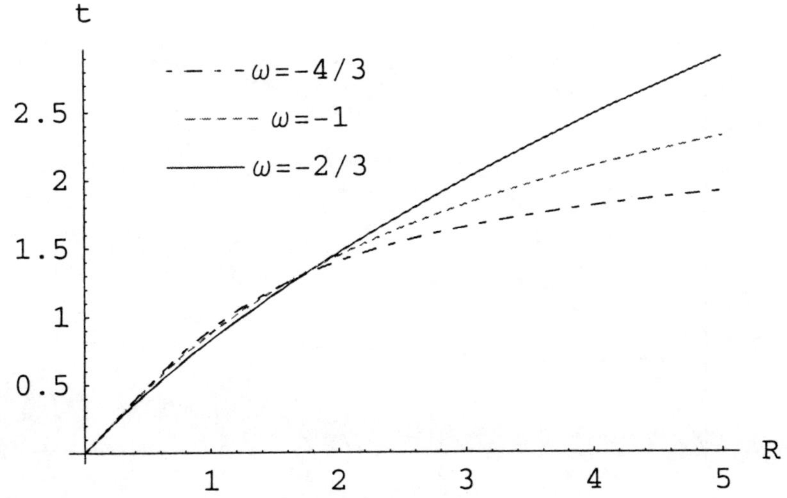

Fig. 2. Plot of the t vs. R for $\omega = -\frac{2}{3}, -1$ and $-\frac{4}{3}$.

than it is in Eq. (16):

$$q = \frac{1}{2}\left[\frac{1}{1+3\omega} + \frac{s^*}{a(1+3\omega)}R^{-5-3\omega}\right]^{-1}. \tag{17}$$

In Eq. (17), there is the definite effect of the wormhole on the acceleration of the universe. When we rewrite Eq. (15) in terms of H_0^2,

$$\frac{H^2}{H_0^2} = \Omega_{\mathrm{de}} R^{-3(1+\omega)} + \Omega_{\mathrm{w}} R^{-2}. \tag{18}$$

Here, Ω_{de} and Ω_{w} are the ratios of the dark energy density and the wormhole density to the critical density.

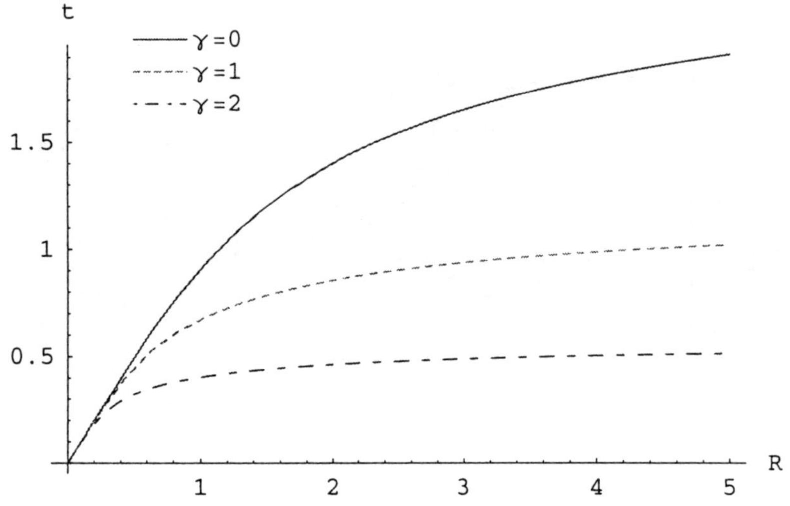

Fig. 3. Plot of the t vs. R for three ratios of $\Omega_{\rm de}/\Omega_{\rm w} = 10^{\gamma}$ when $\omega = -\frac{4}{3}$.

In the early universe, $\omega = 1/3$, the radiation-dominated era, and the first term depends on R^{-4}. Next is the matter-dominated era, $\omega = 0$, and the first term depends on R^{-3}. If we integrate this, we obtain the time dependence of the scale factor R as

$$H_0\sqrt{\Omega_{\rm w}}t = R \times {}_2F_1\left(\frac{1}{k}, \frac{1}{2}, \frac{1}{2} + 1, -\frac{\Omega_{\rm de}}{\Omega_{\rm w}}R^k\right), \tag{19}$$

where $k = -(1 + 3\omega)$. Here, ${}_2F_1(....)$ is the hypergeometric function. Fig. 2 shows the R vs. t curves for $\omega = -2/3, -1$, and $-4/3$. As we see, the scale factor R diverges at finite time for the case of $\omega = -4/3$, the phantom energy only. This is called the 'Big Rip' [9]. The other solutions for $\omega = 1/3$

and 0 are

$$H_0\sqrt{\Omega_{\rm w}}t = \sqrt{\frac{\Omega_{\rm de}}{\Omega_{\rm w}} + R^2}, \quad \text{for} \quad \omega = \frac{1}{3}, \tag{20}$$

$$H_0\sqrt{\Omega_{\rm w}}t = R \times \log \left| R + \sqrt{\frac{\Omega_{\rm de}}{\Omega_{\rm w}} R + R^2} \right|, \quad \text{for} \quad \omega = 0. \tag{21}$$

Fig. 3 shows the $R-t$ relation for $\gamma = 0, 1, 2$, where $\Omega_{\rm de}/\Omega_{\rm w} = 10^\gamma$. In this figure, R diverges at finite time in all cases; however, the larger $\Omega_{\rm w}$, the later the big rip time. The wormhole sector is shown to hold the 'Big Rip' given by the phantom energy when $\Omega_{\rm w} \gg \Omega_{\rm de}$. If we transform Fig. 3 so as that R is the vertical axis and t is horizontal axis, there is a finite time to be divergence of $R(t)$. In the case of wormhole with a small density, the phantom energy cause the big rip without any obstacle. Consequently, our result shows that the wormhole may delay the big rip time.

3. Conclusion

In this paper, we studied the accelerating cosmological model with a traversable wormhole in the center. We showed that the wormhole played the role of the curvature even though the universe is flat. The acceleration will not be directly affected by the wormhole while the velocity will be changed by the wormhole. However, a wormhole effect appears in the deceleration parameter. When we examine the exact solution to the scale factor, we found that a large wormhole could delay the 'Big Rip'.

4. Acknowledgments

The participation at ICGA7 and presentation of this work were supported in part by APCTP.

References

1. M. S. Morris and K. S. Thorne, Am. J. Phys. **56**, 395 (1988); Sean A. Hayward, Sung-Won Kim, Hyunjoo Lee, J. Korean Phys. Soc. **42**, S31 (2003).
2. Sung-Won Kim, Gravitation & Cosmology **6**, 337 (2000).
3. Sung-Won Kim, J. Korean Phys. Soc. **42**, S94 (2003).
4. A. G. Riess *et al.* Astron. J. **116**, 1009 (1998).
5. For example, P. F. Gonzalez-Diaz, Phys. Rev. Lett. **93**, 071301 (2004); V. Faraoni and W. Israel, Phys. Rev. D **71**, 064017 (2005).
6. E. Babichev, V. Dokuchaev, and Yu. Eroshenko, Phys. Rev. Lett. **93**, 021102 (2004).

7. S. Sushkov, Phys. Rev. D **71**, 043520 (2005); F. S. N. Lobo, Phys. Rev. D **71**, 084011 (2005); O. B. Zaslavskii, Phys. Rev. D **72**, 061303 (2005).
8. Sung-Won Kim, Phys. Rev. D **53**, 6889 (1996).
9. R. R. Caldwell, Phys. Lett. B **545**, 23 (2002).

COSMIC BOSE EINSTEIN CONDENSATION – INDUCED INFLATION AND EARLY FORMATION OF BLACK HOLES

TAKESHI FUKUYAMA

Department of Physics, Ritsumeikan University,
Kusatsu Shiga, 525-8577, Japan
E-mail: fukuyama@se.ritsumei.ac.jp

MASAHIRO MORIKAWA

Department of Physics, Ochanomizu University,
2-1-1 Otsuka, Bunkyo, Tokyo,112-8610, Japan
E-mail: hiro@phys.ocha.ac.jp

We do not know the basic ingredient of the universe. A model is proposed in which dark energy is identified as Bose-Einstein condensation of some boson field. A global cosmic acceleration cause by such condensation and the rapid local collapses into black holes of the condensation are examined. We propose a novel mechanism of inflation due to the steady flow of condensation, and the early formation of highly non-linear objects.

1. Introduction

The whole history and structure of the Universe is so far quantitatively well understood but the qualitative understanding is premature. Although the standard ΛCDM model perfectly explains the temperature fluctuations in the sky and the power spectrum of density fluctuations, they are restricted mainly in the linear stage of density fluctuations. On the other hand in the non-linear stage of density fluctuations, there are many peculiarities such as the tremendous amount of quasars which have shone in the early stage of the Universe, omnipresent black holes at the center of each galaxy, extremely mild inflation (accelerated expansion) at present, etc. Especially it is almost impossible to build such many black holes in the early stage only from the gravitational attraction of the ordinary matter.

The above facts naturally suggest a new type of instability associated with a phase transition which might occur in the entire universe in the dark age, the period between the photon decoupling and the present. Moreover such unusual existence of dark matter, dark energy and black holes may be connected with each other; the dark matter and the dark energy may be the same boson field but different existence form and their instability might have built many black holes.

We now propose such cosmological scenario in which the above phase transition and the instability are naturally deduced [1,2]. The starting point to introduce such model is the relevant role of scalar fields in various cosmological models presently studied. It generally provides with unusual coherence of matter and accelerated expansion. The most familiar mechanism for the realization of such scalar field would be the quantum mechanical condensation of boson field which may dominate the universe. This is well known as the Bose Einstein condensation (BEC) from the early days of quantum mechanics and is widely realized in laboratories.

The nature of BEC is quite different from the classical existence form such as gas, liquid, solid and plasma; it is usually described by the non-linear Schrödinger equation (GP-equation).

$$i\hbar\frac{\partial\psi}{\partial t} = -\frac{\hbar^2}{2m}\Delta\psi + V\psi + g\left|\psi\right|^2\psi \tag{1}$$

where $\psi(\vec{x},t)$is the condensate wave function, $V(\vec{x})$ is the potential, and $g = 4\pi\hbar^2 a/m$, and a is the s-wave scattering length. This should be used in the relativistic form

$$\frac{\partial^2\Phi}{\partial t^2} - \Delta\Phi + m^2\Phi - \lambda(\Phi^*\Phi)\Phi = 0 \tag{2}$$

for our cosmological purpose. We choose $\lambda < 0$ in order to induce instability.

How the BEC is possible in the real Universe? This quantum condensation initiates when the thermal de Broglie wave length, i.e. the quantum extension of each particle, overlaps with each other; i.e. the temperature drops below the critical temperature

$$T_{\rm cr} = \frac{2\pi\hbar^2 n^{2/3}}{km} \tag{3}$$

which is proportional to the two-thirds of the number density n. On the other hand the non-relativistic matter behaves as in the same way ($\propto n^{2/3}$) in the expanding universe. The above facts determine the upper limit of the boson mass m from the present value of matter density. It turns out to be about 2eV.

2. Gradual process of BEC

Being a phase transition, the time evolution of BEC process cannot be described simply by a fundamental Lagrangian. In this respect the present model is essentially different from the ordinary interacting dark matter models. We introduce a phenomenological condensation rate Γ which controls the speed of condensation from the gas component. Moreover, the instability of BEC is represented by the abrupt collapse of BEC when the BEC density exceeds some critical value. Then the whole set of equations of motion in our model will be

$$\begin{aligned} H^2 = \left(\tfrac{\dot{a}}{a}\right)^2 &= \tfrac{8\pi G}{3c^2}\left(\rho_g + \rho_\phi + \rho_l\right) \\ \dot{\rho}_g &= -3H\rho_g - \Gamma\rho_g \\ \dot{\rho}_\phi &= -6H\left(\rho_\phi - V\right) + \Gamma\rho_g - \Gamma'\rho_\phi \\ \dot{\rho}_l &= -3H\rho_l + \Gamma'\rho_\phi \end{aligned} \tag{4}$$

where $\rho_g, \rho_\phi, \rho_l$ are the energy densities of the boson gas, BEC, and the localized objects which form after BEC collapse. The coefficient Γ' appears only when ρ_ϕ exceeds the critical density and is very large.

There are two relevant regimes in (4). One is the (a) over-hill regime which appears when the condensation flow is larger than the cosmic expansion, and the other is (b) the mild-inflationary regime which appears when the flow is smaller than the expansion.

3. Over-hill regime

In the early stage, the dense boson gas provides relatively fast condensation and the mean field ϕ goes over the potential hill (Fig. 1). This regime corresponds to the fixed point

$$\phi \to \infty, \rho_\phi \to 0, H \to 0, a \to a_* \tag{5}$$

which is singular. However before the universe reaches this singularity, the uniformity of BEC is broken and it collapses to the localized objects.

This is clarified by the linear instability analysis around the solution of Eq.(4). It turns out that the Jeans scale becomes

$$(m/m_{\text{plank}})m \tag{6}$$

where $m_{\text{plank}} \approx 10^{-5}$gris the Plank mass. When the first BEC collapse occurred at around $z = 20$, the mass scale of the localized object becomes about the galaxy size $10^{11} M_{\odot}$. The detail will be reported elsewhere [3].

After the BEC collapse, the mean field entirely disappears and then the new BEC initiates. If it reaches the critical value, BEC collapses again to form localized objects.

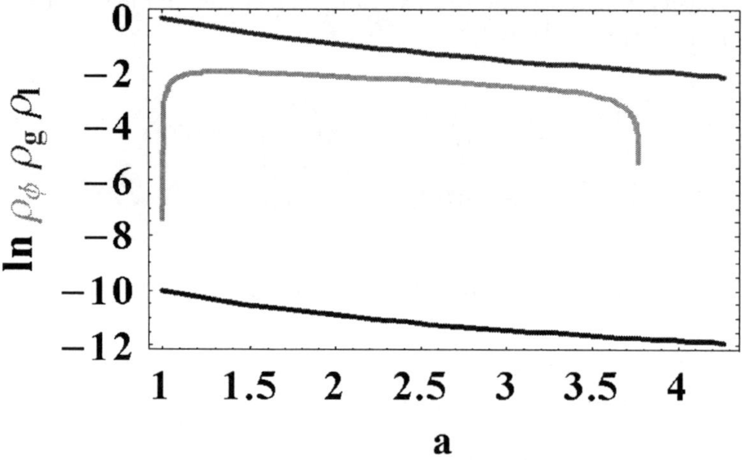

Fig. 1. A numerical simulation of Eq.(4). This is the evolution of energy densities in the over-hill regime. Red and green curves are the bose gas and BEC energy densities, respectively $(\rho_g,\ \rho_\phi)$ as a function of scale factor a, whose unit is arbitrary. Blue curve is the energy density of collapsed BEC. The parameter values are $m^2 = 0.01,\ \lambda = -1,\ \Gamma =$ 0.4and the initial conditions are $\rho_l^{\text{initial}} = 10^{-10},\ \rho_g^{\text{initial}} = 1,\ \phi^{\text{initial}} = \dot{\phi}^{\text{initial}} =$ 10^{-4} in the unit of $8\pi G/\left(3c^2\right) = 1$.

4. Mild-inflationary regime

After several BEC collapses, the boson gas energy density would be consumed to form BEC. Then the condensation flow becomes small and the mean field cannot go over the potential hill. Thus the flow and the potential force finally balance with each other and the condensation ceases at that point. Because it is the middle of the potential bottom and the hill-top, the dominated finite vacuum energy there causes accelerated expansion. This should correspond to now we live. This regime corresponds to the fixed point (Fig. 2)

$$\phi \to \phi_*, H \to H_*, \rho \to 0, \dot{\phi} \to 0 \tag{7}$$

which is the de Sitter space.

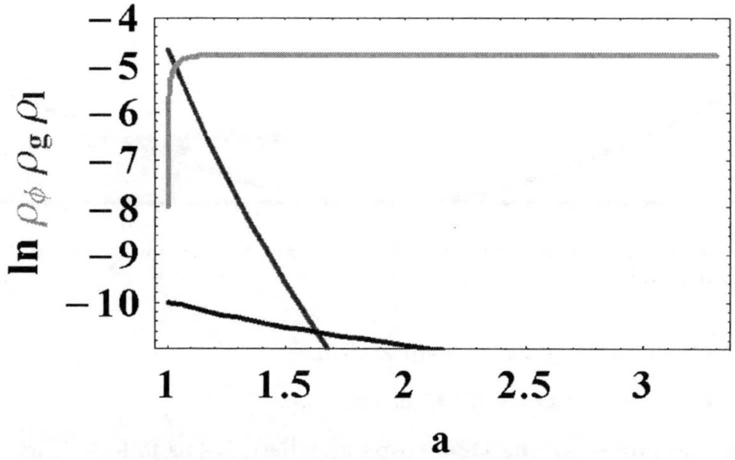

Fig. 2. The same as Fig. 1 but in the mild inflationary regime.

This inflation itself is supported by the vacuum energy as usual, but the vacuum energy is maintained by the balance of the condensation glow and the potential force.

In this occasion, we would like to emphasize that this regime is a novel type of inflation (Fig. 3). This inflation is supported by a balance of the condensation speed ($\Gamma\rho_g$), which promotes the condensation ϕ, and the potential force ($\dot{\phi}V'$), which prohibits the condensation ϕ. Actually at the fixed point, the exact balance

$$\dot{V} = \Gamma\rho_g \tag{8}$$

is established. This can be checked by dividing both sides of the last equation in Eq.(4) by $\dot{\phi}$ and applying the fixed point condition Eq.(7). Though the both sides of Eq.(8) exponentially reduce to zero, the balance itself is kept automatically[a].

[a]This exponentially reducing amplitude of the balance may lead to the asymptotic insta-

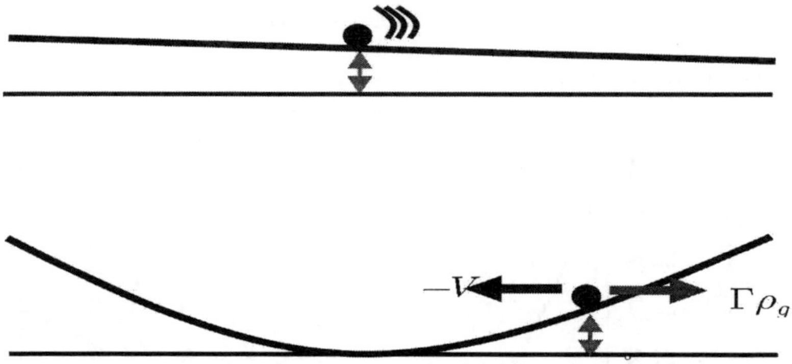

Fig. 3. Comparison of the ordinary inflation (above) and the novel inflation (below) in schematic pictures.

5. BEC cosmology and observations

5.1. *Whole process of BEC cosmology*

The whole process of the BEC cosmology becomes as follows (Fig. 4).

In this case, four over-hill regimes are followed by the final inflation. This behavior, i.e. a final inflation after multiple over-hill regimes, is robust in our model.

5.2. *Log-z periodicity*

In the above demonstration in Fig. 4, BEC collapses take place almost in periodic in the logarism of redshift. Actually the collapsing time $z = 17.0, 10.9, 7.1, 4.1$make an almost log-periodic sequence.

This log-z periodicity is an interesting property of our model. In the over-hill regime, the boson gas density behaves as $\rho_g \propto e^{-\Gamma t}$ because the condensation speed is high. Then the density ρ_g is simply transformed into ρ_ϕ almost without cosmic dilution. On the other hand, BEC collapse when ρ_ϕ reaches the critical value $\rho_\phi^{\rm cr} = V_{\rm max} = m^4/(-2\lambda)$. Then the condensation energy density approximately behaves as

bility of the inflationary regime and the autonomous termination of this regime, provided any small external perturbations.

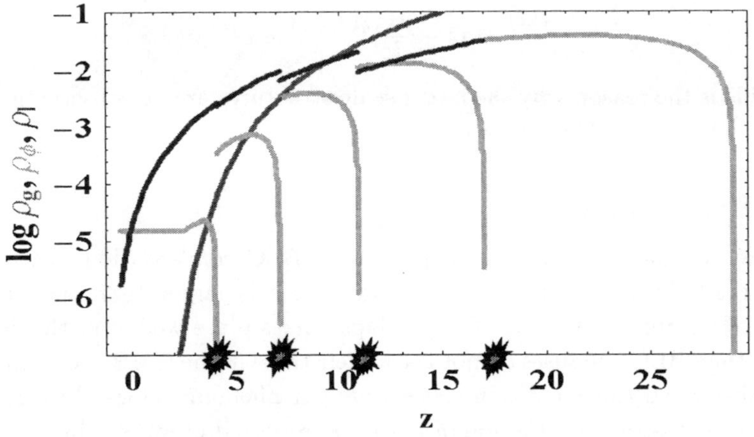

Fig. 4. The whole process of BEC cosmology. This is essentially the combination of Figs. 1 and 2. The bur marks indicate the epochs of BEC collapse, whose sequence turns out to be periodic in the redshift z.

$$\rho_\phi(t) \approx [\rho_g(t_0) - \rho_g(t)]_{\mathrm{mod}\ \rho_\phi^{\mathrm{cr}}} \approx \left[\rho_g(t_0)\left(1 - e^{-\Gamma t}\right)\right]_{\mathrm{mod}\ \rho_\phi^{\mathrm{cr}}} \tag{9}$$

This is the reason for the log-z periodicity.

5.3. *Large scale structure*

Let us further examine why the power spectrum is modified only in the smallest scale. As is shown in Ref. [1], the condensation collapses in a very short time although the evolution speed is huge. Therefore the larger scale fluctuations simply do not have time to grow. Suppose the condensation collapses at the scale $l = a/k \equiv \tilde{k}^{-1} \ll H^{-1}$. Then, for the power law expansion $a \approx t^{2/3}$, the density fluctuation in k-mode behaves as

$$\delta_k \propto a^{2\tilde{k}/H} = t^{4\tilde{k}/(3H)}. \tag{10}$$

The time duration of the collapse, i.e. the time necessary for the condensation to disappear, would be $\Delta t \approx l/c$. Note that all the fluctuations are very unstable only within this time. During this time duration, the linear fluctuation of the scale l can grow

$$\frac{\delta\,(t+\Delta t)}{\delta\,(t)} = (1+\frac{H}{\tilde{k}})^{4\tilde{k}/(3H)} \approx e^{4/3} \approx 3.8 \tag{11}$$

This is the reason why the larger scale structures are not observationally affected by the BEC collapses.

5.4. *CMB isotropy*

We are considering a very slow process of BEC condensation due to the almost adiabatic dynamics. If this time scale is larger than the photon decoupling time, or the first BEC collapse takes place well after the decoupling, then BEC collapses do not leave their trace in the CMB spectrum. On the other hand this condition constraints our phenomenological parameter Γ. The microscopic evaluation of it will be reported elsewhere [3].

6. Conclusions

We conclude our paper by showing a schematic picture of the dynamical sequence of our model (Fig. 5). Initially the boson gas behaves as the cold dark matter, in which the condensate is accumulated very slowly. The condensate behaves as dark energy. After the energy density of the condensate reaches some critical value, it collapses into localized objects such as black holes and the surrounding hot gas, which behave as cold dark matter. This condensation-collapse period repeats multiple times in general and finally the condensation speed balances with the potential force. This balance makes the mean field static and a novel type of inflation initiates. Thus the dark matter turns into dark energy and then into dark matter again.

There are many issues for the improvement of our model. First of all, we have to estimate the structures formed by the BEC collapse, the size and mass, etc. We also have to evaluate the associated gravitational wave, and to identify the boson field. It is most likely that the boson is a fermion pair. For example, neutrinos may have a chance to for a boson. Some of these elaborations of BEC cosmology will be reported soon [3].

References

1. M. Nishiyama, M. Morita, and M. Morikawa, in the Proceedings of the XXXIXth Rencontres de Moriond 143 (2004) [astro-ph/0403571].
2. T. Fukuyama and M. Morikawa, in print, Progr. Theor. Phys. **115** (2006).
3. T. Fukuyama , M. Morikawa and T. Tatekawa, in preparation.

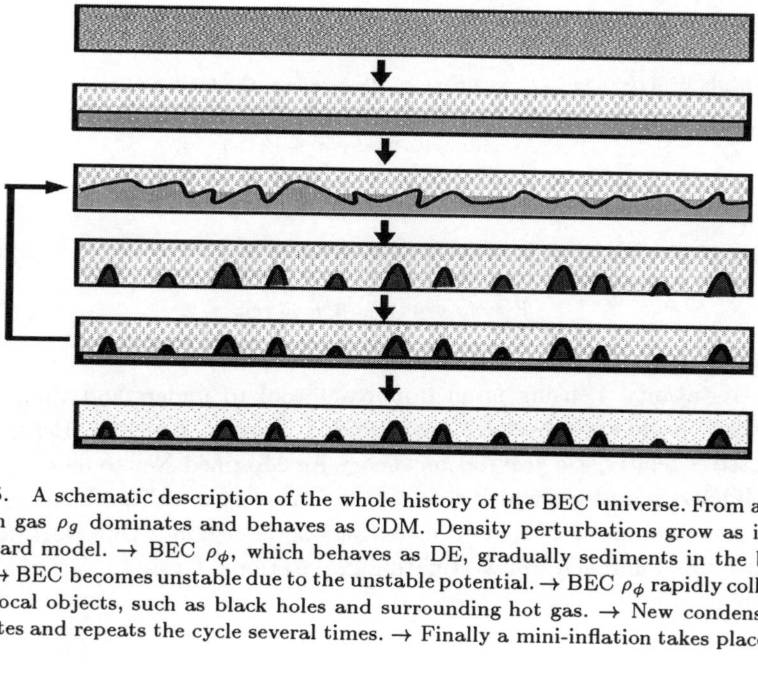

Fig. 5. A schematic description of the whole history of the BEC universe. From above, Boson gas ρ_g dominates and behaves as CDM. Density perturbations grow as in the standard model. → BEC ρ_ϕ, which behaves as DE, gradually sediments in the boson gas. → BEC becomes unstable due to the unstable potential. → BEC ρ_ϕ rapidly collapses into local objects, such as black holes and surrounding hot gas. → New condensation initiates and repeats the cycle several times. → Finally a mini-inflation takes place.

Gravitational Lensing in TeVeS

Mu-Chen Chiu
Department of Physics, National Central University,
Chung-Li, Taiwan 320, R.O.C.
E-mail: chiumuchen@gmail.com

Chung-Ming Ko
Institute of Astronomy, Department of Physics and Center for Complex System,
National Central University, Chung-Li, Taiwan 320, R.O.C.
E-mail: cmko@astro.ncu.edu.tw

Yong Tian
Department of Physics, National Taiwan University,
Taipei, Taiwan 106, R.O.C.
E-mail: yonngtian@gmail.com

Gravitational Lensing is an important tool to understand the "missing mass" problem, especially for Modified Gravity. Recently, Bekenstein proposed a relativistic gravitation theory for Modified Newtonian Dynamics (MOND) paradigm which resolves the "missing mass" problem well on abnormal dynamical behaviors in extragalactic region . Our work follow Bekenstein's approach to investigating gravitational lensing to get theoretical prediction.

1. Introduction

According to Newtonian gravitational theory, the relation between the rotation velocity v and radius r should be $v \propto r^{-1/2}$ in circular motion. In the solar system, it is applied well. However, it fails on galaxies systems. From observation, the plot of rotation curve becomes asymptotically flat. Beside this, Tully-Fisher relation [1]can't be explained by Newtonian gravitational theory. There are enormous evidence showing the discrepancy increasing with the improved telescope. The type of them is from spiral galaxies to elliptical galaxies and the size is from galaxies to cluster. These unusual dynamical behaviors may indicate either the failure of Newtonian gravitational theory or the "missing mass" in galactic regions. Moreover, from gravitation lensing, the observation deflection angel is also larger than expected in General Relativity (GR). This means it is not merely abnormal dynamical behaviors because lensing is some characteristic feature of relativistic gravitation theory.

There is an interesting scheme called Modified Newtonian Dynamics (MOND)[2, 3, 4] for explaining the "missing mass" problem up to a amazing degree in Ref. 5. It assumes that Newton's second law needs to be modified when the acceleration is very small:

$$\tilde{\mu}(|\mathbf{a}|/\mathfrak{a}_0)\mathbf{a} = -\boldsymbol{\nabla}\Phi_{\mathrm{N}}. \tag{1}$$

Here Φ_{N} is the usual Newtonian potential of the visible matter and $\mathfrak{a}_0 \approx 1 \times 10^{-10}$ m s^{-2} from the empirical data, such as the Tully-Fisher relation and rotation curves [6]; $\tilde{\mu}(x) \approx x$ for $x \ll 1$, $\tilde{\mu}(x) \to 1$ for $x \gg 1$. In the solar system where accelerations are strong compared to $\mathfrak{a}_0$, the formula (1) is just Newton's second law $\mathbf{a} = -\boldsymbol{\nabla}\Phi_{\mathrm{N}}$.

However, as a theory MOND is not complete and lacks a relativistic scheme. The progress toward a relativistic gravitational theory is not smooth and easy [7, 8]. Many of these usually violate causality. Two decades later, the invention by Bekenstein, named Te*Ve* S (Tensor-Vector-Scalar) [9] finally succeeded. Now, Many relativistic features such as lensing effects can be tested rigorously[10, 11, 12]

In this paper, we only introduce the concept of Te*Ve* S and gravitational lensing effects of a point-mass model in it. For more details, please see Ref. 10.

2. Te*Ve* S

The conventions in this paper are the metric signature +2 and units with $c = 1$.

Te*Ve* S contains three dynamical fields: an Einstein metric $g^{\mu\nu}$, a scalar field ϕ and a timelike 4−vector field $\mathfrak{U}_\alpha$ with a normalization condition $g^{\alpha\beta}\mathfrak{U}_\alpha\mathfrak{U}_\beta = -1$. In addition, there is a non-dynamical field scalar field σ.

2.1. *Disformal Metric Relation*

If there is no dark matter in the extragalactic region, the abnormal lensing effects may mean that the concept of geodesics is not usual as GR. We may need the physical metric $(-\tilde{g})^{\mu\nu}$ is deviated from $g^{\mu\nu}$ to describe the geodesic equation of matter or light. In Te*Ve* S, $(-\tilde{g})^{\mu\nu}$ is the same as in

Sander's stratified theory in Ref. 13:

$$\tilde{g}_{\alpha\beta} = e^{-2\phi}(g_{\alpha\beta} + \mathfrak{U}_\alpha \mathfrak{U}_\beta) - e^{2\phi}\mathfrak{U}_\alpha \mathfrak{U}_\beta, \tag{2}$$

2.2. *Actions of TeVeS*

The geometric action S_g is still the same as the Einstein-Hilbert action and describe the dynamics of the metric $g^{\mu\nu}$:

$$S_g = (16\pi G)^{-1} \int g^{\alpha\beta} R_{\alpha\beta}(-g)^{1/2} d^4x. \tag{3}$$

In the scalar action, there is no kinetic terms of σ. The equation of motion of σ is an algebraic relation with $h^{\alpha\beta}\phi_{,\alpha}\phi_{,\beta}$. The action of the two scalar fields is taken to be

$$S_s = -\frac{1}{2} \int \left[\sigma^2 h^{\alpha\beta}\phi_{,\alpha}\phi_{,\beta} + \frac{1}{2} G\ell^{-2}\sigma^4 F(kG\sigma^2)\right](-g)^{1/2} d^4x, \tag{4}$$

where $h^{\alpha\beta} \equiv g^{\alpha\beta} - \mathfrak{U}^\alpha \mathfrak{U}^\beta$ and F is a dimensionless function used to produce the non-relativistic MONDian dynamic behaviors. k is a dimensionless constants and ℓ is a constant length.

The vector action is with the form

$$S_v = -\frac{K}{32\pi G} \int \left[g^{\alpha\beta} g^{\mu\nu} \mathfrak{U}_{[\alpha,\mu]} \mathfrak{U}_{[\beta,\nu]} - 2(\lambda/K)(g^{\mu\nu}\mathfrak{U}_\mu \mathfrak{U}_\nu + 1)\right](-g)^{1/2} d^4x, \tag{5}$$

where $\mathfrak{U}_{[\alpha,\mu]} = \mathfrak{U}_{\alpha,\mu} - \mathfrak{U}_{\mu,\alpha}$. Here, λ is a spacetime dependent Langrange multiplier enforcing the normalization condition of $\mathfrak{U}^\alpha$.

The matter action is taken to be

$$S_m = \int \mathcal{L}(-\tilde{g})^{1/2} d^4x, \tag{6}$$

where $\mathcal{L}$ is a Lagrangian density for the fields and a functional of the physical metric and its derivatives.

3. Lensing Effects in TeVeS

In this paper,we have investigated the GL phenomenon under the approximations and presuppositions of

(*i*) The GL lens is assumed to be static, spherically symmetric and following the thin lens formalism, (*ii*) The motions of light rays are described in the framework of a Schwarzschild lens (i.e. a point mass model),

(*iii*) The revised physical metric is obtained by adding a positive scalar field into the potential of the standard Schwarzschild metric in symmetric coordinates,

Under these presuppositions, we find that when θ is larger than θ_0 ($\equiv r_0/D_L$; the MOND length scale), the reduced deflection angle α will approach to a constant for a given mass. This special prediction is a feature of GL in *TeVe* S, which is similar to that obtained by Ref. 15 with an intuitional approach. This is no surprise because just like in GR, the deflection of photons is simply twice the deflection of a massive particle with the speed of light in *TeVe* S, and thus [15] started from the correct premise. However, there is still something that was unknown before the appearance of *TeVe* S. For a static spherically symmetric spacetime, the case $\mu \ll 1$, which yields a relation [See Ref. 10 Eq.(29)], is valid only when $|\boldsymbol{\nabla}\Phi| \ll (4\pi/k)^2\mathfrak{a}_0$. This condition is valid in the MOND regime when $|\boldsymbol{\nabla}\Phi|$ goes up to a couple of orders above $\mathfrak{a}_0$, or equivalently ϱ_0 is around one order of magnitude below r_0 [9].

We should address that the criteria of mass discrepancy for GL effects in *TeVe* S consists with that of stellar dynamics. In other words, only when $\varrho_0 \gg kr_0/4\pi$, the "missing mass" shall appear. This corresponds to the demarcation of the high surface brightness (HSB) and low surface brightness LSB galaxies from the dynamical analysis in Ref. 5.

When we apply the deflection angle law to magnification, we find that in *TeVe* S the difference in the magnifications of the two images in the point mass model depends on the lens mass and source positions, and is always larger than one. This differs from traditional gravitational lensing, which says that the difference must always be one. Tens of thousands of the multiple images lensings found by the Sloan Digital Sky Survey (SDSS) [14] might be applied to check this prediction. For microlensing, light curves in *TeVe* S at deep MOND regime also differ from that in GR. To observe the discrepancy, the sources have to be located about $z_s \geq 1$ [15].

Concerning time delay, the result is even more exotic. Add an arbitrary positive scalar field into the primary Schwarzschild metric [9] the resultant contributions of the scalar field will reduce rather than enhance the potential time delay. Unfortunately, this phenomenon is unmeasurable in GL systems. What we can determine is only the time delay between two images produced by the same source. Therefore, the opposite feature of the time delay in *TeVe* S, which is the same for all images, will be canceled out, and only those parts contributed from the deflection potential can be observed.

Even though the opposite effect on the potential time delay due to the scalar field can not be measured in GL system, the time delay between two images offers another constraint on the needed mass in GR and *TeVe* S, which usually differs from that given by the deflection angle. Actually the mass ratio obtained from these two approaches will be the same, otherwise the mass ratio given by time delay would always be smaller than that by deflection angle. In one word, the MOND and CDM paradigm are not mutually alternatives in a GL system, when we consider deflection angle and time delay at the same time.

Acknowledgments

We thank J.M. Nester for stimulating discussions on the modified gravitational theories and comments on the whole work. We also thank Chi Yuan, Tzi-Hong Chiueh, Kin-Wang Ng and Pisin Chen for the comments. We appreciate the referee for pointing out the potential observation of microlensing. M.C. Chiu and C.M. Ko are supported in part by the National Science Council of Taiwan, by grant NSC-93-2112-M-008-017. Y. Tian is partially supported by the grant NSC93-2112-M-002-025 of the National Science Council. Y. Tian wishes to thank Prof. W-Y. P. Hwang for the supervision during which this research was carried out.

References

1. Tully R.B. & Fisher J.R. 1977, AA, **54**, 661
2. Milgrom, M. 1983a, ApJ, **270**, 365
3. Milgrom, M. 1983b, ApJ, **270**, 371
4. Milgrom, M. 1983c, ApJ, **270**, 384
5. Sanders, R.H. & McGaugh, S.S. 2002, ARAA, **40**, 263
6. Sanders, R.H. & Verheijen, M.A.W. 1998, APJ, **503**, 97
7. Bekenstein, J.D. & Milgrom, M. 1984, ApJ, **286**, 7
8. Bekenstein, J.D. 1988, Physics Letters B, **202**, 497
9. Bekenstein, J.D. 2004, PrD, **70**, 083509
10. Chiu M.-C., Ko C.-M., & Tian Y. 2006, ApJ, **636**, 565
11. Zhao H., Bacon, D. J., Taylor, A. N., & Horne, K. 2006, MNRAS, **368**, 171
12. Chen D.-M & Zhao H. 2006, Submitted to Astrophys.J.Suppl.(astro-ph/0606506)
13. Sanders, R.H. 1997, ApJ, **480**, 492
14. Stoughton, C et al. 2002, AJ ,**123**, 485
15. Mortlock, D.J. & Turner, E.L. 2001, MNRAS **327**, 557

AN ISSUE TO THE COSMOLOGICAL CONSTANT PROBLEM

R TRIAY*

Centre de Physique Théorique *

CNRS Luminy Case 907, 13288 Marseille Cedex 9, France

E-mail: triay@cpt.univ-mrs.fr

On geometrical grounds, the cosmological constant problem turns out to be an artifact due to the unfounded link of this fundamental constant to vacuum energy density of quantum fluctuations.

1. Introduction

The *cosmological constant problem* (CCP) is an enigma of modern cosmology among the particles physics and cosmology communities. The most comprehensive contribution to this problem can be found in [29]. It is a matter of fact that its origin is intimately related to the status of the cosmological constant Λ. It was first assumed as a free parameter in the field equations with the aim of accounting for a static Universe and then rejected because a cosmological expansion was observed subsequently. Such an issue to the cosmological problem has provided us with (authority and/or simplicity) arguments in favor of $\Lambda = 0$ until acceleration of the cosmological expansion could not be avoided for the interpretation of recent cosmological data. However, the related estimate does not agree by hundred orders of magnitude with its expected value as obtained from quantum field theories[5,40,20] by interpreting $\Lambda/8\pi G$ as vacuum energy density of quantum fluctuations. The aim of the present contribution is to understand this problem on geometrical grounds.

2. Status of the cosmological constant

The reason why the status of the cosmological constant Λ has long been discussed[38,11,6,30] has an historical origin, which still contributes to a so-

*Unité Mixte de Recherche (UMR 6207) du CNRS, et des universités Aix-Marseille I, Aix-Marseille II et du Sud Toulon-Var. Laboratoire affilié à la FRUMAM (FR 2291).

ciological debate at the present time. The key point is that either Λ is an *universal constant*[35], as it is clearly established in General Relativity (GR), as similarly as Newton constant of gravitation G, or it is associated to a particles/fields contribution to gravitation.

2.1. *Historical status of* Λ

For solving the cosmological problem, Einstein's goal was to obtain the gravitational field of a static universe, as it was supposed to be at that epoch. Similarly to the necessary modification of Poisson's equation for describing a uniform static distribution of dust in Newtonian gravity[a], Λ was assumed in the gravitational field equations accordingly to GR. With Mach's principle in mind (origin of inertia), a consistent cosmological solution describing a spatially closed universe[9] was derived, named as Einstein's model. A decade later, Friedmann's model[14] was used by G. Lemaître[18] for pointing out the cosmological expansion from Hubble's law[16], when Einstein's model was shown to be unstable[8b]. This summarises very briefly the state of the art as recorded in contemporary textbooks[43]. Henceforth, Friedmann's model with $\Lambda = 0$ was preferred because of Einstein's definite renouncement from the point of view of "logical economy"[10], what became the *Standard world model*[48]. His confession[c] to G. Gamov[15] stands probably for the historical reason why Λ was wrongly understood as a free parameter in GR theory of gravitation by the majority of observational cosmologists until recently. Although, $\Lambda \neq 0$ has been unsuccessfully envisaged in the 60's for explaining observations[17,22,23,33,28]. Nowadays, it is generally believed that $\Lambda \sim 2h^2\,10^{-56}\,\mathrm{cm}^{-2}$ (where $h = H_\circ/100\,km\,s^{-1}\,Mpc^{-1}$) is required for interpreting the CMB temperature fluctuations[32,19,34,1] and for accounting of Hubble diagram of SN[24,25,31,26]. Such a necessity was evident[d] two decades earlier from statistical investigations based on quasars and galaxiess[2,3,12,13,2,44,41].

[a]Such an approach has been used by Neumann (1896) in Newtonian theory, see R.C. Tolman[43].

[b]*i.e.* in addition of suffering from a fine tuning problem on the values of Λ and the specific density of energy of gravitational sources, any irregularity in their distribution causes either a collapse or an expansion.

[c]*loc. cit.* : "the biggest blunder of my life".

[d]It was ignored because not representative of the general consensus at that epoch, what is typical of the present days scholastic attitude.

2.2. *Geometrical status of* Λ

On geometrical grounds, *Principle of General Relativity* applied to gravity provides us with a unique interpretation of Λ. The gravitational field and its sources are characterized respectively by the metric tensor $g_{\mu\nu}$ on the space-time manifold V_4 and by a *vanishing divergence* stress-energy tensor $T_{\mu\nu}$. The gravitational field equations must be invariant with respect to the action of diffeomorphism group of V_4[35,36], and therefore they read

$$T_{\mu\nu} = -A_0 F^{(0)}_{\mu\nu} + A_1 F^{(1)}_{\mu\nu} + A_2 F^{(2)}_{\mu\nu} + \ldots \tag{1}$$

where $F^{(n)}_{\mu\nu}$ stands for a covariant tensor of degree $2n$, defined by means of metric tensor $g_{\mu\nu}$ and its derivatives, and A_n for a *coupling constant*, its value is estimated from observations. The $n = 0, 1$ order terms are uniquely defined

$$F^{(0)}_{\mu\nu} = g_{\mu\nu}, \quad F^{(1)}_{\mu\nu} = S_{\mu\nu} = R_{\mu\nu} - \frac{1}{2} R g_{\mu\nu} \tag{2}$$

where $R_{\mu\nu}$ is the Ricci tensor and R the scalar curvature, whereas $F^{n\geq 2}_{\mu\nu}$ must be derived from additional principles.

Schwarzschild solution of Eq. (1) enables us to identify $A_{n=0,1}$ with Newton approximation, what provides us with modified Poisson equation[35]

$$\mathrm{div}\tilde{\mathrm{g}} = -4\pi \mathrm{G}\rho + \Lambda \tag{3}$$

where $\vec{g}$ stands for the gravitational acceleration field due to sources defined by a specific density ρ, and the following identification of constants

$$\mathrm{G} = \frac{1}{8\pi A_1}, \qquad \Lambda = \frac{A_0}{A_1} \tag{4}$$

which shows their common status of *universal constant*.

3. Modeling gravitational structures

The space-time geometry is constrained by the presence of gravitational sources by means of tensor $T_{\mu\nu}$ in Eq. (1), each right hand terms contributes for describing the geometry within its effective scale. A dimensional analysis of field equations provides us with an estimation of their corresponding magnitudes, what is useful for modelling gravitational structures. Moreover, Newtonian approximation of gravitational field at large scale provides us with a simple way for interpreting Λ effect.

3.1. *Dimensional analysis*

According to GR, the speed of the light $c = 1$ (*i.e.* time can be measured in unit of length[e] $1\text{s} = 2.999\,792\,458\,10^{10}\,\text{cm}$) and then $\text{G} = 7.4243 \times 10^{-29}\,\text{cm}\,\text{g}^{-1}$. Let us choose units of mass and of length[f], herein denoted respectively by M and L. The correct dimensional analysis of GR sets the covariant metric tensor to have the dimension $[g_{\mu\nu}] = L^2$, and thus $[g^{\mu\nu}] = L^{-2}$, $[R_{\mu\nu}] = 1$ and $[R] = L^{-2}$. Since the specific mass density and the pressure belong to T^{μ}_{ν}, one has $[T_{\mu\nu}] = ML^{-1}$. Hence, according to Eq. (1), the dimensions of A_n are the following

$$[A_0] = ML^{-3}, \quad [A_1] = ML^{-1}, \quad \ldots [A_n] = ML^{2n-3} \tag{5}$$

which shows their relative contributions for describing the gravitational field with respect to scale. Namely, the larger their degree n the smaller their effective scale[g]. Equivalently, the estimation of A_0 demands observational data located at scale larger than the one for A_1, *etc....* This is the reason why the Λ effect is not discernible at small scale but requires cosmological distances.

3.2. *Newtonian gravity up to cosmological scales*

The observations show that gravitational structures within scales of order of solar system can be described by limiting the expansion solely to Einstein tensor $S_{\mu\nu}$, when cosmology requires also the first term. The transition scale between A_0 and A_1 is of order of $1/\sqrt{\Lambda} \sim 7h^{-1}$ Gyr. Although GR is preferred for investigating the dynamics of cosmic structures, Newton approximation given in Eq. (3) provides us with an easier schema for realizing the Λ effect. Hence, the acceleration field due to gravity around a point mass m reads

$$\vec{g} = \left(-\text{G}\frac{m}{r^3} + \frac{\Lambda}{3}\right)\vec{r} \tag{6}$$

Since $\Lambda > 0$, the gravity force is attractive at distance $r < r_\circ$ and repulsive at $r > r_\circ$ with a critical distance

$$r_\circ = \sqrt[3]{3m\text{G}/\Lambda} \tag{7}$$

where the gravity vanishes.

[e] This is the reason why any statement on the variation of c is meaningless in GR.
[f] Only two fundamental units can be chosen, the third one is derived.
[g] In other words, the contribution of A_0 dominates at scale larger than the one of A_1, *etc...*

4. The cosmological constant problem

Let us assume :

(1) quantum fluctuations are sources of the gravitational field by means of the following stress-energy tensor[h]

$$T_{\mu\nu}^{\rm vac} = \rho_{\rm vac}\, g_{\mu\nu}, \qquad \rho_{\rm vac} = \hbar k_{\max} \tag{8}$$

in the field equations Eq.(1), where $k_{\max}$ stands for the ultraviolet momentum cutoff up to which the quantum field theory is valid[5], one has

$$\rho_{\rm vac}^{EW} \sim 2\,10^{-4}, \quad \rho_{\rm vac}^{QCD} \sim 1.6\,10^{15}, \quad \rho_{\rm vac}^{Pl} \sim 2\,10^{89} \tag{9}$$

in units of $\mathrm{g\,cm^{-3}}$. The reason why such an estimation is not unique comes from the perturbative aspect of the theory for describing the quantum world, what can be understood as a weakness of this approach.

(2) the cosmological term interprets as the contribution of quantum fluctuations

$$\rho_\Lambda = \frac{\Lambda}{8\pi \mathrm{G}} \tag{10}$$

The difficulty of above hypotheses is that $\rho_{\rm vac}$ differs from

$$\rho_\Lambda \sim h^2\, 10^{-29}\, \mathrm{g\,cm^{-3}} \tag{11}$$

as obtained from astronomical observations, by 25–118 orders of magnitude. Such an enigma is named the *cosmological constant problem*. Other estimations of this quantum effect from the viewpoint of standard Casimir energy calculation scheme[49] provide us with discrepancies of $\sim$ 37 orders of magnitude[7]. A similar problem happens when

$$\Lambda_{\rm vac} = 8\pi \mathrm{G} \rho_{\rm vac} \tag{12}$$

is interpreted as a cosmological constant. Indeed, if the quantum field theory which provides us with an estimate of $\rho_{\rm vac}$ is correct then the distance from which the gravity becomes repulsive in the sun neighborhood ranges from $r_0^{EW} \sim 2\,10^{-2} h^{-2/3}$ a.u. down to $r_0^{Pl} \sim 3\,10^{-11} h^{-2/3}$ Å depending on the quantum field theory, see Eq. (7). Obviously, such results are not consistent with the observations.

[h] The usual picture which describes the vacuum as an isotropic and homogenous distribution of gravitational sources with energy density $\rho_{\rm vac}$ and pressure $p_{\rm vac} = -\rho_{\rm vac}$ (although this is not an equation of state) is not clear and not necessary for the discussion.

Another version of the cosmological constant problem points out a fine tuning problem. It consists on arguing on the smallness of $\Lambda = \Lambda_{\text{vac}} + \Lambda_\circ$, interpreted as an effective cosmological constant, where $\Lambda_\circ$ stands for a bare cosmological constant.

4.1. *Understanding the recession of galaxies*

The observations show that the dynamics of the cosmological expansion after decoupling era agrees with Friedmann-Lemaître-Gamov solution. It describes an uniform distribution of pressureless matter and CMB radiation with a black-body spectra, the field equations are given by Eq. (1) with $n \leq 1$. The present values of related densities are $\rho_{\text{m}} = 3h^2\,10^{-30}\ \text{g}\,\text{cm}^{-3}$ (dark matter included) and $\rho_{\text{r}} \sim 5h^2\,10^{-34}\ \text{g}\,\text{cm}^{-3}$. Their comparison to the expected vacuum energy density ρ_{vac} shows that if quantum fluctuations intervene in the dynamics of the cosmological expansion then their contribution prevails over the other sources (by 26–119 orders of magnitude today). Such an hypothesis provides us with a vacuum dominate cosmological expansion since primordial epochs. Therefore, one might ask whether such disagreements with observations can be removed by taking into account higher order $n \geq 2$ terms in Eq. (1). With this in mind, let us describe the dynamics of structures at scales where gravitational repulsion ($\Lambda > 0$) is observed. Since the values of universal constants G and Λ are provided by observations, it is more convenient to use adapted units of time l_g and of mass m_g defined as follows

$$l_g = 1/\sqrt{\Lambda} \sim h^{-1}\,10^{28}\ \text{cm}, \qquad m_g = 1/(8\pi G\sqrt{\Lambda}) \sim 4h^{-1}\,10^{54}\ \text{g} \tag{13}$$

herein called *gravitational units*. They are defined such that the field equations read in a normalized form

$$T_{\mu\nu} = -g_{\mu\nu} + S_{\mu\nu} + A_2 F^{(2)}_{\mu\nu} + \dots \tag{14}$$

i.e. $A_0 = A_1 = 1$, where the stress-energy tensor $T_{\mu\nu}$ accounts for the distribution of gravitational sources. It is important to note that, with gravitational units, Planck constant reads

$$\hbar \sim 10^{-120} \tag{15}$$

Indeed, such a tiny value as *quantum action unit* compared to $\hbar = 1$ when quantum units are used instead, shows clearly that Eq. (14) truncated at order $n \leq 1$ is not adapted for describing quantum physics[45,46]. This is the main reason why it is hopeless to give a quantum status to Λ[41]. As approximation, because of dimensional analysis described above, the contribution of higher order terms being the more significant as the density is large, Eq. (14) can be splited up with respect to scale into two equations systems. The first one corresponds to terms of order $n < 2$ (the usual Einstein equation with Λ) and the second one

$$T_{\mu\nu}^{\text{vac}} = A_2 F_{\mu\nu}^{(2)} + \dots \tag{16}$$

stands for the field equations describing the effect of quantum fluctuations on the gravitational field at an appropriated scale (quantum), interpreted as correction of the RW metric $g_{\mu\nu}$. The identification of constants A_n (*e.g.*, $A_2 = \hbar$) and the derivation of tensors $F_{\mu\nu}^{(n)}$ with $n \geq 2$ requires to model gravitational phenomena at quantum scale, see *e.g.*[42,39]. Unfortunately, the state of the art does not allow yet to provide us with a definite answer for defining the right hand term of Eq. (16), see *e.g.*[27].

5. Conclusion

Seeking for the contribution of quantum fluctuations into the cosmological constant is a motivation that inherits from the previous attitude consisting on rejecting the cosmological term. On the other hand, to rescale the field equations for describing the cosmological expansion prevents us to assume the vacuum acting as a cosmological constant. As a consequence, one understands that such an interpretation turns to be the origin of the cosmological constant problem. In other words, such a problem is the price to pay for identifying Λ_{vac} to the cosmological constant. Because the understanding of quantum gravity is still an ongoing challenge, the correct field equations describing the contribution to gravity of quantum fluctuations are not yet established. However, the dimensional analysis shows that the related gravitational effects are expected at small (quantum) scales and do not participate to the general expansion of the universe according to observations.

References

1. A. Benoit *et al.*, *Astron. Astrophys.*, **399**, L25 (2003)
2. G. Bigot, H.H.Fliche, R. Triay, *Astron. Astrophys.*, **206**, 1 (1988)
3. G. Bigot, R. Triay, in *The quest for the fundamental constants in Cosmology*, 9th Moriond Astrophysics meeting, eds J. Audouze and J. Trân Thanh Vân, Editions Frontières (1989)
4. S.M. Carroll, W.H. Press, E.L. Turner, *Annu. Rev. Astron. Astrophys.* **30**, 499 (1992)

5. S.M. Carroll, *Living Reviews in Relativity*, **4**,1 (2001)
6. J.C. Charlton, M.S. Turner, *Astrophys. J.*, **313**,495 (1987)
7. I.O. Cherednikov, *Acta Phys. Pol. B*, **33** (2002)
8. A.S. Eddington, *Month. Not. Roy. Astron. Soc.*, **90**, 668 (1930)
9. A. Einstein, *Sitz. Preuss. Akad. wiss. Phys-Math*, **142**, 87 (1917)
10. A. Einstein, *Sitz. Preuss. Akad. wiss. Phys-Math*, **235**,37 (1931); Appendix to the 2nd edn. of *The Meaning of Relativity*,(1945) on the "cosmological problem"; reprinted in (all later editions) Princeton University Press (1955), p.109
11. J.E. Felten, R. Isaacman *Rev. Mod. Phys.*, **58**,689 (1986)
12. H.H. Fliche, J.M. Souriau, *Astron. & Astrophys.*, **78**, 87 (1979)
13. H.H. Fliche, J.M. Souriau, R. Triay, *Astron. & Astrophys.*, **108**, 256 (1982)
14. A. Friedmann, *Z. Phys.*, **10**,377 (1922); *ibid.* **21**,326 (1924)
15. G. Gamow, in *My world line*, Viking Press, New York (1970), p.44
16. E.P. Hubble, *Proc. Nat. Acad. Sci.*, **15**, 168 (1927)
17. N. Kardashev, *Astrophys. J.*, **150**, L135 (1967)
18. G. Lemaître, *Ann. Soc. Sci. Bruxelles.*, **47 A**, 49 (1927); *Month. Not. Roy. Astron. Soc.*, **91**, 483 (1931)
19. C.B. Netterfield *et al.*, *Astron. J.*, **571**, 604 (2002)
20. T. Padmanabhan, *Phys. Rep.*, **380**,235 (2003)
21. J.V. Peach, *Astrophys. J.*, **159**, 753 (1970)
22. V. Petrosian, E.E., Salpeter, P. Szekeres *Astrophys. J.*, **147**, 1222 (1967)
23. V. Petrosian, E.E., Salpeter, *Astrophys. J..*, **151**, 411 (1968)
24. J.V. Perlmutter *et al.*, *Nature*, **391**, 51 (1998)
25. J.V. Perlmutter *et al.*, *Astrophys. J.*, **483**, 565 (1999)
26. A.J. Riess *et al.*, *Astron. J.*, **116**, 1009 (1998)
27. C. Rovelli, *Int. Jour. Mod. Phys. D*, **12**,1509 (2003)
28. M. Rowan-Robinson, *Month. Not. Roy. Astron. Soc.*, **141**, 445 (1968)
29. V. Sahni, A. Starobinsky, *Int. Jour. Mod. Phys. D*, **9**, 373 (2000)
30. A. Sandage, *Ann. Rev. Astron. Astrophys.*, **26**, 561 (1988)
31. P.B. Schmidt *et al.*, *Astrophys. J.*, **507**, 46 (1998)
32. J.L. Sievers *et al.*, *Astrophys. J.*, **591**, 599 (2003)
33. I. Shklovsky, *Astrophys. J.*, **150**, L1 (1967)
34. D.N. Spergel *et al.*; *Astrophys. J. Suppl.*, **148**, 175 (2003)
35. J.M. Souriau, *Géométrie et relativité* (Hermann, Paris, 1964)
36. J.M. Souriau, *Ann. Inst. H. Poincaré*, **XX** (4), 315 (1974)
37. J.M. Souriau, in Géométrie symplectique et physique mathématique, *Colloques Internationaux CNRS*, **237**, 59 (1974)
38. J.M. Souriau, in *Journées Relativistes – 1976*, ULB Bruxelles (1977)
39. K.S. Stelle, *Phys. Rev. D*, **16**,953 (1977)
40. M. Straumann, in *1rst Sem. H. Poincaré.* Paris (2002)
41. J.M. Souriau, R. Triay, *Grav. & Cosm.*, **3**,(9),51 (1997)
42. G.'t Hooft, M. Veltmann, *Ann. Inst. H. Poincaré*, (A) Phys. Th. **XX** (1), 69 (1974)
43. R.C. Tolman, in *Relativity Thermodynamics and Cosmology*, Oxford at the Clarendon Press (1934), p.344

44. R. Triay, in *Large Scale Structures and Peculiar Motions in the Universe*, eds. D.W. Latham and L.N. da Costa, ASP conference series (1989)
45. R. Triay, in *The Cosmological Model*, 37th Moriond Astrophysics meeting, Les Arcs, Editions Frontière (2002)
46. R. Triay, in Proceedings of *the Tenth Marcel Grossmann Meeting on General Relativity*, p. 2266 (Rio de Janeiro, Brazil 2003) . Edit. M. Novello & S.P. Bergliaffa, Ser. Edit. R. Ruffini. World Scientific (2006)
47. R. Triay, *Int. Journ. Mod. Phys. D*, **14**10,1667 (2005)
48. S. Weinberg, *Gravitation and Cosmology*, Ed. John Wiley & Sons Inc.
49. Ya. B. Zel'dovich, *JETP Lett.*, **6**, 316 (1967)
50. Ya. B. Zel'dovich, I.D. Novikov, in *The Structure and Evolution of the Universe*, Vol. 2, Ed. G. Steigman, The University of Chicago press (1983)

Astrophysics

Black Hole Mass Determination for Blazars*

J. H. Fan[1], J.Li[1], J.L. Zhou[1], T.X. Hua[1], Y. X. Wang[2], J. H. Yang[3]

1. Center for Astrophysics, Guangzhou University, Guangzhou 510006, China

2. College of Science and Trade, Guangzhou University, Guangzhou 511442, China

3. Department of Physics and Electronics Science, Hunan University of Arts and Science, Changde 415000, China

We determined the central black hole mass (M) for 25 γ-ray-loud blazars using their available variability timescales. In this method, the absorption effect depends on the γ-ray energy, emission size and property of the accretion disk. Using the intrinsic γ-ray luminosity as a fraction λ of the Eddington luminosity, $L_{\gamma}^{in} = \lambda L_{Ledd.}$ and the optical depth equal to unity, we can determine the upper limit of the central black hole masses. We found that the black hole masses range between $10^7 M_{\odot}$ and $10^9 M_{\odot}$ when $\lambda = 0.1$ and 1.0 are adopted. For the black hole mass there is no clear difference between BLs and FSRQs, which suggests that the central black hole masses do not play an important role in the evolutionary sequence of blazars or there is no evolution between BLs and FSRQs.

1. Introduction

The EGRET instrument at CGRO has detected many blazars (i.e. flat spectrum radio quasars (FSRQs) and BL Lacertae objects (BLs)). Blazars emit most of their bolometric luminosity in γ-rays (E > 100 MeV) (Hartman et al. 1999). Many γ-ray emitters are also superluminal radio sources (von Montigny et al. 1995). These objects share some common properties, such as luminous γ-ray emission and strong variability in the γ-ray and other bands on timescales from hours to days (see below). These facts suggest that γ-ray emission in blazars is likely arise from a jet. To explain its observational

*This work is supported by the National 973 project (NKBRSF G19990754), the National Science Fund for Distinguished Young Scholars (10125313), the National Natural Science Foundation of China (10573005), and the Fund for Top Scholars of Guangdong Provience (Q02114). We also thank the financial support from the Guangzhou Education Bureau and Guangzhou Science and Technology Bureau..

properties, a beaming (black hole + accretion disk + jet) model has been proposed. In the beaming model, a supermassive black hole is surrounded by an accretion disk. Many authors have tried to estimate the masses using different methods (see Barth et al. 2002; McLure, & Dunlop 2001; Rieger & Mannheim, 2000, 2003; Yuan et al. 2004, Yuan 2006, Fan 2005).

Since there is a large number of soft photons around the central black hole, it is generally believed that the escape of high energy γ-rays from the AGN depends on the $\gamma - \gamma$ pair production process. Therefore, the opacity of $\gamma - \gamma$ pair production in γ-ray-loud blazars can be used to constrain the basic parameters. Becker & Kafatos (1995) have calculated the γ-ray optical depth in the X-ray field of an accretion disk and found that the γ-rays should preferentially escape along the symmetry axis of the disk, due to the strong angular dependence of the pair production cross section. The phenomenon of $\gamma - \gamma$ "focusing" is related to the more general issue of $\gamma - \gamma$ transparency, which sets a minimum distance between the central black hole and the site of γ-ray production (Becker & Kafatos 1995, Zhang & Cheng 1997). So, the γ-rays are focused in a solid angle, $\Omega = 2\pi(1-cos\Phi)$, suggesting that the apparent observed luminosity should be expressed as $L_\gamma^{obs} = \Omega D^2 F_\gamma^{obs}(> 100MeV)$, where F_γ^{obs} and D are observed γ-ray energy flux and luminosity distance respectively. The observed γ-rays from an AGN require that the jet almost points towards us and that the optical depth τ is not greater than unity. The γ-rays are from a solid angle, Ω, instead of being isotropic. In this sense, the non-isotropic radiation, absorption and beaming (boosting) effects should be considered when the properties of a γ-ray-loud blazars are discussed. In addition, the variability time scale may carry the information about the γ-ray emission region. These considerations require a new method to estimate the central black hole mass and other basic parameters of a γ-ray-loud blazar, which is the focus of the present work.

$H_0 = 75$ km s^{-1} Mpc^{-1}, and $q_0 = 0.5$ are adopted throughout the paper.

2. Mass estimation method and result

2.1. *Method*

Here we describe our method of estimating the basic parameters, namely, the central black hole mass (M), the boosting factor (or Doppler factor) (δ), the propagation angle (Φ) and the distance along the axis to the site of the γ-ray production (d) for γ-ray-loud blazars with short timescale variabilities (see Cheng et al. 1999 and Fan 2005 for detail). To do so, we consider a two-temperature disk. The γ-ray observations suggest that the γ-rays are strongly boosted. From the high energy γ-ray emission we know that the optical depth of γ-γ pair production should not be larger than unity. In addition, the observed short-time scale gives some information about the size of emitting region. This can be used to constrain the basic parameters of a γ-ray-loud blazar as in the following.

Based on the papers by Becker & Kafatos (1995) and Cheng, Fan, and Zhang(1999), and the considerations of *Optical depth, Time scale and the site of $\gamma-$ray production, the $\gamma-$Ray luminosity*, and *the optical depth mimimum*, we can finally get four relations,

$$\frac{d}{R_g} = 1.73 \times 10^3 \frac{\Delta T_D}{1+z} \delta M_7^{-1}$$

$$L_{iso}^{45} = \frac{\lambda 2.52\ \delta^{\alpha_\gamma+4}}{(1-cos\Phi)(1+z)^{\alpha_\gamma-1}} M_7$$

$$9 \times \Phi^{2.5}(\frac{d}{R_g})^{-\frac{2\alpha_X+3}{2}} + kM_7^{-1}(\frac{d}{R_g})^{-2\alpha_X-3} = 1$$

$$22.5\Phi^{1.5}(1-cos\Phi) - 9 \times \frac{2\alpha_X+3}{2\alpha_\gamma+8}\Phi^{2.5} sin\Phi$$

$$-\frac{2\alpha_X+3}{\alpha_\gamma+4} kM_7^{-1} A^{-\frac{2\alpha_X+3}{2}} (1-cos\Phi)^{-\frac{2\alpha_X+3}{2\alpha_\gamma+8}} sin\Phi = 0 \qquad (1)$$

in which, there are four basic parameters. So, for a source with available data in the X-ray and γ-ray bands, the masses of the central black holes, M_7, the Doppler factor, δ, the distance along the axis to the site of the γ-ray production, d, and the propagation angle with respect to the axis of the accretion disk, Φ, can be derived from Eq. (1), where $R_{ms} = 6R_g$, $R_0 = 30R_g$, and E_γ= 1GeV are adopted.

2.2. *Results*

For a sample of 25 sources(Table 1), The data are from a paper by Fan (2005), X06 is from a paper by Xie et al. (2006). The black hole masses obtained in this work are in the range of $10^7 M_\odot$ to $10^9 M_\odot$, $(0.57 \sim 60) \times 10^7 M_\odot$ for $\lambda = 1.0$ and $(0.87 \sim 95.0) \times 10^7 M_\odot$ for $\lambda = 0.1$.

If we consider BLs and FSRQs separately, the distribution of the mass upper limits is not very show much different, their average masses are log$M$ = 8.05 ± 0.54 for FSRQs, and log$M$ = 8.1 ± 0.46 for BLs. There is no difference in black hole mass between BLs and FSRQs.

3. Discussion and Conclusion

In AGNs, the central black hole plays an important role in the observational properties. It may also shed some light on the evolution. There are several methods for black hole mass determinations although consensus has not been reached. In the present work, we proposed a method to estimate the central black hole mass. The central black hole masses are constrained by the optical depth of the $\gamma-\gamma$ pair production. This method can be used to determine the central black hole mass of high redshift gamma-ray sources.

BLs and FSRQs are two subclasses of blazars. From the observational point of view, except for the emission line properties (the emission line strength in FSRQs is strong while that in BL is weak or invisible), other observational properties are very similar between them (Fan 2002). Somebody proposed that there is an evolutionary process between BLs and FSRQs with the strong emission line FSRQs evolving into non-emission line BLs. If this is the case, there should be a tendency that the black hole masses in BLs are greater than those in FSRQs. However, it is not the case based on our estimation of central black hole masses. In the case of black hole mass, there is no clear difference between BLs and FSRQs, which suggests that the central black hole masses do not play an important role in the evolutionary sequence of blazars or there is no evolution process between BLs and FSRQs.

In this paper, the optical depth of a γ-ray travelling in the field of a two-temperature disk and beaming effects have been used to determine the central mass, M, for 25 γ-ray-loud blazars with available short time-scales. The masses obtained in the present paper are in the range of $10^7 M_\odot$ to $10^9 M_\odot$ for the whole sample. In the case of black hole mass, there is no clear difference between BLs and FSRQs, which suggests that the central black hole masses do not play an important role in the evolutionary sequence of blazars or there is no evolution process between BLs and FSRQs.

References

. Barth, A.J., Ho, L.C., Sargent, W.L.W., 2002, ApJL, 566, L13
. Becker P., Kafatos, M. 1995, ApJ, 453, 83
. Cheng K.S., Fan, J.H., Zhang, L., 1999, A&A, 352, 32
. Fan J.H. 2005, A&A, 436, 799
. Fan J.H. 2003, ApJL, 585, L23
. Hartman, R. 1999, ApJ, 121, 79
. McLure, R. J. & Dunlop, J. S., 2001, MNRAS, 327, 199
. Rieger, F. M. & Mannheim, K., 2000, A&A, 359, 948
. Rieger, F. M. & Mannheim, K., 2003, A&A, 397, 121

. von Montigny C., Bertsch D.L., Chiang J. et al. 1995, ApJ 440, 525
. Xie G.Z., et al. 2006, AJ, 131, 1210
. Yuan, Y.F., et al. 2004, ApJ, 606, 1112
. Yuan Y.F. 2006, in this proceeding volume
. Zhang L., Cheng K.S. 1997, ApJ 475, 534

Name (1)	z (2)	*ID* (3)	f_{1KeV} (4)	Ref (5)	α_X (6)	Ref (7)	F (8)	α_γ (9)	ΔT_D (10)	Ref (11)	M_1 (12)	$M_{0.1}$ (13)
0208-512	1.003	Q	0.61	C97	1.04	C97	9.1	0.69	134.4	S96	60.9	95.0
0219+428	0.444	B	1.56	Fo98	1.6	Fo98	0.25	1.01	30.0	DG	19.73	29.80
0235+164	0.94	B	2.5	M96	1.01	M96	0.65	1.85	72	M96	36.75	53.69
0420-014	0.915	Q	1.08	Fo98	0.67	C97	0.64	1.44	33.6	DG	10.98	16.43
0458-020	2.286	Q	0.1	DG	0.67	C97	0.68	1.45	144.0	DG	31.48	47.2
0521-365	0.055	B	1.78	DG	0.68	DG	0.32	1.63	72	DG	31.1	46.44
0528+134	2.07	Q	0.65	C97	0.54	C97	3.08	1.21	24.	DG	4.52	6.97
0537-441	0.894	B	0.81	C97	1.16	C97	2.0	1.0	16.	H96	10.46	15.96
0716+714	0.3	B	1.35	Fo98	1.77	Fo98	0.46	1.19	1.92	DG	2.22	3.28
0735+178	0.424	B	0.248	Fo98	1.34	Fo98	0.30	1.6	28.8	DG	20.03	29.37
0829+046	0.18	B	1.07	DG	0.67	C97	0.34	1.47	24.	DG	12.14	18.2
0836+710	2.17	Q	0.819	Fo98	0.42	Fo98	0.33	1.62	24.	DG	1.67	2.53
1101+384	0.031	B	37.33	Fo98	2.10	Fo98	0.27	0.57	1.92	DG	1.5	2.31
1156+295	0.729	Q	0.8	C97	0.86	C97	1.63	0.98	1.06	X06	8.0	10.7
1219+285	0.102	B	0.42	C97	1.30	C97	0.53	0.73	1.71	X06	7.82	10.1
1226+023	0.158	Q	12.07	Fo98	0.81	Fo98	0.09	1.58	24	C88	8.9	13.27
1253-055	0.537	Q	2.43	H96b	0.68	H96b	2.8	1.02	12.	K93	6.57	10.2
1510-089	0.361	Q	0.718	Fo98	0.90	Fo98	0.49	1.47	57.6	DG	28.03	41.9
1622-297	0.815	Q	0.08	M97	0.67	C97	17.	0.87	4.85	M97	5.44	8.51
1633+382	1.814	Q	0.42	C97	0.53	C97	0.96	0.86	16.	M93	3.43	5.42
1652+399	0.033	B	10.1	C97	1.60	C97	0.32	0.68	6.	Q96	4.01	6.27
2155-304	0.117	B	0.058	U97	1.25	U97	0.34	0.56	3.3	Ch99	3.26	5.11
2200+420	0.07	B	1.84	P96	1.31	P96	1.71	0.68	3.2	B97	11.0	15.95
2230+114	1.04	Q	0.486	Fo98	0.67	C97	0.51	1.45	48	P88	15.33	22.95
2251+158	0.859	Q	1.08	Fo98	0.62	Fo98	1.16	1.21	1.92	DG	0.57	0.87

Black Holes as the Central Engines for Astrophysical Sources *

Y.-F. Yuan and J.-M. Shi
*Center for Astrophysics, University of Science and Technology of China,
Hefei, Anhui 230026, P.R. China, E-mail: yfyuan@ustc.edu.cn*

The extraction of gravitational potential energy from matter which accretes into the deep potential well of black holes is believed to be the central engine of several important energetic sources in the Universe, such as active galactic nucleus (AGNs), gamma ray bursts (GRBs), X-ray binaries (XRBs), and so on. In this talk, I give a brief review of our recent works on black hole astrophysics. These works include: the possible observational properties of the afterglow of ultra-luminous quasars; the observed profile of an emission line from the conical jets around rotating a black hole; the temporal evolution of the hyperaccretion flow under beta equilibrium; and the relationship between the lack of type I x-ray bursts in XRBs and the evidence for black hole event horizons.

1. Afterglow of Ultra-luminous Quasars

Quasars were discovered in the early 1960's, the amazing characteristics of quasars are their huge luminosity ($\simeq 10^{46}$erg s^{-1}) and fast variations ($\simeq$ hours). It was quickly conceived that accretion of their surrounding matter onto supermassive black holes with $M \simeq 10^9 M_\odot$ are the central engine of quasars. [1,2,3] Now it is generally believed that a supermassive black hole (SMBH) with $M \simeq 10^6 - 10^9 M_\odot$ residing in the galactic center is the essential ingredient of the unified model for AGNs. Recently, based on the methods of stellar kinematics and gas dynamics, more and more solid evidence of the existence of SMBHs in the centers of galaxies are accumulating. [4] The discovery of the currently known highest redshift quasar ($z = 6.4$) from the Sloan Digital Sky Survey indicates that there are already SMBHs with $M > 10^9 M_\odot$. [5] The origin and evolution of SMBHs are becoming one

*This work is supported by the Program for New Century Excellent Talents in University, and the National Natural Science Foundation (10233030,10573016).

of the highlights in astrophysics. It is still unclear what is the relationship between the origin/evolution of SMBHs and the cosmological ionization and galaxy formation. [6]

Quasars represent a brief phase in the life-cycle of most massive galaxies. The lifetime of quasars remains uncertain, but they are generally thought to be short-lived, with estimated ages of $\sim 10^7$ to 10^9 years. [7,8] While the quasar is active, it provides a source of energy to heating infalling gas. If quasars are ultraluminous, shining brighter than the Compton-limit luminosity, their radiation heats the surrounding baryon gas to the Compton temperature, forming Compton spheres extending to the Strömgren radius of Fe^{26+}/He^{2+}. [9] After the quasars shut off, we predicted [9] that their "afterglow" can be detected through three signatures: (1) an extended X-ray envelope, with a characteristic temperature of $\sim 3 \times 10^7$ K; (2) Lyα and Lyβ lines and the $K-$edge of Fe^{26+}; and (3) nebulosity from hydrogen and helium recombination emission lines.

We show that the afterglows of ultraluminous quasars form relic Compton shperes that could be detected by *Chandra*, the planned *XEUS* mission, and ground-based optical telescopes. The luminosity and size of quasar afterglows can be used to constrain the lifetime of quasars, a key constraint on quasar and galaxy formation. [9]

2. Dynamics Evidence of Existence of Black Holes

The ASCA X-ray satellite first observed the broad iron Kα line at 6.4KeV which is thought to arise from X-ray irradiation of the cold surface layers of the accretion disk in Seyfert 1 galaxies. This line is intrinsically narrow in frequency, but the observed line profile of Fe Kα exhibits is so broad and asymmetric (red wing), which is corresponding to the Doppler motions with relativistic speeds ($\sim 0.3c$) and the gravitational redshift at the inner radius of 6 Schwarzschild radii. [10,11] The observations of broad Fe Kα line provides the most convincing evidence to date for the existence of SMBHs.

On the other hand, blueshifted and extremely strong X-ray emission lines have been detected recently in both AGNs and stellar black hole candidates (BHCs). It is argued that these lines might be from the relativistic outflow in black hole systems. [12] Motivated by the observations, we first studies the observed profile of an emission line from relativistic outflows around rotating black holes, simply assuming the jets as pencil beams. It was found that the observed line emission shows a doubly peaked profile at large viewing angles for extreme Kerr black holes. From the qualitatively analysis and numerical calculations, it was known that the low frequency peak is from the outer region of jets, while the high frequency peak is from the inner region. [13]

Later on, we improved our investigation by studying the emission line from a more realistic, cone-shaped jet, and we found that an intrinsically narrow line is basically singly peaked, the high frequency peak from the inner region disappears or weakens significantly. [12] This result is quite different from our previous results and emphasizes the difference between the observed line profile in the jet case and those in the disc case in which double peaks are frequently observed in the previous theoretical investigations. [14,15] In our opinion, the weakening or disappearance of the high frequency peak is due to the simplified assumption on the geometrical structure of jets. Roughly speaking, the differential emitting area of the jets is proportional to its radius, therefore, the innermost the position of the region of emission the fewer the photons emitted.

3. Central Engine of Gamma Ray Bursts

Gamma ray bursts is one of the most violent events in the Universe: during their explosion, $\simeq 10^{51} - 10^{53}$ergs is released in 0.1-1000 seconds. According to their durations (T_{dur}), GRBs are classified into two groups: short ($T_{dur} \leq 2$ s) and long bursts ($T_{dur} \geq 2$ s). [16] Multi-wavelength follow-up observations indicate long bursts are associated with the "collapsar" model in which a massive, rotating star collapses into a black hole, while short bursts invoke the merger of two neutron stars or a neutron star and a black hole. After the merger of a compact binary, or the collapse of a massive star, a dense, hot torus likely forms around a newly born black hole. The accretion of the torus onto black hole power the GRB fireball. The typical mass accretion rates are about $0.1 - 10 M_{\odot}/s$. [17,18,19] The most efficient cooling mechanism for the dense torus is provided by neutrino emission, then the annihilation of the emitted neutrinos might lead to the fireball generation (even though it occurs on timescales that are typically too short to provide the bulk of the GRB energy (e.g. Ref. 20).

This kind of hyperaccretion models have been discussed by many authors. A 1-dimensional, time-dependent disk model was presented in Ref. 21. Our current work improves on previous, 1-D time-dependent modeling of an hyperaccreting torus at several levels. [22] In particular, we introduce a detailed treatment of the equation of state, and calculate self-consistently the chemical equilibrium in the gas that consists of helium, free protons, neutrons and electron-positron pairs. We compute the chemical potentials of the species, as well as the electron fraction, throughout the disc. Another important addition compared to our previous work is the inclusion of the cooling term due to the photodisintegration of Helium. The presence of this term can substantially affect the energy balance and, it may eventually lead to an instability in the flow. The identification of instabilities in GRB accretion disks has lately become a topic of much interest in light of the recent *Swift* observations of flares in the early afterglow light curves. These flares could indeed be produced if the disk were to break up into rings which would then accrete on their viscous timescales. [23] Our calculations [22] have shown that such conditions can indeed be realized in a hyperaccreting torus.

4. Does the Event Horizon of Black Holes Exist?

Theoretical investigation indicated that the maximum mass of a neutron star is about $3.2M_{\odot}$. [24] If the masses of the compact objects in XRBs exceed the maximum mass of a neutrons star, they are observationally selected as black hole candidates (BHCs). So far, eighteen excellent BHCs have been discovered in this way. However, the existence of black holes remains open to doubt until other conceivable options are excluded. Q-stars is one of the possibilities. The maximum mass of Q-stars could be greater than $10M_{\odot}$, but Q-stars have hard surfaces. Can one show that such objects are ruled out as BHCs? One way to do this is to demonstrate that BHCs do not have hard surfaces. It is argued that BHCs should exhibit Type I X-ray bursts if they have surfaces. [25,26,27]

This still leaves open the possibility that BHCs may be made up of some kind of exotic dark matter with which normal gas does not interact. That is, the dark matter may be "porous" and permit accreting gas to fall through and to collect at the center. The dark matter and the fermionic gas would then behave as two independent fluids that interact only via gravity. [28] We refer to these objects as fermion-fermion stars and boson-fermion stars, respectively, where the first half of the name refers to the nature of the dark matter, and the second "fermion" in each name corresponds to the nucleonic gas component.

In order to check whether the BHCs could be fermion-fermion stars and boson-fermion stars, we consider models with a dark mass of $10M_{\odot}$ and a range of gas mass from $10^{-6}M_{\odot}$ to nearly $1M_{\odot}$, and analyze the bursting properties of the models when they accrete gas. We show that all the models would experience thermonuclear Type I X-ray bursts at appropriate mass accretion rates. For a range of accretion rates from a tenth of Eddington upto almost the Eddington rate, bursts occur reasonably frequently and have substantial fluences. These bursts would be hard to miss. Since no Type I bursts have been reported from black hole candidates,[29,30] the models are ruled out. The case for identifying black hole candidates in XRBs as true black holes is thus strengthened.

References

1. Ya. B. Zel'dovich and I. D. Novikov, *Sov. Phys. Dokl.* **158**, 811 (1964)
2. E. E. Salpeter, *Astrophys. J.* **140**, 796 (1964)
3. D. Lynden-Bell, *Nat.* **223**, 690 (1969)
4. L. Ho and J. Kormendy, in P. Murdin ed. *Enc. of Astron. & Astrophys.*, Inst. of Phys. Pub., (2000)
5. X. Fan, et al., *Astron. J.* **122**, 2833 (2001)
6. R. Barkana and A. Loeb, *Phys. Rep.* **349**, 125 (2001)
7. P. Jakobsen, et al., *Astron. Astrophys.* **397**, 891 (2003)
8. P. Martini and D.P. Schneider, *Astrophys. J. Letts.* **597**, L109 (2003)
9. J.-M. Wang, Y.-F. Yuan and L. Ho, *Astrophys. J. Letts.* **625**, L5 (2005)
10. A. C. Fabian, et al., *PASP* **112**, 1145 (2000)
11. C. S. Reynolds and M. A. Nowak, *Phys. Rep.* **377**, 389 (2003)
12. Y.-F. Yuan, *Astrophys. Space Sci.* **297**, 423 (2005)
13. J.-M. Wang, Y.-Y. Zhou, Y.-F. Yuan et al., *Astrophys. J.* **544**, 381 (2000)
14. A. C. Fabian, et al., *Mon. Not. R. Astron. Soc.* **238**, 729 (1989)
15. A. Loar, *Astrophys. J.* **376**, 90 (1991)
16. C. Kouveliotou, et al., *Astrophys. J. Lett.* **413**, L101 (1993)
17. T. Di Matteo, R. Perna and R. Narayan, *Astrophys. J.* **579**, 706 (2002)
18. Y.-F. Yuan, *Phys. Rev.* **D72**, 013007 (2005)
19. J.-M. Shi, Master thesis, USTC (2005)
20. S. Setiawan, M. Ruffert and H.-Th Janka, *Mon. Not. R. Astron. Soc.* **352**, 753 (2004)
21. A. Janiuk, R. Perna, T. Di Matteo and B. Czerny, *Mon. Not. R. Astron. Soc.* **355**, 950 (2004)
22. A. Janiuk, Y.-F. Yuan, R. Perna and T. Di Matteo, in T. Bulik, B. Rudak, G. Madejski, eds. *Astrophysical Sources of High Energy Particles and Radiation*, AIP, (2005)
23. R. Perna, P. Armitage and B. Zhang, *Astrophys. J. Lett.* **636**, L29 (2006)
24. C. E. Rhoades and R. Ruffini, *Phys. Rev. Lett.* **32** 324 (1974)
25. R. Narayan, et al., *Astrophys. J.* **308**, 110 (1997)
26. M. R. Garcia, et al., *Astrophys. J. Lett.* **553**, L47 (2001)
27. R. Narayan, *Prog. of Theore. Phys. Suppl.* **155**, 263 (2004)
28. Y.-F. Yuan, R. Narayan and M. Rees, *Astrophys. J.* **616**, 1112 (2004)
29. D. Tournear, et al., *Astrophys. J.* **595**, 1058 (2003)
30. J. E. McClintock, R. Narayan, G. B. Rybicki, **615**, 402 (2004)

The Third Cloud over General Relativity – Anomalous Acceleration of Pioneer 10 and 11 and its Possible Explanation *

Xu/Chongming and Wu/Xuejun
Purple Mountain Observatory, Nanjing 210008, China
Department of Physics, Nanjing Normal University, Nanjing 210097, China
E-mail:cmxu@pmo.ac.cn

In this paper, we review the problem of the anomalous acceleration of Pioneer 10 and 11 in the region of the deep space (> 10AU), which is un-modelled sunward constant acceleration $(8.74 \pm 1.33) \times 10^{-8}\mathrm{cm/s^2}$. The anomalous acceleration has been discovered for long time ago (1992) by Anderson et al., but recently it is confirmed further more after excluding almost all of known conventional physical errors. The Pioneer anomaly is so called the third dark cloud over the general relativity besides the dark matter and the dark energy. We also briefly introduce several published explanation on the Pioneer anomaly by means of conventional physics or new physics, but none of them is fully successful or commonly acceptable in our point view. A new plan to launch a mission to explore the Pioneer anomaly is mentioned in the paper. In the second part, we initiate the possible influence from the anomalous acceleration on the orbit motion of eight planets as a perturbation (pericenter-shift and radiu-shift) is calculated by us. To compare with the present observational precision, such kind of influence definitely does not exist for the bounded orbit motion of eight planets. Therefore we believe that, if the weak equivalence principle is correct everywhere, it is impossible to explain the Pioneer anomaly by means of any revising gravitational theory based on the general relativity.

1. Introduction

In the first part of this paper, we introduce the recent hot topic: the anomalous acceleration of Pioneer 10 and 11 in the region of the deep space which

*This work is supported by the National Natural Science Foundation of China (Grant No. 10273008)

is un-modelled sunward constant acceleration $(8.74 \pm 1.33) \times 10^{-8}\text{cm/s}^2$ [1,2,3,4]. The unknown anomalous acceleration of Pioneer is called the "Pioneer anomaly (PA)". The PA is so called the third dark cloud over the general relativity (first suggested by J. Ehlers (2005)[5]) besides the dark matter and the dark energy.

In the section 2, we will describe the radio-metric datum from Pioneer 10 and 11 since their launching (1972 and 1973) and other spacecraft (e.g. Galileo spacecraft and Ulysses spacecraft). The error from the datum is also mentioned. After excluding almost all of known conventional physical errors, it is believable that PA represents a real phenomenon (perhaps a new physics). More than two dozens of explanation on PA by means of conventional physics or new physics [2,4,6], but none of them is successful or commonly acceptable in our point view. A new plan to launch a mission to explore PA[3,4] is described also.

In the second part of this paper (section 3), we calculate the possible influence on the orbit motion of eight planets caused by the anomalous acceleration, while PA is taken as perturbation in the calculation. Our calculation, especially shift of pericenter, show the accumulated effects. But such kind of perturbation of the orbit motion does not find in the high precise ephemerides (EPM2004)[7].

Some conclusion remark is made at the last section. We believe that if the weak equivalence principle is correct everywhere, it is almost impossible to explain PA by means of any revising gravitational theory base on the general relativity. Therefore PA is really the third cloud over the general relativity.

2. Pioneer Anomaly

The Pioneer 10 & 11 missions are launched on Mar. 2, 1972 and Dec. 4, 1973 respectively. The Pioneer 10 & 11 pass hyperbolic escape orbits close to the plane of the ecliptic to opposite direction in the solar system. The analysis of the radio metric tracking data from the Pioneer 10 & 11 spacecraft has consistently indicated the existence of a constant un-modelled Doppler frequency shift of $\dot{\nu} \sim (5.99 \pm 0.01) \times 10^{-9}\text{Hz/s}$. The frequency shift can be understood as a constant sunward acceleration of $a_p = (8.74 \pm 1.33) \times 10^{-8}\text{cm/s}^2$. The datum are taken between 1987-1998 (11.5 years, 20055 datum) for Pioneer 10 (which was 40 AU to 70.5 AU distant from the sun) and between 1987-1990 (3.75 years, 19198 datum) for Pioneer 11 (22.4 to 31.7 AU).

Additionally, the anomalous acceleration exist not only in the Pioneer 10 & 11, but also in the Galileo spacecraft (launched in 1989) and in Ulysses spacecraft (launched in 1990). Their anomalous accelerations are

$$a_p = (8 \pm 3) \times 10^{-8}\text{cm/s}^2 \quad \text{(for Galileo spacecraft)}\,,$$
$$a_p = (12 \pm 3) \times 10^{-8}\text{cm/s}^2 \quad \text{(for Ulysses spacecraft)}\,.$$

Because Galileo and Ulysses spacecraft are influenced strongly by the solar radiation pressure, the measurement of the anomalous acceleration a_p may not be so reliable. Only when the distance is larger than 20 AU (where the solar pressure negligible), the measurement of a_p might be more reliable.

The analysis of datum (1987-1994) for Pioneer 10 is independently treated by Markwardt[8] $a_p = (8.60 \pm 1.34) \times 10^{-8}\text{cm/s}^2$. The error budget for Pioneer 10 has been discussed in detail[2,3]. The systematics comes from three parts (all of following unit is 10^{-8}cm/s^2):

1. Systematics generated external to the spacecraft (including: solar radiation pressure and mass (± 0.01, bias + 0.03); solar wind ($\pm < 10^{-5}$); solar corona (± 0.02); electro-magnetic Lorentz forces ($\pm < 10^{-4}$); influence of the Kuiper belt's gravity (± 0.03); influence of the Earth orientation (± 0.001); mechanical and phase stability of DSN (Deep Space Network) antennae ($\pm < 0.001$); phase stability and clocks ($\pm < 0.001$); DSN station location ($\pm < 10^{-5}$); troposphere and ionosphere ($\pm < 0.001$)).

2. On-board generated systematics (including: radio beam reaction force (± 0.11, bias +1.10); RTG (Radioisotope Thermoelectric Generators) heat reflected off the craft (± 0.55, bias -0.55); differential emissivity of the RTGs (± 0.85); non-isotropic radiative cooling of the spacecraft (± 0.48); expelled Helium produced within the RTGs (± 0.16, bias +0.15); gas leakage (± 0.56); variation between spacecraft determination (± 0.17, bias +0.17)).

3. Computational systematics (including: numerical stability of least-squares estimation (± 0.02); accuracy of consistency and model tests (± 0.13); mismodeling of maneuvers (± 0.01); mismodeling of the solar corona (± 0.02); annual/diurnal terms (± 0.32))

The total error and bias are estimated at ± 1.33 and ± 0.90 (10^{-8}cm/s^2). After counting all above error, the PA is believable and seems a real physical phenomenon which has been explained neither conventional physics nor engineering parameters of the spacecraft.

There are few dozens models (both within conventional physics and new physics which include unknown mass distribution in the solar system, local effect of the cosmic expansion, Schwarzschild solution with an expanding boundary, variable cosmological term Λ, gravitational constant G changed with time, NGT (Non symmetric Gravitational Theory), Momd (Modified Newtonian Dynamics), Bach's conformal gravity $\cdots$ and so on (references are in [2,4,6]). But non of them is commonly acceptable, because a successful theory has to explain one more unknown another problem (such as dark matter problem).

Some new plan to explore PA are suggested, e.g. Cosmic Vision[3] and Pioneer Anomaly Explorer[4]. The scientific objectives of new missions are:

- to investigate the origin of the Pioneer Anomaly with an improvement by a factor of 1000;
- to improve spatial, temporal and directional resolution;
- to test Newtonian gravity potential at large distances;
- to discriminate amongst candidate theories explaining a_p;
- to study the deep-space environment in the outer solar system;
- to improve limits on the extremely low-frequency gravitational radiation.

We expect a new mission to explore PA in a period of 10 years.

3. Orbit motion of planets perturbed by a_p

Normally there is no influence from a_p on the orbit motion to be considered. In Refs.[2,4] if we took the anomalous acceleration a_p as perturbation, then the perturbation of radial distance $\Delta r \sim -ra_r/a_N$ where a_N is the Newtonian radial acceleration. In the case of Earth-Mars $\Delta r \sim -21$km to -76 km, but in Viking mission (1976) the measurement accuracy from radio ranging is about 12 m[9]. Therefore if such kind of perturbation existed, it would be found already, then such kind of the perturbation does not exist at all. Our point view is that to estimate the influence of a_p on the orbit of planets by means of the perturbation of the semi-major axis and the data from Viking mission is not enough, since a_p is sunward and constant. If we consider a circular orbit as an extreme example, the contribution from a_p would not change the figure of the orbit, only change the radius or equivalently to the total solar mass increasing up to $1.35 \times 10^{-7} M_\odot$. To calculate the pericenter shift $\Delta\phi$ is important. Later we will understand net orbital effects for the semi-major axis to be zero.

In this section, we consider a_p as a constant radial acceleration to perturb the orbit motion of eight planets. The perturbation are linearly independent of other orbital perturbations. Since for Ulysses spacecraft (5.4 AU) and Galileon spacecraft (5.2 AU) the anomalous acceleration also exist, we assume the perturbation from constant sunward a_p for whole solar system, not only outer solar system[10,11a]. The Newtonian perturbation theory is applied for simplification. We take the general perturbation formulae for a test particle (or planet) as following (to see Eq.(4.54) of ref.[12]).

$$\dot{a} = \frac{2}{n\sqrt{1-e^2}}\left[e \sin f S + \frac{p}{r}\mathcal{T}\right], \tag{1}$$

$$\dot{e} = \frac{\sqrt{1-e^2}}{na}[\sin f S + (\cos f + \cos E)\mathcal{T}], \tag{2}$$

$$\frac{di}{dt} = \frac{r \cos u}{na^2\sqrt{1-e^2}}W, \tag{3}$$

$$\dot{\Omega} = \frac{r \sin u}{na^2\sqrt{1-e^2}\sin i}W, \tag{4}$$

$$\dot{\phi} = \frac{\sqrt{1-e^2}}{nae}\left[-\cos f S + \left(1+\frac{r}{p}\right)\sin f\mathcal{T}\right] - \frac{r \cot i \sin u}{na^2\sqrt{1-e^2}}W, \tag{5}$$

$$\dot{M}_0 = \left(\frac{1-e^2}{nae}\cos f - \frac{2r}{na^2}\right)S - \frac{1-e^2}{nae}\left(1+\frac{r}{p}\right)\sin f\mathcal{T}, \tag{6}$$

where a is the semi-major axis, e the eccentricity, i the inclination, Ω the longitude of the ascending node, ϕ the argument of pericenter and $M_0 = M - \int ndt$, M the mean anomaly and $n = 2\pi/T$ (T is the orbital period of the planet (test particle)), f the true anomaly counted from pericenter, $p = a(1-e^2)$ the semi-latus rectum of the Keplerian ellipse, $u = \phi + f$, S the radial projection of the perturbing acceleration, $\mathcal{T}$ and W the transverse (inside plane) and the normal (outside plane) projections of the perturbing acceleration (in the present situation $\mathcal{T} = W = 0$).

Considering the constant perturbing acceleration $S(= a_p)$, Eqs.(1)–(6) become

$$\dot{a} = \frac{2e}{n\sqrt{1-e^2}}\sin f S, \tag{7}$$

$$\dot{e} = \frac{\sqrt{1-e^2}}{na}\sin f S, \tag{8}$$

$$\frac{di}{dt} = 0, \tag{9}$$

$$\dot{\Omega} = 0, \tag{10}$$

$$\dot{\phi} = -\frac{\sqrt{1-e^2}}{nae}\cos f S, \tag{11}$$

$$\dot{M}_0 = \left(\frac{1-e^2}{nae}\cos f - \frac{2r}{na^2}\right)S. \tag{12}$$

[a] we just read ref.[11] when we prepared this proceeding paper

The secular effects can be got from average over one orbital period T. As we know from Eqs.(9) and (10), i (inclination) and Ω (longitude of the ascending node) are not perturbed by S. The shifts of the semi-major axis and the eccentricity over one orbital revolution vanish (from Eqs. (7) and (8)): $\overline{\frac{da}{dt}} = \overline{\frac{de}{dt}} = o$ as mentioned above. We also have

$$\overline{\frac{d\phi}{dt}} = \frac{S\sqrt{1-e^2}}{na}, \tag{13}$$

$$\overline{\frac{dM_0}{dt}} = -\frac{3S}{na} \tag{14}$$

where "—" means average over one orbital period. We do not calculate the datum from the Pluto since up to now the Pluto does not run a full orbital revolution during our observing on it. Datum from all of eight planets are calculated and shown in the Table:

Table 1. Table of perturbation caused by a_p.

	a	e	T	$\frac{d\phi}{dt}$	$\Delta\phi$	$\dot{a}$	Δa
Mercury	0.467	0.2056	0.24	$7.2\times^{-14}$	-9.6	$-4.41\times10^{-2}\sin f$	0
Venus	0.7282	0.0668	0.615	$3.62\times^{-13}$	-15.7	$-3.60\times10^{-2}\sin f$	0
Earth	1.0167	0.0167	1	$1.72\times^{-12}$	-18.7	$-1.46\times10^{-2}\sin f$	0
Mars	1.666	0.0934	1.88	$3.50\times^{-13}$	-21.3	$-1.54\times10^{-1}\sin f$	0
Jupiter	5.455	0.0484	11.867	$1.31\times^{-12}$	-41.3	$-5.30\times10^{-1}\sin f$	0
Saturn	10.07	0.0560	29.46	$1.52\times^{-12}$	-55.4	$-1.44\sin f$	0
Uranus	20.1	0.0461	84	$2.64\times^{-12}$	-70.5	$-3.39\sin f$	0
Neptune	30.4	0.010	164.8	$1.58\times^{-11}$	-102.9	$-1.44\sin f$	0

But such kind of perturbation on the shift of pericenter does not find in ephemerides (EPM2004)[7]. Therefore our conclusion is that a_p does not exist in the orbital motion of planets. Although the conclusion is the same as in Refs.[2,4], but we use more strict way.

4. Conclusion

(1) The PA is a real physical phenomenon and has not been explained either by means of conventional physics or traditional engineering parameters of the spacecraft. The existence of PA should be interested in for the fundamental physics.

(2) PA does not exist in the orbital motion of planets. If we recognize the weak equivalence principle correct everywhere, it is almost impossible to

explain in terms of any revising gravitational theory based on the general relativity. Therefore PA really challenge to the general relativity. In our opinion, maybe the transmission formula of the electromagnetic wave in the gravitational field would be changed.

Acknowledgments

Author would like to thank Prof. Yi/Zhao-Hua and Prof. S.G. Turyshev for their enlightening discussion.

References

1. J. D. Anderson, P. A. Laing, E. L. Lau, A. S. Liu, M. M. Nieto and S. G. Turyshev, *Phys. Rev. Lett.* **81**, 2858 (1998).
2. J. D. Anderson, P. A. Laing, E. L. Lau, A. S. Liu, M. M. Nieto and S. G. Turyshev, *Phys. Rev.* **D65**, 082004 (2002).
3. S. G. Turyshev, M. M. Nieto and J. D. Anderson, *arXiv: gr-qc/***0503021** (2005).
4. H. Dittus, S. G. Turyshev, C. Lämmerzahl, et al., *arXiv: gr-qc/***0506139v1** (2005).
5. S. G. Turyshev, "*Pioneer anommaly*", Colloquium of Dept. of Phys., Univ. of Bremen (Gemany), June 2, 2005.
6. " *Laser, Clooks and Drag-Free: Key Technologies for Future Space Exploration Testing Einstein's General Relativity in Space*" held at ZARM, Univ. of Bremen (Germany), May 30-June 1, 2005.
7. E. V. Pitjeva, *Solar System Research* **39**, 176 (2005).
8. C. Markwardt, *arXiv: gr-qc/***0208046v** (2002).
9. R. D. Reasenberg, et al., *Astrophys. J. Lett.* **234**, L219 (1979).
10. G. L. Page, D. S. Dixon annd J. F. Wallin, *arXiv: astro-ph/***0504367v3** (2005).
11. L. Iorio, *arXiv: gr-qc/***0601055v1** (2006).
12. Z. H. Yi, *Fundamental Celestial Mechanics*, Nanjing Univ. Pub., 1993 (in Chinese).

General Formula for Comparison of Clock Rates – Applications to Cosmos and Solar System*

Xu/Chongming and Wu/Xuejun
Purple Mountain Observatory, Nanjing 210008, China
Department of Physics, Nanjing Normal University, Nanjing 210097, China
E-mail:cmxu@pmo.ac.cn

Erwin Brüning
School of Mathematical Sciences, University of KwaZulu-Natal,
Durban 4000, South Africa

In this paper we deduce a quite general formula which allows the relation of clock rates at two different space time points to be discussed. In the case of a perturbed Robertson-Walker metric, our analysis leads to an equation which includes the Hubble redshift, the Doppler effect, the gravitational redshift and the Rees-Sciama effects. In the case of the solar system, when the 2PN metric is substituted into the general formula, the comparison of the clock rates on both the earth and a space station could be made. It might be useful for the discussion on the precise measurements on future ACES and ASTROD.

1. Introduction

A spatial experiment (to obtain an accuracy of order 10^{-16} in fractional frequency) named ACES (Atomic Clock Ensemble in Space) mission [1] is scheduled to be launched in 2006 by ESA (European Space Agency). Also, ASTROD (Astro-dynamical Space Test of Relativity using Optical Devices) [2] has been proposed, and the desired accuracy of the optical clock is about 10^{-17}. These plan need 2PN level on the comparison of clock rates. The precision of 2PN level on the comparison of clock rates has been discussed[3] by means of world function, but the calculation of the world function is not

*This work is supported by the National Natural Science Foundation of China (Grant No. 10273008)

easy, and they discuss the frequency shift only in the field of an axisymmetric rotating body, not in a general case. Therefore we deduce a general formula in a different way. Our general formula can also be applied to cosmos and easily extended to an even higher order (higher than 2PN level). We hope our general formula might be useful in ACES and ASTROD, and it might be also useful for the precise measurement of Hannay effect [4]

Many formulae have been suggested for the comparison of clock rates at different positions, based on certain simplifying assumptions. The change of the clock rates can be related to: the relativistic Doppler effect, the gravitational redshift, the Hubble redshift, the Rees-Sciama effect[5] and so on. The physical conditions causing all these effects may be all present. In early 90's these effects have been combined into one equation (see Eq.(6) in Ref.[6]) in first order approximation. Since all of terms are the level of the first order approximation, the coupling terms do not exist, and the authors do not deduce the equation through an exact method. Accordingly, a comprehensive approach, starting from first principles, is needed in which the physical conditions for all these effects are taken into account at the same time. Such an approach should lead us to a synthetic formula which reflects all these effects in a compact way and which should provide additional information, due to possible interactions which could not be incorporated in the isolated approaches for the individual effects.

2. General formula

In a global coordinates (ct, x^i), a source A moves with a velocity v_A^i and a receiver B with a velocity v_B^i. The clock rates in A and B are directly related with their own proper time $\Delta\tau_A$ and $\Delta\tau_B$ ($\frac{\Delta\tau_A}{\Delta\tau_B} = \frac{\Delta\tau_A}{\Delta t_A}\frac{\Delta t_A}{\Delta t_B}\frac{\Delta t_B}{\Delta\tau_B}$), Since $-c^2 d\tau^2 = ds^2$, we have

$$\Delta t_a = \frac{\Delta\tau_a}{\sqrt{-[g_{00}(a) + 2g_{0i}(a)v_a^i/c + g_{ij}(B)v_a^i v_a^j/c^2]}}, \tag{1}$$

where $a = A, B$ means point A or B, $g_{\mu\nu}$ is the global metric. As abbreviation we introduce

$$G_a = -(g_{00}(a) + 2g_{0i}(a)v_a^i/c + g_{ij}(a)v_a^i v_a^j/c^2)\,,.$$

Δt, being an integrable coordinate time interval, has unique meaning throughout space. One of the main purpose of our paper is to calculate the relation between Δt_A and Δt_B by means of a "calculus of differences". Assuming that, at t_{A_1} a source A emits a first pulse at position

$A_1(x^i_{A_1})$, then a receiver B received the first pulse at position $B_1(x^i_{B_1})$ at time t_{B_1}. A second pulse is emitted at $A_2(x^i_{A_2})$ and t_{A_2}, which is received by B at $B_2(x^i_{B_2})$ and t_{B_2}. Then $t_B = t_A + \int_A^B F(t, x^i)dx$ where we have defined $F(t, x^i) \equiv \left[-g_{0i}\frac{dx^i}{dx} - \sqrt{(g_{0i}g_{0j} - g_{00}g_{ij})\frac{dx^i}{dx}\frac{dx^j}{dx}}\right]/(cg_{00})$ and $dx^2 \equiv \delta_{ij}dx^i dx^j$. According to the "calculus of differences", we have $\Delta t_B = \Delta t_A + \Delta\left[\int_A^B F(t, x^i)dx\right]$. In a linear approximation in "calculus of differences"

$$\Delta\int_A^B F dx = \int_{x(B)}^{x(B+\Delta B)} F dx - \int_{x(A)}^{x(A+\Delta A)} F dx + \int_{x(A)}^{x(B)} \Delta F dx\,. \quad (2)$$

$\Delta x(A)$ and $\Delta x(B)$ are given by $\Delta x(a) \equiv x(a+\Delta a) - x(a) = \frac{\mathbf{k}_a}{|k_a|}\cdot\frac{d\mathbf{x}}{dx}\Big|_a \Delta x = \frac{\mathbf{k}_a}{|k_a|}\cdot \mathbf{v}_a\Delta t_a$. Here $a = A, B$, $\mathbf{k}_A$ is the wave vector at point a of the light signal. After calculating all the terms and considering $\int_{x(A)}^{x(B)} \frac{\partial F}{\partial x^i}\Delta x^i dx \ll \int_{x(A)}^{x(B)} \frac{\partial F}{\partial t}\Delta t dx$, the general formula is obtained finally as

$$\frac{\Delta\tau_B}{\Delta\tau_A} = \sqrt{\frac{G_B}{G_A}}\left(\frac{1 - F(A)\frac{\mathbf{k}_A\cdot\mathbf{v}_A}{|k_A|}}{1 - F(B)\frac{\mathbf{k}_B\cdot\mathbf{v}_B}{|k_B|}}\right) + \frac{\sqrt{G_B}}{\Delta\tau_A\left(1 - F(B)\frac{\mathbf{k}_B\cdot\mathbf{v}_B}{|k_B|}\right)}\int_{x(A)}^{x(B)}\left(\frac{\partial F}{\partial t}\Delta t + \frac{\partial F}{\partial x^i}\Delta x^i\right)dx\,. \quad (3)$$

Eq.(3) is our main result. Based on this equation we can calculate the comparison of the clock rate between two arbitrary points of space time. The equation is dependent on the metric and the path of the null geodesic line. The metric can be solved from the field equation in different situations (the energy momentum distribution and boundary conditions) while the path of the null geodesic line can be obtained from null geodesic equation.

For a moving source in Minkowski metric, our formula reduces to the Doppler effect in the special relativity. Again considering a static gravitational field (e.g. Schwarzschild metric), in which both source and receiver without moving ($\mathbf{v}_A = \mathbf{v}_B = 0$), the general form then becomes the equation of gravitational redschift shown in normal textbook of gravity.

3. Application in Cosmos and in Solar System

Consider a linearly simplest perturbed Robertson-Walker metric of the form

$$ds^2 = -c^2\left(1 - \frac{2w}{c^2}\right)dt^2 + \left(1 + \frac{2w}{c^2}\right)\frac{R^2\delta_{ij}dx^i dx^j}{\left(1 + \frac{k}{4}r^2\right)^2}\,, \quad (4)$$

where the gravitational potential $w = w(t, x^i)$ is assumed to be a small quantity. We assume the Doppler effect caused only by the motion of the source, then $\mathbf{v}_B = 0$ (also possible $\mathbf{v}_A = 0$). The velocity of the source A is $v^i_A = R(t_A)dx^i_A/dt$. $F(A)$, G_A, and G_B can be calculated easy and the integral in Eq.(3) can be derived as

$$I = \frac{\Delta t_A}{R(t_A)} \int_{x(A)}^{x(B)} \left(\dot{R}(t) + 2\frac{R(t)}{c^2}\frac{\partial w}{\partial t} \right) dt$$
$$= \Delta t_A \left[\left(\frac{R(t_B) - R(t_A)}{R(t_A)} \right) + \frac{2}{c^2 R(t_A)} \int_{x(A)}^{x(B)} R(t) \frac{\partial w}{\partial t} dt \right] . \qquad (5)$$

We finally obtain

$$\frac{\Delta\tau_B}{\Delta\tau_A} = \left[1 + \frac{1}{c^2} \left(w(t_A, x^i_A) - w(t_B, x^i_B) \right) + \frac{v_A^2}{c^2} \right]$$
$$\left\{ \frac{R(t_B)}{R(t_A)} - \frac{\mathbf{k}_A \cdot \mathbf{v}_A}{c|k_A|} + \frac{2}{c^2 R(t_A)} \int_{x(A)}^{x(B)} R(t) \frac{\partial w}{\partial t} dt \right\} , \qquad (6)$$

where $\frac{1}{c^2}\left(w(t_A, x^i_A) - w(t_B, x^i_B)\right)$ is the contribution from the normal gravitational redshift; $(\mathbf{k}_A \cdot \mathbf{v}_A)/(c|k_A|)$ and v_A^2/c^2 are the Doppler effect and transverse Doppler effect (or relativistic Doppler effect) respectively; $R(t_B)/R(t_A)$ just contributes to Hubble redshift; the last term is related to Rees-Sciama effect. We thought that our general formula allows to derive the Birkinshaw-Gull effect [7] too, if we use suitable perturbation functions for $w(t, x^i)$ and $w_j(t, x^i)$.

Our formula (3) is of substantial generality and accordingly allows a great number of fruitful applications. A important application probably is to the solar system since there we have a chance to do measurement and thus by comparison with measurements our formula can be tested. Here, we provide a fundamental formula with the precision $O(4)$, but we have not expanded the potential and vector potential by means of multiple moments that has to be done in the practical problems. In near future high precision measurement will be done up to 2PN level, thus allowing the coupling term (i.e., the term connecting the gravitational redshift, the Doppler redshift and so on) to be measured. Our scheme, offers the possibility for this if an appropriate assumptions about the metric are used. Accordingly, in this section we start from DSX formalism [8] and its extension [9]. First we evaluate the terms G_A and G_B for the extended DSX metric, considering the terms $F(A)\mathbf{k}_A \cdot \mathbf{v}_A/|k_A|$ and $F(B)\mathbf{k}_B \cdot \mathbf{v}_B/|k_B|$, the integral (the second term) in Eq.(3) can be evaluated by the median method, i.e., $\int_A^B \frac{\partial F}{\partial t} \Delta t dx = \Delta\bar{t} \int_A^B \frac{\partial F}{\partial t} dx$, where $\Delta\bar{t}$ is the median value, $\Delta\bar{t} = \eta \Delta t_A$, η is a introduced parameter which is closed to 1. Gathering all evaluations done, we finally obtain a general formula for the comparison of clock rates in the solar system, on the 2PN level of precision:

$$
\begin{aligned}
\frac{\Delta\tau_B}{\Delta\tau_A} &= 1 + \left\{ \frac{1}{c^2}\Big(w(A) - w(B)\Big) + \frac{1}{2c^2}\left(v_A^2 - v_B^2\right) - \left(\frac{\mathbf{k}_A \cdot \mathbf{v}_A}{c|k_A|} - \frac{\mathbf{k}_B \cdot \mathbf{v}_B}{c|k_B|}\right) \right. \\
&\left. -\frac{1}{c^2}\left(\frac{\mathbf{k}_B \cdot \mathbf{v}_B}{|k_B|}\right)\left(\frac{\mathbf{k}_A \cdot \mathbf{v}_A}{c|k_A|} - \frac{\mathbf{k}_B \cdot \mathbf{v}_B}{c|k_B|}\right)\right\} \\
&+\frac{1}{c^3}\left\{\Big(w(B) - w(A)\Big)\left(\frac{\mathbf{k}_B \cdot \mathbf{v}_B}{c|k_B|} + \frac{\mathbf{k}_A \cdot \mathbf{v}_A}{c|k_A|}\right) + 2w(A)\left(\frac{\mathbf{k}_B \cdot \mathbf{v}_B}{c|k_B|} - \frac{\mathbf{k}_A \cdot \mathbf{v}_A}{c|k_A|}\right)\right. \\
&\left.-\frac{1}{2}\Big(v^2(B) - v^2(A)\Big)\left(\frac{\mathbf{k}_B \cdot \mathbf{v}_B}{c|k_B|} - \frac{\mathbf{k}_A \cdot \mathbf{v}_A}{c|k_A|}\right) + \left(\frac{\mathbf{k}_B \cdot \mathbf{v}_B}{c|k_B|}\right)^2\left(\frac{\mathbf{k}_B \cdot \mathbf{v}_B}{c|k_B|} - \frac{\mathbf{k}_A \cdot \mathbf{v}_A}{c|k_A|}\right)\right\} \\
&+\frac{1}{c^4}\left\{\frac{1}{2}\Big(w(B) - w(A)\Big)^2 + \frac{1}{2}\Big(10w(A) - w(B)\Big)v_A^2 - \frac{1}{2}\Big(w(A) + 6w(B)\Big)v_B^2\right. \\
&+\frac{1}{8}\Big(3v_A^4 - 2v_A^2v_B^2 - v_B^4\Big) + 4\Big(w_i(B)v_B^i - w_i(A)v_B^i - w_i(B)\frac{k_B^i}{|k_B|}\frac{\mathbf{k}_B \cdot \mathbf{v}_B}{c|k_B|} \\
&+w_i(A)\frac{k_A^i}{|k_A|}\frac{\mathbf{k}_A \cdot \mathbf{v}_A}{c|k_A|}\Big) + \left(\frac{\mathbf{k}_B \cdot \mathbf{v}_B}{c|k_B|}\right)^2\Big(3w(B) + w(A) + \frac{1}{2}(v_A^2 - v_B^2)\Big) \\
&\left.-2\left(\frac{\mathbf{k}_B \cdot \mathbf{v}_B}{c|k_B|}\right)\left(\frac{\mathbf{k}_A \cdot \mathbf{v}_A}{c|k_A|}\right)\Big(w(A) + w(B)\Big) + \left(\frac{\mathbf{k}_B \cdot \mathbf{v}_B}{c|k_B|}\right)^3\left(\frac{\mathbf{k}_B \cdot \mathbf{v}_B}{c|k_B|} - \frac{\mathbf{k}_A \cdot \mathbf{v}_A}{c|k_A|}\right)\right\} \\
&+\frac{2\eta}{c^2}\int_A^B \frac{\partial w}{\partial t}dt + O(5)\,. \qquad (7)
\end{aligned}
$$

To sum up our discussion we can say that Formula (7) 'contains' the Doppler effect, transverse Doppler effect (relativistic Doppler effect), gravitational redshift and their complete coupling effects to 2PN level in the solar system. In addition there is a term which is the integral of the rates of change of the scalar potential along the null geodetic line from source A to receiver B. This is probably the most interesting result in our paper. Hopefully this integral term and the coupling effects can be tested in the future with a deep space explorer and are confirmed.

References

1. A. Spallicci, et al., *Class. Quantum Grav.* **14**, 2971 (1997).
2. W. T. Ni, S. Shiomi and Liao/ A-C, *Class. Quantum Grav.* **21**, S641 (2004).
3. B. Linet and P. Teyssandier, *Phys. Rev.* D **66** 024045 (2002).
4. J. H. Hannay, *J. Phys. A: Math. Gen.* **18**, 221 (1985).
5. M. Rees and D. W. Sciama, *Nature* **217**, 511 (1968).
6. E. Martinez-Gonzalez, J. L. Sanz and J. Silk, *Astrophys. J.* **355**, L5-L9 (1990).
7. M. Birkinshaw and S. F. Gull, *Nature* **302**, 315 (1983).
8. T. Damour, M. Soffel and C. Xu, *Phys. Rev.* D **45**, 1012 (1992).
9. C. Xu and X. Wu, *Chin. Phys. Lett.* **20**, 195 (2003).

Searching for sub-millisecond pulsars: A theoretical view *

R. X. Xu (Xu/Renxin)
School of Physics, Peking University, Beijing 100871, China
E-mail: r.x.xu@pku.edu.cn

Sub-millisecond pulsars should be triaxial (Jacobi ellipsoids), which may not spin down to super-millisecond periods via gravitation wave radiation during their lifetimes if they are extremely low mass bare strange quark stars. It is addressed that the spindown of sub-millisecond pulsars would be torqued dominantly by gravitational wave radiation (with braking index $n \simeq 5$). The radio luminosity of sub-millisecond pulsars could be high enough to be detected in advanced radio telescopes. Sub-millisecond pulsars, if detected, should be very likely quark stars with low masses and/or small equatorial ellipticities.

1. Introduction

Historically, the very idea of "*gigantic nucleus*" (neutron star) was first given by Landau more than 70 years ago, was soon involved in supernova study of Baade and Zwicky, and seems to be confirmed after the discovery of radio pulsars. Although this is a beautiful story, no convincing work, either theoretical from first principles or observational, has really proved this idea that pulsars are normal neutron stars, since nucleon (neutron or proton) was supposed to be point-like elementary particles in Landau's time but not now. An idea of quark stars, which are composed by free quarks rather than free nucleus, was suggested as more and more sub-nucleon phenomena were understood, especially the asymptotic freedom nature of strong interaction between quarks; and it is found that no problem exists in principle if pulsars are quark stars. Therefore, astrophysicist without bias should think equally neutron and quark stars to be two potential models for the nature of pulsar-like stars[1].

*This work is supported by National Scientific Foundation of China (10573002) and the Key Grant Project of Chinese Ministry of Education (305001).

One of the most important problems could then be differentiating neutron and quark stars in the new millennium astrophysics. Many likely ways to identifying a quark star are proposed[1], but the search of sub-millisecond pulsars would be an excellent key experiment. The reason for this is very intuitive and almost model-independent. Normal neutron stars are gravitationally confined, and thus can not spin with periods being less than a break period of $\sim 0.5 M_1^{1/2} R_6^{-3/2}$ ms (mass: $M_1 = M/M_\odot$, radius: $R_6 = R/10^6$ cm), but quark stars are chromatically confined, without limitation by the Kepler frequency. Additionally, a much high spin frequency may not damped significantly if a quark star has a very low mass. Therefore, sub-millisecond pulsars should be expected to be detected if pulsars are (low-mass) quark stars, with a rapid spin at birth, but could not exist if they are normal neutron stars.

We are viewing the implications of possibly future discovery of sub-millisecond pulsars in this paper, with (solid) quark star models to be focused for the nature of pulsar-like compact objects. The formation, evolution, and magnetospheric activity of sub-millisecond pulsars are also discussed. Because the pulse profiles of radio pulsars are highly modulated ($\sim 100\%$), with very small pulse-widths (a few $\sim 10^\circ$), the timing precision could only be high enough to uncover sub-millisecond pulsars in radio band. We focus thus radio pulsars here.

2. Spindown of sub-millisecond pulsars

2.1. *Spin periods limited by gravitational wave emission*

Fluid pulsars. In Newtonian theory, a rapidly rotating fluid Maclaurin spheroid is secularly unstable to become a Jacobi ellipsoid, which is non-axisymmetric, if the ratio of the rotational kinetic energy to the absolute value of the gravitational potential energy $T/|W| > 0.1375$. For a Maclaurin spheroid with homogenous density ρ, the ratio $T/|W| = 0.1375$ results in an eccentricity $e = 0.81267$ (or $c/a = \sqrt{1-e^2} \simeq 0.6$) and thus in a critical frequency $\Omega_c \sim 5.6\times 10^3$/s (i.e., spin period $P_c \sim 1.1$ ms) if $\rho = 4\times 10^{14}$ g/cm^3. Sub-millisecond pulsars could then be Jacobi ellipsoids, to be triaxial. In the general relativistic case[2,3], gravitational radiation reaction amplifies an oscillation mode, and it is then found that the critical value of $T/|W|$ for the onset of the instability could be much smaller than 0.1375 for neutron stars with mass of $\sim M_\odot$. This sort of non-axisymmetric stellar oscillations will inevitably result in gravitational wave radiation, and put limits on the spin periods.

A kind of oscillation mode, socalled r-mode, is focused on in the literatures[4,5,6]. The r-mode oscillation is also called as the Rossby waves that are observed in the Earth's ocean and atmosphere, the restoring force of which is the Coriolis force. This instability may increase forever if no dissipation occurs. Therefore, whether the instability can appear and how much the oscillation amplitude is depend on the interior structure of pulsars, which is a tremendously complicated issue in supranuclear physics.

The Kepler frequency of low-mass bare strange stars could be approximately a constant,

$$\Omega_0 = \sqrt{\frac{GM}{R^3}} = 1.1 \times 10^4 \ \ \mathrm{s}^{-1}, \tag{1}$$

where the average density is taken to be $\sim 4 \times 10^{14}$ g/cm^3. Correspondingly, the spin period $P_0 = 2\pi/\Omega_0 \simeq 0.6$ ms $< P_c$. It is found by Xu[7] that the gravitational wave emissivity of quark stars is mass-dependent. The r-mode instability could not occur in fluid bare strange stars with radii being smaller than ~ 5 km (or mass of a few $0.1 M_\odot$) unless these stars rotates faster than the break frequency (in fact, a more effective gravitational wave emission mode occurs if $P <\sim P_0$, see Eq.(5)). These conclusions do not change significantly in the reasonable parameter-space of bag constant, strong coupling constant, and strange quark mass.

Some recent observations in X-ray astronomy could hint the existence of low-mass bare strange stars[8]. The radiation radii (of, e.g., 1E 1207.4-5209 and RX J1856.5-3754) are only a few kilometers (and even less than 1 km). No gravitational wave emission could be detected from such fluid stars even they spin only with a period of ~ 1 ms.

Solid pulsars. A protoquark stars should be in a fluid state when their temperatures are order of 10 MeV, but would be solidified as they cool to very low temperatures[9,?]. Assuming the initial ellipticity of a solid quark star keeps the same as that of the star just in its fluid phase, strain energy has to develop when a solid quark star spins down. Quake-induced glitches of the quark star occur when the strain energy reaches a critical value[10], and we thus suggest that the stellar ellipticity would be approximately determined by the conventional Maclaurin spheroids (for $P \gg 1$ ms)

$$\varepsilon(P) \simeq \frac{5\Omega^2}{8\pi G\rho} \simeq 3 \times 10^{-3} P_{10\mathrm{ms}}^{-2}, \tag{2}$$

where the spin period $P = 2\pi/\Omega = P_{10\mathrm{ms}} \times 10$ ms, provided that the density $\rho \simeq 4 \times 10^{14}$ g/cm^3 is a constant.

A pulsar must be non-axisymmetric in order to radiate gravitationally. A triaxial pulsar, with deformation ellipticity ϵ_e in its equatorial plane, or a wobbling pulsar, either freely or forcedly, may thus radiate gravitational waves, the frequency of which is 2Ω for the former but is $\Omega + \Omega_{\rm prec}$ (the precession frequency $\Omega_{\rm prec}$ is orders of magnitude smaller than Ω) for the later. This wave results in a perturbed metric, which is order of h_0 being given by[11],

$$h_0 = \frac{128\pi^3 G\rho_0}{15c^4} \cdot \frac{R^5}{dP^2}(\epsilon_e \text{ or } \epsilon\theta) \approx 2.8 \times 10^{-20} R_6^5 d_{\rm kpc}^{-1} P_{\rm 10ms}^{-2}(\epsilon_e \text{ or } \epsilon\theta), \quad (3)$$

where approximations $I \simeq 0.4MR^2$ and $M \simeq 4\pi R^3\rho_0/3$ are applied for solid quark stars in the right equation, the pulsar's distance to earth is $d = d_{\rm kpc} \times 1$ kpc, θ is the wobble angle.

For normal neutron stars, ϵ_e and $\epsilon\theta$ are supported by crustal shear stress and magnetic pressure. However for solid quark stars, this mechanisms may not work due to a relatively negligible magnetic and Coulomb forces. Nevertheless, glitches of solid quark stars could also produce bumps, with a maximum ellipticity[12],

$$\epsilon_{\rm max} \sim 10^{-3}(\frac{\sigma_{\rm max}}{10^{-2}})R_6^{-6}(1 + 0.084R_6^2)^{-1}, \quad (4)$$

where $\sigma_{\rm max}$ is the stellar break strain. This ellipticity is larger for low-mass quark stars due to weaker gravity. This maximum ellipticity could be much smaller than the ellipticity of Maclaurin spheroids with $P <\sim 1$ ms (sub-millisecond pulsars). Therefore, for the sake of simplicity, we assume that the real ellipticity of a solid quark star could be $\varepsilon(P)$, due to stress releases through star-quake induced glitches[10], in the discussion below.

LIGO is sensitive to hight frequency waves, which recently puts upper limits on h_0 for 28 known pulsars through the second LIGO science run[13]. The upper limits are order of 10^{-24}, which means approximately an limit of $\epsilon R_6^5 < 10^{-4}$. Only three normal pulsars are targeted; others are millisecond pulsars. The upper limits of masses and radii for millisecond pulsars are constrained[7] by the second LIGO science run. Specially, the radius of the fastest rotating pulsar, PSR B1937+21, could be smaller than ~ 2 km if its wobble angle θ is between 1° and 10°.

Energy of gravitational wave is not as instinctive as the perturbed metric, h_0. Nonetheless, for triaxial sub-millisecond pulsars, the luminosity of gravitational waves radiation from a solid pulsar could be obtained[14], and the total rotation energy loss via gravitational and magnetospheric (photons and particles) emission is

$$-I\Omega\dot{\Omega} = \frac{32G\Omega^6 I^2 \epsilon_e^2}{5c^5} + \frac{2}{3c^3}\mu_m^2 M^2 \Omega^4, \tag{5}$$

where μ_m is the magnetic momentum per unit mass. It is worth mentioned that the work relevant to rapid rotating Jacobi ellipsoids was done by many authors[15,16,17,18,19]. The ratio of the term due to gravitational wave and that due to magnetospheric activity in Eq.(5) is

$$f_r = \frac{192GR_{\rm eff}^4 \epsilon_e^2}{125c^2\mu_m^2}\Omega^2 \simeq 4.5 \times 10^{11} R_{\rm effkm}^4 \epsilon_e^2 \mu_{m-6}^{-2} P_{\rm ms}^{-2}, \tag{6}$$

where an effective radius $R_{\rm eff} = R_{\rm effkm} \times (1\ {\rm km})$ is defined through $I = 2MR_{\rm eff}^2/5$, $\mu_m = \mu_{m-6} \times (10^{-6}\ {\rm G\ cm^3\ g^{-1}})$, $P = 2\pi/\Omega = P_{\rm ms} \times (1\ {\rm ms})$. It is evident from Eq.(6) that the braking torqued by gravitational wave dominates the spindown of a sub-millisecond pulsar unless $\epsilon_e \ll 1$ and/or $R_{\rm eff} \ll 1$ km. The braking index for gravitational wave emission is $n \equiv \Omega\ddot{\Omega}/\dot{\Omega}^2 = 5$ from Eq.(5), and this feature of $n > 3$ could be evidence for gravitational wave in return. Therefore, gravitational wave radiation alone could result in an increase of the rotation period of sub-millisecond pulsars,

$$\dot{P} = \frac{K}{P^3}, \quad \text{with } K \equiv \frac{512\pi^4 GI\epsilon_e^2}{5c^5} \simeq \frac{4096\pi^5 G}{75c^5}\rho\epsilon_e^2 R_{\rm eff}^5, \tag{7}$$

where the pulsar mass is approximated by $M \simeq 4\pi R_{\rm eff}^3 \rho/3$.

The solution of P from Eq.(7) is $P^4 = P_i^4 + 4Kt$ (P_i denotes the initial spin period), which is shown in Fig. 1 for different parameters of ϵ_e and $R_{\rm eff}$ ($\rho = 4 \times 10^{14}$ g/cm^3 is assumed). We see that the period increase rate is proportional to $(\epsilon_e^{2/5} R_{\rm eff})^5$, a much small value of ϵ_e and/or a low mass could make it possible that sub-millisecond pulsars exist in the Universe. Since $K \propto \rho R_{\rm eff}^5 \sim MR_{\rm eff}^2$, P would increase much faster for normal neutron stars than that for quark stars because of the sharp difference between the mass-radius relations of these two kinds of stars, if the equatorial ellipticities, ϵ_e, are almost the same. The origin of ϵ_e could be due to the secular instability of Maclaurin spheroid (to become Jacobi ellipsoid finally) before solidification, or the glitch-induced bumpy surface after solidification. Detailed calculation on ϵ_e is necessary and essential for us to know the spindown features of sub-millisecond pulsars as well as to estimate the gravitational wave strength emitted from such stars. It is worth noting that quark nuggets (stars) born during cosmic QCD phase separation may exist as super-millisecond pulsars due to spindown in Hubble time ($\sim 10^{10}$ years) although they might still keep sub-millisecond spin if their values of K is small enough.

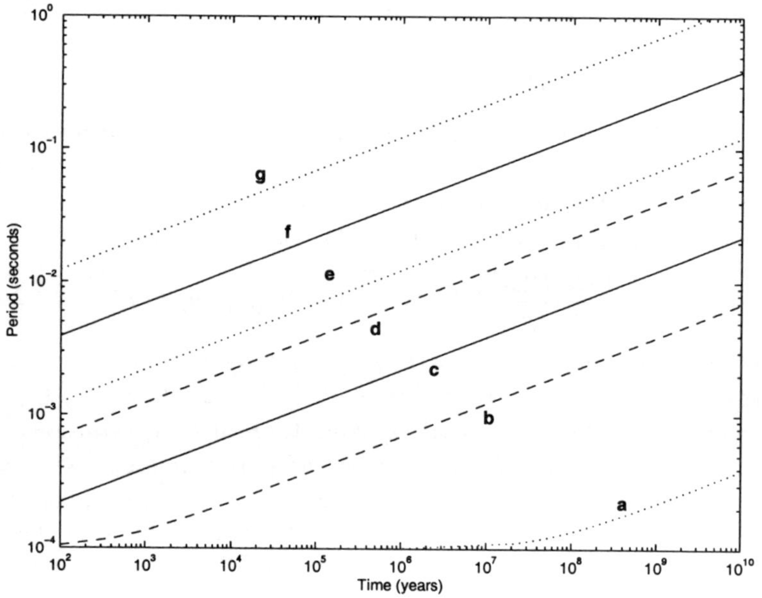

Fig. 1. Gravitational-wave-radiation-induced period evolution of sub-millisecond pulsars with an initial period $P_i = 0.1$ ms for different parameters of equatorial ellipticity, ϵ_e, and effective radius, $R_{\rm eff}$, but a fixed density $\rho = 4 \times 10^{14}$ g/cm^3. The lines are labelled from "a" to "g". "a": $\epsilon_e = 10^{-3}$, $R_{\rm eff} = 0.01$ km; "b": $\epsilon_e = 10^{-3}$, $R_{\rm eff} = 0.1$ km; "c": $\epsilon_e = 0.01$, $R_{\rm eff} = 0.1$ km; "d": $\epsilon_e = 0.1$, $R_{\rm eff} = 0.1$ km; "e": $\epsilon_e = 10^{-3}$, $R_{\rm eff} = 1$ km; "f": $\epsilon_e = 0.01$, $R_{\rm eff} = 1$ km; "g": $\epsilon_e = 0.1$, $R_{\rm eff} = 1$ km.

2.2. *Propeller-torqued spindown*

Besides spindown mechanisms due to gravitational wave radiation and to magnetospheric activity, another one could also work when a sub-millisecond pulsar is in a propeller phase. Propeller torque acts on a pulsar through magnetohydrodynamical (MHD) coupling at magnetospheric boundary with Alfvén radius, and accreted matter inward has to been ejected at the boundary. The matter going outward gains kinematic energy during this process, and the pulsar losses then its rotation energy.

A real difficulty to estimate quantitatively propeller-torqued spindown is to know the accretion rate, $\dot{M}$. The accreted matter could be either the debris captured during the birth of pulsars or the inter-stellar medium. Such accretion details, including the MHD coupling, have not been known with certainty yet.

3. Magnetospheric activity of sub-millisecond pulsars

The potential drop in the open-field-line region is essential for the magnetospheric activity of sub-millisecond pulsars. In case of approximately constant $\mu_{\rm m}$, the potential drop between the center and the edge of a polar cap can be expressed as[8],

$$\phi = \frac{64\pi^3}{3c^2}\bar{B}\mu_m R_{\rm eff}^3 P^{-2} \simeq 2.2\times 10^{13}({\rm volts})\ \mu_{m-6} R_{\rm effkm}^3 P_{\rm ms}^{-2}, \tag{8}$$

where the bag constant $\bar{B} = 60$ MeV/fm$^3 \simeq 10^{14}$ g/cm^3 (i.e., $\rho/4$). It is well known that pair production mechanism is a key ingredient for pulsar radio emission. A pulsar is called to be "death" if the pair production condition can not be satisfied. Although a real deathline depends upon the dynamics of detail pair and photon production, the deathline can also be conventionally taken as a line of constant potential drop ϕ. Assuming a critical drop $\phi_c = 10^{12}$ volts, a sub-millisecond pulsar with $P = 0.1$ ms could still be active even its radius is only 0.08 km, in case of $\mu_{m-6} = 1$.

The potential drop in the open field line region would be much higher than that presented in Eq.(8) if the effect of inclination angle is included[20]. Note that this conclusion favors the magnetospheric activity of sub-millisecond pulsars.

Part of the power of the magnetospheric activity is in the electromagnetic emission of radio band. If the radio power accounts for $\eta \approx 10^{-10\sim-5}$ times of the magnetospheric activity[21], the radio luminosity is then, from Eq.(5),

$$L_{\rm radio} = \eta\frac{512\pi^6}{27c^3}\mu_m^2\rho^2 R_{\rm eff}^6 P^{-4} \simeq 1.1\times 10^{32}\eta({\rm erg/s})\ \mu_{m-6}^2 R_{\rm effkm}^6 P_{\rm ms}^{-4}, \tag{9}$$

where $\rho = 4\times 10^{14}$ g/cm^3. This power is in the same order of that of normal radio pulsars observed[21] even the radius of sub-millisecond pulsars is less than 1 km. Although sub-millisecond pulsars could be radio loud, one needs a very short sampling time, and has to deal with then a huge amount of data in order to find a sub-millisecond pulsar. Due to its large receiving area and wide scanning sky, the future radio telescope, *FAST* (five hundred meter aperture spherical telescope), to be built in Yunnan, China, might uncover sub-millisecond radio pulsars.

4. Conclusions and discussions

We show that sub-millisecond pulsars should be in Jacobi ellipsoidal figures of equilibrium. It is addressed that the spindown of sub-millisecond pulsars would be torqued dominantly by gravitational wave radiation, and that such pulsars may not spin down to super-millisecond periods via gravitation wave radiation during their lifetimes if they are extremely low mass bare strange quark stars. It is possible, based on the calculation of Fig. 1, that isolated super-millisecond pulsars could be quark nuggets (stars) born during cosmic QCD phase separation (via spindown in Hubble timescale).

The radio luminosity of sub-millisecond pulsars could be high enough to be recorded in advanced radio telescopes (e.g., the future *FAST* in China). Sub-millisecond pulsars would not be likely to be normal neutron stars. It could then be a clear way of identifying quark stars as the real nature of pulsars to search and detect sub-millisecond radio pulsars.

Where to find sub-millisecond radio pulsars? This is a question related to how sub-millisecond pulsars origin. Actually, a similar question, which was listed as one of Lorimer-Kramer's 13 open questions in pulsar astronomy[21], is still not answered: How are isolate millisecond pulsars produced? More further issues are related: Should sub-millisecond pulsars be in globular clusters? Can millisecond and sub-millisecond pulsars form during cosmic QCD separation? How to estimate an initial period of quark star in this way? Could AIC (accretion-induced collapse) of white dwarfs produce pulsars with periods < 10 ms? A recent multi-dimensional simulations of AIC was done[22] in the normal neutron stars regime, but a quark-star version of AIC simulation is interesting and necessary. Another interesting idea is: could low mass quark stars form during the fission of a progenitor quark star? (Quark matter produced in this way might be chromatically charges?). All these ideas are certainly interesting, and could *not* be ruled out simply and quickly by first principles.

Acknowledgments

The author would like to thank many stimulating discussions with the members in the pulsar group of Peking University.

References

1. Xu R. X. 2006, Pulsars and Quark stars, in: The proceedings of the 2005 Lake Hanas International Pulsar Symposium, Chin. Jour. Astron. & Astrophys., in press (astro-ph/0512519)
2. Chandrasekhar S. 1970, Phys. Rev. Lett., 24, 611
3. Friedman J. L., Schutz B. F. 1978, ApJ, 222, 281
4. Andersson N. 1998, ApJ, 502, 708
5. Friedman J. L., Morsink, S. M. 1998, ApJ, 502, 714

6. Lindblom L., Owen B. J., Morsink S. M. 1998, Phys. Rev. Lett., 80, 4843
7. Xu R. X. 2006, Astropart. Phys., 25, 212 (astro-ph/0511612)
8. Xu R. X. 2005, Mon. Not. Roy. Astron. Soc., 356, 359 (astro-ph/0402659)
9. Xu R. X. 2003, ApJ, 596, L59 (astro-ph/0302165)
10. Zhou A. Z., Xu R. X., Wu, X. J., Wang, N. 2004, Astropart. Phys., 22, 73 (astro-ph/0404554)
11. Thorne K. 1995, Gravitational waves from compact bodies, in: Compact stars in binaries, Eds. J. van Paradijs, E. P. J. van den Heuvel, E. Kuulkers, Kluwer Academic Publishers, p.153 (gr-qc/9506084)
12. Owen B. J. 2005, Phys. Rev. Lett., 95, 211101 (astro-ph/0503399)
13. Abbott B., et al. (LIGO Scientific Collaboration) 2005, Phys. Rev. Lett., 94, 181103
14. Weinberg S., *Gravitation and Cosmology: principles and applications of the general theory of relativity*, John Wiley, Chap. 10, 1972
15. Kochhar R. K., Sivaram C. 1983, ApSS, 96, 447
16. Lai D., Shapiro S. L. 1995, ApJ, 442, 259
17. Gondek-Rosinska D., Gourgoulhon E. 2002, Phys. Rev. D66, 044021
18. Amsterdamski P., Bulik T., Gondek-Rosinska D., Kluzniak W., 2002, A&A 381, L21
19. Gondek-Rosinska D., Gourgoulhon E., Haensel P. 2003, A&A 412, 777
20. Yue Y. L., Cui X. H., Xu R. X. 2006, preprint (astro-ph/0603468)
21. Lorimer D. R., Kramer M., *Handbook of pulsar astronomy*, Cambridge Univ. Press, 2005
22. Dessart L., Burrows A., Ott C., Livne E., Yoon S. C., Langer N. 2006, ApJ, in press (astro-ph/0601603)

Quantum Gravity

Low-energy quantum gravity*

Michael A. Ivanov
Physics Dept.,
Belarus State University of Informatics and Radioelectronics,
6 P. Brovka Street, BY 220027, Minsk, Republic of Belarus.
E-mail: ivanovma@gw.bsuir.unibel.by.

If gravitons are super-strong interacting particles and the low-temperature graviton background exists, the basic cosmological conjecture about the Dopplerian nature of redshifts may be false. In this case, a full magnitude of cosmological redshift would be caused by interactions of photons with gravitons. Non-forehead collisions with gravitons will lead to a very specific additional relaxation of any photonic flux. It gives a possibility of another interpretation of supernovae 1a data - without any kinematics. A quantum mechanism of classical gravity based on an existence of this sea of gravitons is described for the Newtonian limit. This mechanism needs graviton pairing and "an atomic structure" of matter for working it. If the considered quantum mechanism of classical gravity is realized in the nature, then an existence of black holes contradicts to Einstein's equivalence principle. In this approach the two fundamental constants - Hubble's and Newton's ones - should be connected between themselves. Every massive body would be decelerated due to collisions with gravitons that may be connected with the Pioneer 10 anomaly.

1. Introduction

In this contribution, I would like to describe a very unexpected possibility to consider gravity as a very-low-energy stochastic process. I enumerate those discoveries and observations which may support this my opinion. 1. In 1998, Anderson's team reported about the discovery of anomalous acceleration of NASA's probes Pioneer 10/11 [1]; this effect is not embedded in a frame of the general relativity, and its magnitude is somehow equal to $\sim Hc$, where H is the Hubble constant, c is the light velocity. 2. In the

*A full 20-page version of this contribution is available as hep-th/0510270 v3

same 1998, two teams of astrophysicists, which were collecting supernovae 1a data with the aim to specificate parameters of cosmological expansion, reported about dimming remote supernovae [2,3]; the one would be explained on a basis of the Doppler effect if at present epoch the universe expands with acceleration. This explanation needs an introduction of some "dark energy" which is unknown from any laboratory experiment. 3. In January 2002, Nesvizhevsky's team reported about discovery of quantum states of ultra-cold neutrons in the Earth's gravitational field [4]. Observed energies of levels (it means that and their differences too) in full agreement with quantum-mechanical calculations turned out to be equal to $\sim 10^{-12}$ eV. If transitions between these levels are accompanied with irradiation of gravitons then energies of irradiated gravitons should have the same order - but it is of 40 orders lesser than the Planck energy.

2. Passing photons through the graviton background [5]

Due to forehead collisions with gravitons, an energy of any photon should decrease when it passes through the sea of gravitons. From another side, none-forehead collisions of photons with gravitons of the background will lead to an additional relaxation of a photon flux, caused by transmission of a momentum transversal component to some photons. It will lead to an additional dimming of any remote objects, and may be connected with supernova dimming. We deal here with the uniform non-expanding universe with the Euclidean space, and there are not any cosmological kinematic effects in this model. Then average energy losses of a photon with an energy E on a way dr will be equal to [5]: $dE = -aEdr$, where a is a constant. If a *whole* redshift magnitude is caused by this effect, we must identify $a = H/c$, where c is the light velocity, to have the Hubble law for small distances. The expression is true if the condition $\bar{\epsilon} << E(r)$ takes place. Photons with a very small energy may lose or acquire an energy changing their direction of propagation after scattering. Early or late such photons should turn out in the thermodynamic equilibrium with the graviton background, flowing into their own background. Perhaps, the last one is the cosmic microwave background. Photon flux's average energy losses on a way dr due to non-forehead collisions with gravitons should be proportional to $badr$, where b is a new constant of the order 1. These losses are connected with a rejection of a part of photons from a source-observer direction. We get for the factor b (see [6]): $b \simeq 2.137$. Both redshifts and the additional relaxation of any photonic flux due to non-forehead collisions of gravitons with photons lead in our model to the following luminosity distance D_L : $D_L = a^{-1} \ln(1 +$

$z) \cdot (1+z)^{(1+b)/2} \equiv a^{-1} f_1(z)$, where $f_1(z) \equiv \ln(1+z) \cdot (1+z)^{(1+b)/2}$. In Figure 3 of [7], the graph of f_1 is shown; observational data (82 points) are taken from Table 5 of [8]. The predictions fit observations very well for roughly $z < 0.5$. It excludes a need of any dark energy to explain supernova dimming. Discrepancies between predicted and observed values of $\mu_0(z)$ are obvious for higher z.

It follows from a universality of gravitational interaction, that not only photons, but all other objects, moving relative to the background, should lose their energy, too, due to such a quantum interaction with gravitons. We get for the body acceleration $w \equiv dv/dt$ by a non-zero velocity: $w = -ac^2(1-v^2/c^2)$. For small velocities: $w \simeq -Hc$. If the Hubble constant H is equal to $2.14 \cdot 10^{-18} s^{-1}$ (it is the theoretical estimate of H in this approach), a modulus of the acceleration will be equal to $|w| = 6.419 \cdot 10^{-10}\ m/s^2$, that has the same order of magnitude as a value of the observed additional acceleration $(8.74 \pm 1.33) \cdot 10^{-10} m/s^2$ for NASA probes [1].

3. Gravity as the screening effect

It was shown by the author [9] that screening the background of super-strong interacting gravitons creates for any pair of bodies both attraction and repulsion forces due to pressure of gravitons. For single gravitons, these forces are approximately balanced, but each of them is much bigger than a force of Newtonian attraction. If single gravitons are pairing, an attraction force due to pressure of such graviton pairs is twice exceeding a corresponding repulsion force if graviton pairs are destructed by collisions with a body. In such the model, the Newton constant is connected with the Hubble constant that gives a possibility to obtain a theoretical estimate of the last. We deal here with a flat non-expanding universe fulfilled with super-strong interacting gravitons; it changes the meaning of the Hubble constant which describes magnitudes of three small effects of quantum gravity but not any expansion or an age of the universe.

If masses of two bodies are m_1 and m_2 (and energies E_1 and E_2), $\sigma(E_1, \epsilon)$ is a cross-section of interaction of body 1 with a graviton with an energy $\epsilon = \hbar\omega$, where ω is a graviton frequency, $\sigma(E_2, \epsilon)$ is the same cross-section for body 2. Then the following attractive force will act between bodies 1 and 2 :

$$F_1 = \frac{1}{3} \cdot \frac{D^2 c (kT)^6 m_1 m_2}{\pi^3 \hbar^3 r^2} \cdot I_1,$$

where $I_1 = 5.636 \cdot 10^{-3}$. When $F_1 \equiv G_1 \cdot m_1 m_2 / r^2$, the constant G_1 is equal to:

$$G_1 \equiv \frac{1}{3} \cdot \frac{D^2 c(kT)^6}{\pi^3 \hbar^3} \cdot I_1.$$

By $T = 2.7\ K$: $G_1 = 1215.4 \cdot G$, that is three order greater than the Newton constant, G. But if single gravitons are elastically scattered with body 1, then our reasoning may be reversed: the same portion of scattered gravitons will create a repulsive force $F_1^{'}$ acting on body 2 and equal to $F_1^{'} = F_1$. So, for bodies which elastically scatter gravitons, screening a flux of single gravitons does not ensure Newtonian attraction. But for black holes which absorb any particles and do not re-emit them, we will have $F_1^{'} = 0$. It means that such the object would attract other bodies with a force which is proportional to G_1 but not to G, i.e. Einstein's equivalence principle would be violated for them. This conclusion stays in force for the case of graviton pairing, too.

To ensure an attractive force which is not equal to a repulsive one, particle correlations should differ for *in* and *out* flux. For example, single gravitons of running flux may associate in pairs [9].

If running graviton pairs ensure for two bodies an attractive force F_2, then a repulsive force due to re-emission of gravitons of a pair alone will be equal to $F_2^{'} = F_2/2$. It follows from that the cross-section for *single additional scattered* gravitons of destructed pairs will be twice smaller than for pairs themselves (the leading factor $2\hbar\omega$ for pairs should be replaced with $\hbar\omega$ for single gravitons). For pairs, we introduce here the cross-section $\sigma(E_2, <\epsilon_2>)$, where $<\epsilon_2>$ is an average pair energy with taking into account a probability of that in a realization of flat wave a number of graviton pairs may be equal to zero, and that not all of graviton pairs ride at a body. We get for graviton pairs: $<\epsilon_2>= 2\hbar\omega(1-P(0,2x))\bar{n}_2^2 \exp(-\bar{n}_2) \cdot P(0,x)$. Then a force of attraction of two bodies due to pressure of graviton pairs, F_2 will be equal to:

$$F_2 = \frac{8}{3} \cdot \frac{D^2 c(kT)^6 m_1 m_2}{\pi^3 \hbar^3 r^2} \cdot I_2,$$

where $I_2 = 2.3184 \cdot 10^{-6}$. The difference F between attractive and repulsive forces will be equal to: $F \equiv F_2 - F_2^{'} = \frac{1}{2} F_2 \equiv G_2 \frac{m_1 m_2}{r^2}$, where the constant G_2 is equal to: $G_2 \equiv \frac{4}{3} \cdot \frac{D^2 c(kT)^6}{\pi^3 \hbar^3} \cdot I_2$. If one assumes that $G_2 = G$, then it follows that by $T = 2.7K$ the constant D should have the value: $D = 0.795 \cdot 10^{-27} m^2/eV^2$.

We can establish a connection between the two fundamental constants,

G and H :

$$H = (G \frac{45}{64\pi^5} \frac{\sigma T^4 I_4^2}{c^3 I_2})^{1/2} = 2.14 \cdot 10^{-18}\ s^{-1},$$

or in the units which are more familiar for many of us: $H = 66.875\ km \cdot s^{-1} \cdot Mpc^{-1}$. This value of H is in the good accordance with the majority of present astrophysical estimations [2], but it is lesser than it follows from the observed value of anomalous acceleration of Pioneer 10 [1].

4. Some cosmological consequences of the model

If the described model of redshifts is true, what is a picture of the universe? It is interesting that in a frame of this model, every observer has two own spheres of observability in the universe (two different cosmological horizons exist for any observer). One of them is defined by maximum existing temperatures of remote sources - by big enough distances, all of them will be masked with the CMB radiation. Another, and much smaller, sphere depends on their maximum luminosity - the luminosity distance increases with a redshift much quickly than the geometrical one. The ratio of the luminosity distance to the geometrical one is the quickly increasing function of z : $D_L(z)/r(z) = (1+z)^{(1+b)/2}$, which does not depend on the Hubble constant. An outer part of the universe will drown in a darkness. Some other possible cosmological consequences of an existence of the graviton background were described in [9,7].

5. Conclusion

If this mechanism is realized in the nature, both the general relativity and quantum mechanics should be modified. Any divergencies, perhaps, would be not possible in such the model because of natural smooth cut-offs of the graviton spectrum from both sides. Gravity at short distances, which are much bigger than the Planck length, needs to be described only in some unified manner.

References

1. J.D. Anderson et al. *Phys. Rev. Lett.* **81,** 2858 (1998); *Phys. Rev.* **D65,** 082004 (2002); [gr-qc/0104064 v4].
2. A.G. Riess et al. *AJ* **116,** 1009 (1998).
3. S. Perlmutter et al. *ApJ* **517,** 565 (1999).
4. V.V. Nesvizhevsky et al. *Nature* **415,** 297 (2002).

5. M.A. Ivanov *General Relativity and Gravitation* **33,** 479 (2001); Erratum: **35,** 939 (2003); [astro-ph/0005084 v2].
6. M.A. Ivanov, [astro-ph-0405083].
7. M.A. Ivanov, [hep-th/0506189].
8. A.G. Riess et al. [astro-ph/0402512] (to appear in *ApJ*).
9. M.A. Ivanov, [gr-qc/0207006].

Mass renormalization at finite temperature and random motion of an electron driven by quantum electromagnetic fluctuations

Hongwei Yu

Department of Physics and Institute of Physics, Hunan Normal University, Changsha, Hunan 410081, China

We discuss the random motion of an electron driven by quantum electromagnetic fluctuations at finite temperature in the Minkowski spacetime and calculate the mean squared fluctuations in the velocity of the electron. We find that at very late times, this random motion leads to a kinetic energy, $e^2/(6m\beta^2)$, for the electron, which may be interpreted as a shift in its rest mass, or mass renormalization. However, our result differs from that of other earlier works on temperature-dependent quantum-electrodynamic corrections to the inertial mass. It is argued that our calculations are gauge invariant while the gauge invariance of other earlier works remains unclear.

1. Introduction

Quantum fluctuation is an intrinsic nature of quantum theory which has profoundly changed our conception of empty space or vacuum. A fundamental feature to be expected of any field which is quantized is the quantum fluctuations. Due to the intrinsic quantum nature and the uncertainty principle, knowing the value of a quantized field means losing all knowledge about its canonical conjugate. As an result, we can not simultaneously precisely determine the strength of a quantum field and its change rate with time. Therefore, there always exist fluctuating quantized fields even in vacuum.

A question then arises naturally as to whether there is any observable manifestations of vacuum fluctuations. In quantum electrodynamics in an unbounded Minkowski spacetime, at least, the effects of electromagnetic vacuum fluctuations upon an electron in such space are usually regarded as unobservable. The divergent parts of the electron self-energy are absorbed by mass and wavefunction renormalizations, and the finite self-energy function can be taken to vanish for real (as opposed to virtual) electrons. Although some physical quantities in quantum field theory, such as energy, are not well-defined in vacuum and we have to use certain renormalization scheme to make them finite, *changes* in the vacuum fluctuations, however, usually exhibit normal behavior and can produce observable effects. A well-known such example is the Casimir effect [1] which arises because of the changes in the electromagnetic vacuum fluctuations induced by the presence of two conducting plane boundaries in vacuum.

Since quantum fields fluctuate in vacuum, one may also expect that test particles under the influence of a fluctuating quantum field will no longer follow fixed classical trajectories in vacuum, but will rather undergo random motion around a mean path in much the same way as a classical stochastic field causes random motion of a test particle. This random motion can be described by the quadratic fluctuations of some quantity characterizing the variation from the mean trajectory, such as velocity and position. Recently, this kind of quantum Brownian (random) motion has been studied in the hope that it may offer another possible experimentally observable effect of vacuum fluctuations in addition to the Casimir effect. In particular, the Brownian (random) motion of a charged particle caused by *changes* in the electromagnetic vacuum fluctuations near a perfectly reflecting plane boundary [2] and in between two plane boundaries [3] has been investigated and the effects have been calculated of the modified electromagnetic vacuum fluctuations due to the presence of the boundary upon the motion of a charged test particle. It has been shown that both the mean squared fluctuations in the velocity and position of the test particle are negative in the case of directions parallel to the boundary. An interpretation has been offered of this as reducing the quantum uncertainty which would otherwise be present. However, the mean squared fluctuations in the normal velocity and position are positive and can be associated, in the case of a single plane, with an effective temperature of

$$T_{eff} = \frac{\alpha}{\pi}\,\frac{1}{k_B m z^2} = 1.7 \times 10^{-6}\left(\frac{1\mu m}{z}\right)^2 K = 1.7 \times 10^{2}\left(\frac{1\mathring{A}}{z}\right)^2 K\ , \quad (1)$$

where k_B is Boltzmann's constant and z is the distance from the boundary.

In the present paper, we would be interested in the random motion of an electron caused by quantum electromagnetic field fluctuations at nonzero temperature (as opposed to zero temperature vacuum fluctuations) in the unbounded flat spacetime, i.e., the random motion driven by quantum fluctuations of a thermal bath of photons.

2. Brownian motion of an electron at finite temperature and electron mass renormalization

Let us now consider the motion of an electron subject to quantum electromagnetic field fluctuations at finite temperature T in the Minkowski (unbounded flat) space. We will use Lorentz-Heaviside units with $c = \hbar = 1$ in our discussions. In the limit of small velocities, the motion of the electron is described by a non-relativistic equation of motion (Langevin equation) with a fluctuating electric force

$$\frac{d\mathbf{v}}{dt} = \frac{e}{m}\,\mathbf{E}(\mathbf{x}, t)\,; \tag{2}$$

assuming that the particle is initially at rest and has a charge to mass ratio of e/m. The velocity of the electron at time t can be calculated as follows

$$\mathbf{v} = \frac{e}{m}\int_0^t \mathbf{E}(\mathbf{x}\ , t)\, dt = \left(\frac{4\pi\alpha}{m^2}\right)^{1/2}\int_0^t \mathbf{E}(\mathbf{x}\ , t)\, dt\ , \tag{3}$$

where α is the fine-structure constant. The mean squared fluctuations in speed in the i-direction can be written as (no sum on i)

$$\langle \Delta v_i^2 \rangle = \frac{4\pi\alpha}{m^2}\int_0^t\int_0^t \langle E_i(\mathbf{x}, t_1)\ E_i(\mathbf{x}, t_2)\rangle_\beta\, dt_1\, dt_2\ , \tag{4}$$

where $\langle E_i(\mathbf{x}, t_1)\ E_i(\mathbf{x}, t_2)\rangle_\beta$ is the renormalized[a] electric field two-point function at finite temperature $T = \frac{1}{k_B\beta}$ and we have used the fact that $\langle E_i\rangle_\beta = 0$. We have, for simplicity, assumed that the distance does not change significantly on the time scale of interest in a time t, so that it can be treated approximately as a constant. If there is a classical, nonfluctuating field in addition to the fluctuating quantum field, then Eq. (4) describes the velocity fluctuations around the mean trajectory caused by the classical field. Note that when the initial velocity does not vanishes, one has to also consider the influence of fluctuating magnetic fields on the velocity dispersion of the test particles. However, it has been shown that this influence is, in general, of the higher order than that caused by fluctuating electric fields and is thus negligible [4].

Let us note that the two point function for the photon field at finite temperature, $D_\beta^{\mu\nu}(x, x') = \langle 0|A^\mu(x)\, A^\nu(x')|0\rangle_\beta$, can be written as an infinite

[a] We adopt the well-established renormalization procedure in quantum field theory in which physical quantities are calculated and supposedly experimentally measured against vacuum.

imaginary-time image sum of the corresponding zero-temperature two-point function, $D_0^{\mu\nu}(x-x')$, i.e.,

$$D_\beta^{\mu\nu}(x,x') = \sum_{n=-\infty}^{\infty} D_0^{\mu\nu}(\mathbf{x}-\mathbf{x}', t-t'+in\beta) \ , \tag{5}$$

where argument x stands for a four-vector, i.e., ($\mathbf{x}, t$). In the Feynman gauge, we have

$$D_0^{\mu\nu}(x-x') = \frac{\eta^{\mu\nu}}{4\pi^2(\Delta t^2 - \Delta\mathbf{x}^2)} \ . \tag{6}$$

Using the above result, we can obtain the components of the renormalized electric field two-point function at finite temperature

$$\begin{aligned}\langle E_x(\mathbf{x},t')E_x(\mathbf{x},t'')\rangle_\beta &= \langle E_y(\mathbf{x},t')E_y(\mathbf{x},t'')\rangle_\beta = \langle E_z(\mathbf{x},t')E_z(\mathbf{x},t'')\rangle_\beta \\ &= \frac{1}{\pi^2}\sum_{n=-\infty}^{\infty}{}' \frac{1}{(\Delta t + in\beta)^4} = \frac{\pi^2}{3\beta^4}\left(2+\cosh\frac{2\pi\Delta t}{\beta}\right)\mathrm{csch}^4\left(\frac{\pi\Delta t}{\beta}\right) - \frac{1}{\pi^2\Delta t^4} \ .\end{aligned} \tag{7}$$

Here a prime means that the $n = 0$ term is omitted in the summation.

Substituting the above results into Eq. (4) and carrying out the integration, we find that the velocity dispersions are given by

$$\begin{aligned}\langle\Delta v_x^2\rangle = \langle\Delta v_y^2\rangle = \langle\Delta v_z^2\rangle &= \frac{e^2}{m^2}\int_0^t\int_0^t \langle E_x(\mathbf{x},t')\ E_x(\mathbf{x},t'')\rangle_\beta \, dt'\, dt'' \\ &= \frac{e^2\mathrm{csch}^2(\frac{\pi t}{\beta})}{18\pi^2 m^2\beta^2 t^2}\left[5\pi^2t^2 + 3\beta^2 + (\pi^2t^2 - 3\beta^2)\cosh\frac{2\pi t}{\beta}\right] \ .\end{aligned} \tag{8}$$

In the high temperature limit, i.e., when $t \gg \beta$, we have

$$\langle\Delta v^2\rangle = \frac{e^2}{3m^2\beta^2} - \frac{e^2}{\pi^2 m^2 t^2} \ . \tag{9}$$

To get a concrete idea of how large t should be in order that the condition $t \gg \beta$ is fulfilled, let us assume that the temperature T is about $\sim 10^2$ Kevin, which can well be considered as high since we are discussing a quantum effect, then the condition becomes $t \gg 5.7 \times 10^{-14} sec.$. This is rather small. It is interesting to note that, for the random motion driven by quantum fluctuations at finite temperature here, no dissipation is needed for $\langle\Delta v_i^2\rangle$ to be bounded at late times in contrast to the random motion due to thermal noise.

At very late times ($t \to \infty$), this random motion leads to a kinetic energy for the electron

$$\Delta_0(T) = \frac{1}{2} m \langle \Delta v^2 \rangle = \frac{e^2}{6m\beta^2} = \frac{2\pi\alpha}{3m\beta^2} \,. \tag{10}$$

This temperature dependent kinetic energy associated with the random motion driven by quantum electromagnetic field fluctuations at finite temperature may be interpreted as a shift in the rest mass of the electron or electron mass renormalization. Thus, the discussion in the present paper seems to offer a transparent physical picture for the origin of the mass renormalization in quantum field theory, that is, the mass renormalization of an electron can be understood as a result of the Brownian motion caused by quantum electromagnetic (vacuum or finite temperature) fluctuations. This is pretty much similar as that the charge renormalization can be regarded as a consequence of the screening effect induced by vacuum polarization. It is worth noting that similar result for $\Delta_0(T)$ has been obtained in other earlier works from different physical prospectives[5,6,7]. However, there seems to be a discrepancy of a factor of two between our result (Eq. (10)) and that of the other works [b]. We would like to point out, however, that calculations performed in the present paper use the electromagnetic field strength, **E**, and are thus gauge invariant, while other works, for example, Refs. [5,6,7] employ a gauge-dependent photon propagator or a Hamiltonian that involves gauge-dependent potentials, **A** [8], and therefore the gauge invariance of their result at least remains unclear. We hope to return to the issue of gauge invariance elsewhere in the future. Meanwhile, it is interesting to note that similar issue of loss of gauge invariance also arises when one tries to calculate the electron self-energy in a spacetime with one periodically compactified spatial dimension [9].

Acknowledgments

This work was supported in part by the National Natural Science Foundation of China under Grants No.10375023 and No.10575035, the Program for NCET (No.04-0784) and the Key Project of Chinese Ministry of Education (No.205110)

[b] See for example, Ref. [6]. Note that un-rationalized Gaussian units were used there as opposed to the rationalized one in the present paper.

References

1. H. B. G. Casimir, Proc. K. Ned. Akad. Wet. **51**, 793 (1948).
2. H. Yu and L.H. Ford, Phys. Rev. D **70**, 065009 (2004).
3. H. Yu and J. Chen, Phys. Rev. D **70**, 125006(2004).
4. M. Tan and H. Yu, Chin. Phys. Lett. **22**, 2165 (2005).
5. G. Barton, J. Phys. B **20**, 879 (1987).
6. G. Barton, Phys. Rev. D **40**, 4096 (1989).
7. J.F. Donoghue, B.R. Holstein, and R.W. Robinett, Ann. Phys. 164, 233(1985); **172**, 483 (1986).
8. It seems that the relation involving the energy density of the normal mode given immediately below (A7) in Ref. [6] has an extra factor of 2. If this factor is dropped, the classical model in the Appendix of Ref. [6] would give a result the same as that of the present paper.
9. T, Yoshimura, Phys. Lett. A **72**, 391 (1979); D.J. Toms, Phys. Lett. A **77** , 303 (1980).

Quantum gravity phenomenology and black hole physics

Yi Ling *

Center for Gravity and Relativistic Astrophysics,
Department of Physics, Nanchang University,
Jiangxi, 330047, China
E-mail: yling@ncu.edu.cn

This report is an extension of the talk given at ICGA7. We present a brief review on quantum gravity phenomenology at first, focusing on the fate of Lorentz symmetry at Planck scale. Then we investigate its possible impacts on the study of black holes physics. We argue that modified dispersion relations may require a modification of Bekenstein entropy bound, leading to a correction to the Bekenstein-Hawking entropy formula for black holes. In particular one specific modified dispersion relation is proposed in the context of doubly special relativity, which changes the picture of Hawking radiation and can prevent black holes from total evaporation.

1. Introduction

1.1. *Searching the signature of quantum gravity*

Recently some observational results suggest that the validity of Lorentz symmetry at extremely high energy level need to be reconsidered. One evidence comes from the measurement of the strength of the electromagnetic interaction[1]. It seems that the fine structure constant is not an exact constant but time dependent, which has been a variation of $\delta\alpha/\alpha = -0.72 \pm 0.18 \times 10^{-5}$ over the past 6-10 billion years. Since the fine structure constant is defined as

$$\alpha = \frac{e^2}{\hbar c}, \tag{1}$$

this peculiar phenomenon invoked a varying speed of light theory and non-standard cosmology with varying c. Moreover, the observed threshold

*Work partially supported by grant (No.10405027, 10205002) of the National Science Foundation of China.

anomalies in ultra high energy cosmic rays point to the possibility that the usual energy-momentum relation or dispersion relations may be modified[2]. For instance, for the reaction

$$p + \gamma_{CMB} \rightarrow p + \pi^0, \tag{2}$$

the usual dispersion relations give rise to the GZK cutoff as

$$E_{GZK} \sim E_p = \frac{m_p m_\pi}{2E_\gamma}. \tag{3}$$

While if dispersion relation is modified as

$$\epsilon^2 \simeq m^2 + p^2 + \lambda(\epsilon, p, \eta), \tag{4}$$

where η is a dimensionless parameter, then the GZK cutoff has the form

$$E_{GZK} \sim E_p = \frac{m_p m_\pi}{2E_\gamma} + \frac{\lambda_{p'} - \lambda_p + \frac{m_p}{m_\pi}\lambda_\pi}{4E_\gamma}. \tag{5}$$

This provides a plausible mechanism to increase the value of the threshold greatly.

It has been proposed that such a modification of dispersion relations is attributed to the semi-classical effect of quantum gravity. In this talk we briefly review some progress in this direction, then turn to study the possible impacts of modified dispersion relations on black hole physics. More details on this topics can also be found in relevant references[3,4].

1.2. *Semi-classical limit of loop quantum gravity*

Currently loop quantum gravity (LQG) as a non-perturbative approach to a quantum theory of gravity involves two fundamental issues. One is dubbed as the dynamical problem, namely how to solve the Hamiltonian constraint in the framework of diffeomorphism-invariant quantum field theory, while the other is on the classical limit of quantum gravity, as any successful theory of quantum gravity should be able to go back to the usual Einstein's general relativity at the classical limit, leading to the emergence of a classical space-time. Recently the second issue has been greatly investigated in various cases. One attractive feature in the picture of semi-classical space time is that the usual energy-momentum relations or dispersion relations may be modified at high energy level. This has been testified in the following three circumstances

(1) *weave states in loop representation*—As far as the author knows, the first hint of modified dispersion relations in the framework of loop quantum gravity appears in the paper by Gambini and Pullin[5] where the

corrections to Maxwell equations are evaluated by considering the action of the Hamiltonian on the weave states which are constructed by a collection of Planck-scale loops.

(2) *Kodama state in loop quantum gravity with a positive cosmological constant* —Loop quantum gravity with a positive cosmological constant contains some distinct and remarkable features such as the existence of Kodama states, and its implications to semi-classical limit of quantum gravity have been investigated and summarized in the paper by Smolin[6]. Through a semi-classical approximation it is found that the effect of quantum gravity is to make the dispersion relations of particles energy dependent.

(3) *Coherent states for quantum gravity* —The complexifier coherent states for quantum general relativity are proposed in the paper by Sahlmann *et al.*[7] and applied to construct a quantum field theory on curved spacetime limit[8]. Non-standard dispersion relations for the scalar and the electromagnetic field are obtained in this context as well.

From all above investigations, there are strong indications that the usual energy-momentum relation or dispersion relations may be modified at the semi-classical level.

2. Modified dispersion relations

Doubly special relativity is proposed as a deformed formalism of special relativity at the semi-classical limit of quantum gravity. It preserves the relativity of inertial frames while at the same time keep Planck energy as an invariant scale, namely a universal constant for all inertial observers[9,10]. This can be accomplished by a non-linear Lorentz transformation in momentum space, which leads to a deformed Lorentz symmetry such that the usual energy-momentum relations or dispersion relations in special relativity may be modified with corrections in the order of Planck length. As pointed out in the paper by Magueijo and Smolin[10], in a DSR framework the modified dispersion relation may be written as

$$\epsilon^2 f_1^2(\epsilon,\eta) - p^2 f_2^2(\epsilon,\eta) = m_0^2, \tag{6}$$

where f_1 and f_2 are two functions of energy from which a specific formulation of boost generator can be defined. We adopt a modified dispersion relation (MDR) by taking $f_1^2 = [1-\eta(l_p E)^n]$ and $f_2^2 = 1$, such that

$$\epsilon^2 = \frac{p^2 + m_0^2}{[1-\eta(l_p\epsilon)^n]}, \tag{7}$$

where Planck length $l_p \equiv \sqrt{8\pi G} \equiv 1/M_p$. If $l_p\epsilon \ll 1$, this modified dispersion relation goes back to the ordinary one

$$\epsilon^2 = p^2 + m_0^2 + \eta(l_p\epsilon)^n(p^2 + m_0^2 + ...). \tag{8}$$

For massless particles it can also be approximately written as

$$\epsilon^2 = p^2 + \eta(l_p\epsilon)^n p^2 + ... \ . \tag{9}$$

3. Impacts on black hole physics

3.1. *Modification of Bekenstein entropy bound*

In this section we present a modified version of Bekenstein entropy bound due to the modification of dispersion relations. We first recall the physical motivation of Bekenstein entropy bound. Consider a region with size R containing energy E. Bekenstein entropy bound claims that the maximum entropy of the system should be bounded as

$$S_{max} \leq \alpha E R. \tag{10}$$

Conventionally the constant α is taken as 2π. We may derive this bound in a heuristic way by employing the uncertainty relations and the standard energy-momentum relation. First, the minimum energy $\epsilon(R)$ of a massless particle localized inside this region is $\epsilon(R) \sim p \sim \delta p \sim 1/R$. Then the maximum number of particles inside this region can be estimated as

$$N_{max} \sim \frac{E}{\epsilon(R)} \sim ER. \tag{11}$$

Thus the upper limit to the entropy should have the form

$$S_{max} \leq \alpha E R, \tag{12}$$

as the number of microstates of particles exponentially increases with the number of particles. Now if the usual dispersion relation is modified as (9), correspondingly the energy ϵ of one particle with position uncertainty $\delta x \sim R$ will approximately satisfy the modified uncertainty relation

$$\epsilon(R) \geq \delta\epsilon \geq \frac{1}{R}[1 + \frac{1}{2}(n+1)\eta(l_p/R)^n]. \tag{13}$$

Repeat above calculation, we find the minimum energy

$$\epsilon(R) \sim \frac{1}{R}[1 + \frac{1}{2}(n+1)\eta(l_p/R)^n], \tag{14}$$

and the maximum number of particles is

$$N_{max} \sim \frac{E}{\epsilon(R)} \sim \frac{ER}{[1+\frac{1}{2}(n+1)\eta(l_p/R)^n]}. \tag{15}$$

Therefore the modified Bekenstein entropy bound for a system would have a form

$$S_{max} \leq \frac{2\pi ER}{[1+\frac{1}{2}(n+1)\eta(l_p/R)^n]}. \tag{16}$$

3.2. *Corrections to black hole temperature and entropy*

In this subsection we present two different ways to obtain the corrections to black hole entropy due to the modification of dispersion relations (7), but for convenience we take $n = 2$ and $\eta = 1$ explicitly in next discussions. Now consider the process that particles with energy ϵ and size δx fall into a black hole. In order to preserve the generalized second law (GSL) of thermodynamics, Bekenstein proposed that the change of the horizon area of the black hole should satisfy the following relation (17)

$$\frac{\delta A}{4L_p^2} \geq 2\pi\epsilon\delta x, \tag{17}$$

such that the total entropy of a black hole and the region outside horizon never decreases[a], where position uncertainty of the particle has been taken as the Schwarzschild radius while falling in the black hole. Correspondingly the minimum increase of entropy is $ln2$. On the other hand, from (9) we may obtain a uncertainty relation between the energy and momentum of a single particle such that

$$\delta\epsilon \sim \frac{\delta p}{(1-\frac{3}{2}l_p^2p^2)}. \tag{18}$$

Identifying $\epsilon \sim \delta\epsilon$ and applying the ordinary uncertainty relation to photons in the vicinity of black hole horizons $\delta p \sim 1/\delta x \sim 1/(4\pi R)$ where R is the radius of Schwarzschild black hole, we may have

$$\delta A_{min} \cong 16\pi G ln2\epsilon\delta x = \frac{4Gln2}{1-3l_p^2/8\pi A}, \tag{19}$$

[a]Here we take the following assumption as the starting point, namely Bekenstein argument (17) still holds without any modification even though the dispersion relation is deformed. Alternatively,we might consider the possible modifications to this inequality and its implications will be discussed elsewhere.

where a calibration factor $2ln2$ is introduced. Thus

$$\frac{dS}{dA} \cong \frac{\delta S}{\delta A_{min}} = \frac{1}{4G}(1 - \frac{3l_p^2}{8\pi A}), \tag{20}$$

and consequently

$$S = \frac{A}{4G} - \frac{3}{4} ln\left(\frac{A}{4G}\right). \tag{21}$$

Above estimation is applicable to large black holes in the case of $l_p\epsilon \ll 1$. However, when $l_p\epsilon \sim 1$ the modified dispersion relation can not be approximately expanded as (8) and we need treat it non-perturbatively.

$$l_p^2\epsilon^4 - \epsilon^2 + (p^2 + m_0^2) = 0. \tag{22}$$

For photons this gives a relation as

$$\epsilon^2 = \frac{1}{2l_p^2}\left[1 - \sqrt{1 - 4l_p^2 p^2}\right]. \tag{23}$$

In this situation we adopt the scheme presented in the paper by Adler *et al.*[11] as an alternative way to calculate the corrections to black hole temperature and entropy. In their paper[11] it is claimed that the characteristic temperature of the black hole is supposed to be proportional to the expectation mean value of photon energy ϵ,namely $\epsilon = T$. Thus we may obtain the temperature of black holes directly from the equation (23) as

$$T = \left[\frac{M_p^2}{2}\left(1 - \sqrt{1 - \frac{4M_p^2}{M^2}}\right)\right]^{1/2}. \tag{24}$$

Now it is straightforward to calculate the black entropy using the first law of thermodynamics. Plugging the temperature into $dM = TdS$, we have

$$S = \frac{1}{2\sqrt{G}} \int_{A_{min}}^{A} (A - \sqrt{A^2 - 8GA})^{-1/2} dA, \tag{25}$$

where $A_{min} = 8G \sim l_p^2/\pi$ is the cutoff corresponding to a black hole with minimum mass $M = 2M_p$. (25) can be integrated out exactly and for large black holes, namely $A \gg 8G$, it approximately has the familiar formula

$$S = \frac{A}{4G} - \frac{1}{2} ln\frac{A}{4G} + \ldots . \tag{26}$$

Comparing with (21) we find the answer is almost the same but the factor in logarithmic term is different, which results from the fact that some approximations for large black holes have been taken into account during the derivation of (21), for instance in (18). In this sense the latter is a more precise approach to obtain corrections to black hole entropy .

3.3. *The picture of Hawking evaporation at Planck scale*

At the end of this section we point out that it is interesting to study the evaporation process of black holes in particular at late times with the use of the Stefan-Boltzmann law. Define $x = M/M_p$, then the evolution of the black hole mass can be described by the equation

$$\frac{dx}{dt} = \frac{-1}{t_f}\left(x - \sqrt{x^2 - 4}\right)^2, \tag{27}$$

where $t_f = 16\pi/(\sigma M_p)$. Solving this equation we easily find that the life of black holes reads as

$$t = t_c - \frac{t_f}{24}\left[x^3 + (x^2 - 4)^{3/2} - 6x\right], \tag{28}$$

where t_c is an integral constant. From above equation it should be noticed that $dx/dt = -4/t_f$ is a finite number at the end $x = 2$ rather than infinity in ordinary case. Furthermore, the heat capacity of black holes is

$$C = \frac{dM}{dT} = -\frac{M^3 T}{M_p^4}\left(1 - \frac{4M_p^2}{M^2}\right)^{1/2}, \tag{29}$$

which becomes vanishing when the black hole mass approaches a nonzero scale, $M = 2M_p$. This implies that a black hole would not evaporate completely but have a remnant with Planck masses. Thus modified dispersion relations provide a mechanism to take black hole remnants as a natural candidate for cold dark matter.

4. Concluding remarks

The modification of dispersion relations can be viewed as a semi-classical effect of quantum gravity. In this paper we discussed its possible impacts on black holes physics. In general, the temperature as well as the entropy of black holes will be corrected due to modified dispersion relations. In particular the mass of black holes is bounded from below at the Planck scale such that the remnant may be treated as a candidate for cold dark matter. We end this paper by a list of some remarks.

(1) The identification of black hole temperature with energy of particles currently only works for Schwarzschild black holes. More general, we expect the surface gravity on horizons might play an important role linking the black hole temperature and modified dispersion relations.

(2) The factor of logarithmic correction to black hole entropy can not be uniquely fixed, but dependent on the specific form of modified dispersion relations. This can be understood as this specific MDR we proposed is only a coarse grained model at semi-classical limit of quantum gravity.
(3) The impact of doubly special relativity on gravity is not taken into account in this paper, but the relevant discussion may be found elsewhere[12,13].

Acknowledgments

I would like to thank the organizers of ICGA7, specially Prof. Chiang-Mei Chen, James M. Nester and Hoi-lai Yu, for invitation and hospitality during the conference. The partial support from the National Center for Theoretical Sciences in Taiwan is also acknowledged.

References

1. Webb.J.K. et al. Phys. Rev. Lett. **87**, 091301(2001).
2. S. R. Coleman and S. L. Glashow, Phys. Rev. D **59**, 116008 (1999).
3. G. Amelino-Camelia, M. Arzano, Y. Ling and G. Mandanici, Class. Quant. Grav. **23**, 2585 (2006) [arXiv:gr-qc/0506110].
4. Y. Ling, B. Hu and X. Li, Phys. Rev. D **73**, 087702 (2006).
5. R. Gambini and J. Pullin, Phys. Rev. D **59**, 124021 (1999), J. Alfaro, H. A. Morales-Tecotl and L. F. Urrutia, Phys. Rev. D **65**, 103509 (2002).
6. L. Smolin, arXiv:hep-th/0209079.
7. H. Sahlmann, T. Thiemann and O. Winkler, Nucl. Phys. B **606**, 401 (2001).
8. H. Sahlmann and T. Thiemann, Class. Quant. Grav. **23**, 909 (2006).
9. G. Amelino-Camelia, Phys. Lett. B **510**, 255 (2001) [arXiv:hep-th/0012238].
10. J. Magueijo and L. Smolin, Phys. Rev. Lett. **88**, 190403 (2002), J. Magueijo and L. Smolin, Phys. Rev. D **67**, 044017 (2003).
11. R. J. Adler, P. Chen and D. I. Santiago, Gen. Rel. Grav. **33**, 2101 (2001), P. Chen and R. J. Adler, Nucl. Phys. Proc. Suppl. **124**, 103 (2003).
12. J. Magueijo and L. Smolin, Class. Quant. Grav. **21**, 1725 (2004).
13. Y. Ling, X.Li and H.Zhang, "Thermodynamics of modified black holes from gravity's rainbow", [arXiv:gr-qc/0512084].

Uniformly Accelerated Detector in (3+1)D Spacetime: From Vacuum Fluctuations to Radiation Flux

Shih-Yuin Lin and B. L. Hu*

Center for Quantum and Gravitational Physics, Institute of Physics, Academia Sinica, Nankang, Taipei 11529, Taiwan and Department of Physics, University of Maryland, College Park, Maryland 20742-4111, USA

We analyze the interaction of a uniformly accelerated detector with a quantum field in (3+1)D spacetime and derive the two-point correlation functions of the detector and of the field separately with full account of their interplay. We find that there does exist a positive radiated flux of quantum nature emitted by the detector in steady state, with a hint of certain features of the Unruh effect. We further verify that only some part of the radiation is conserved with the total energy of the dressed detector. Since this part of the radiation ceases in steady state, the hint of the Unruh effect in late-time radiated flux is actually not directly from the energy flux that the detector experiences in Unruh effect.

1. Introduction

To understand the black hole radiation,[1] Unruh studied a detector theory[2] and found that, even in flat spacetime, a uniformly accelerated detector (UAD) moving in Minkowski vacuum experiences a thermal bath with a temperature $T_U = \hbar a/2\pi c k_B$, where a is the proper acceleration.

In theories of particle detector, the accelerated detector can be a point-like object with internal degree of freedom coupled with a quantum field. It is common knowledge that an accelerated point-charge coupled with electromagnetic(EM) field give rise to EM radiation.[3,4] Hence there has been a speculation that some evidence about the Unruh effect could be found

*Work supported in part by the NSC Taiwan under grant NSC93-2112-M-001-014 and by the US NSF under grant PHY03-00710 and PHY-0426696

in the radiation emitted by a uniformly accelerated detector,[5] even under steady state conditions.

Prior work shows that, at least in (1+1)D, there is no emitted radiation from a linear uniformly accelerated oscillator under equilibrium conditions (steady state and uniform acceleration),[6] while there exists a "polarization cloud" around it.[7] Nevertheless, most experimental proposals on the detection of Unruh effect are designed for the physical four dimensional spacetime, so the chance is still open before a detailed analysis for (3+1) dimensions has been done.

In the following we study the Unruh-DeWitt detector theory in (3+1)D.[8] We use the Heisenberg operator method to obtain the exact evolution of two-point functions of the detector and of the field. Then we calculate the radiated flux through the event horizon for the UAD to see whether there is radiation emitted by UAD in steady state, and whether the evidence of the Unruh effect can be found in that radiation.

2. Unruh-DeWitt Detector Theory

Let us consider a Unruh-DeWitt(UD) detector-field system, whose total action is given by[2,9]

$$S = \int d\tau \frac{m_0}{2}\left[\dot{Q}^2 - \Omega_0^2 Q^2\right] - \int d^4x \frac{1}{2}\partial_\mu \Phi \partial^\mu \Phi$$
$$+\lambda_0 \int d\tau \int d^4x Q(\tau)\Phi(x)\delta^4\left(x^\mu - z^\mu(\tau)\right) \qquad (1)$$

where Q is the internal degree of freedom of the detector, assumed to be a harmonic oscillator with mass m_0 and a (bare) natural frequency Ω_0, τ is the detector's proper time, and $\dot{Q} \equiv dQ(\tau)/d\tau$. Φ is the massless scalar field, and λ_0 is the coupling constant.

Consider the UD detector moving in a prescribed trajectory $z^\mu(\tau)$ in a four-dimensional Minkowski space. The trajectory of the detector is not considered as a dynamical variable, and the motion of the detector is gauged by an external agent. We are thus dealing with a hybrid of quantum field theory of the massless scalar field, quantum mechanics of the detector's intenal degree of freedom, and the classical external agent.

Let the detector go along $z^\mu(\tau) = (a^{-1}\sinh a\tau, a^{-1}\cosh a\tau, 0, 0)$ (i.e., in uniform acceleration), so $U \equiv t - x^1 = 0$ is the event horizon for the detector. Suppose the system is prepared before $\tau = \tau_0$, and the coupling S_I is turned on precisely at the moment τ_0 when we allow all the dynamical variables to begin to interact and evolve. Then, by virtue of the linear coupling in (1), the time evolution of $\hat{\Phi}(\mathbf{x})$ is simply a linear transformation in the phase space spanned by $(\hat{\Phi}(\mathbf{x}), \hat{\Pi}(\mathbf{x}), \hat{Q}, \hat{P})$. For the case with initial operators being the free field operators, one can go further by introduce $\hat{b}^{\dagger}_{\mathbf{k}}$, $\hat{b}_{\mathbf{k}}$, $\hat{a}^{\dagger}$ and $\hat{a}$, which are the creation and annihilation operators defined in free theories for the scalar field and the detector, respectively. Then $\hat{\Phi}(x)$ and $\hat{Q}(\tau)$ can be expressed in the form

$$\hat{\Phi}(x) \sim \sum_{\mathbf{k}} \left[f^{(+)}(x;\mathbf{k})\hat{b}_{\mathbf{k}} + f^{(-)}(x;\mathbf{k})\hat{b}^{\dagger}_{\mathbf{k}} \right] + f^{a}(x)\hat{a} + f^{a*}(x)\hat{a}^{\dagger}, \quad (2)$$

$$\hat{Q}(\tau) \sim \sum_{\mathbf{k}} \left[q^{(+)}(\tau,\mathbf{k})\hat{b}_{\mathbf{k}} + q^{(-)}(\tau,\mathbf{k})\hat{b}^{\dagger}_{\mathbf{k}} \right] + q^{a}(\tau)\hat{a} + q^{a*}(\tau)\hat{a}^{\dagger}. \quad (3)$$

The whole problem therefore can be transformed from solving the Heisenberg equations of motion for the operators into solving c-number functions $f^{s}(x)$ and $q^{s}(\tau)$ with suitable initial conditions.

The UD detector considered here is a quantum mechanical object, which means that there is a natural cutoff on frequency at the energy threshold of detector creations. Thus it is justified to assume that the detector has a finite extent and the retarded Green's function can be regularized accordingly[10]. After the regularization and renormalization, one has

$$(\partial_{\tau}^{2} + 2\gamma\partial_{\tau} + \Omega_{r}^{2})q^{(+)}(\tau;\mathbf{k}) = \frac{\lambda_0}{m_0} f_0^{(+)}(z(\tau);\mathbf{k}), \quad (4)$$

where $f_0^{(+)}(x;\mathbf{k}) \equiv \exp(-i\omega t + i\mathbf{k}\cdot\mathbf{x})$ is the free field solution in Minkowski coordinate. Thus $q^{(+)}$ behaves like a damped harmonic oscillator driven by the vacuum fluctuations of the scalar field, with the damping constant $\gamma \equiv \lambda_0^2/8\pi m_0$ and the renormalized natural frequency Ω_r. Note that (4) is causal and local in τ. Once the form of $f_0^{(+)}$ is given, $q^{(+)}(\tau)$ in (4) is totally determined by the motion of the detector from τ_0 to τ. Hence the response of $q^{(+)}$ is pure kinematics. In contrast, after including the back reaction of the field, the equation of motion for q^{a} looks like the one describing a damped harmonic oscillator without driving force.

3. Evolution of Detector's Two-Point Functions

Suppose the detector is initially prepared in a state that can be factorized into the coherent state $|\, q\,\rangle$ for Q and the Minkowski vacuum $|\, 0_M\rangle$ for the scalar field Φ, i.e., $|\tau_0\,\rangle = |\, q\,\rangle\,|\, 0_M\rangle$. Then the two-point functions of Q will split into two parts, $\langle\, Q(\tau)Q(\tau')\,\rangle = \langle\, q\,|\, q\,\rangle \langle\, Q(\tau)Q(\tau')\,\rangle_{\rm v} + \langle\, Q(\tau)Q(\tau')\,\rangle_{\rm a} \langle 0_M|0_M\rangle$. Here $\langle\, Q(\tau)Q(\tau')\,\rangle_{\rm v}$ can be interpreted as account-

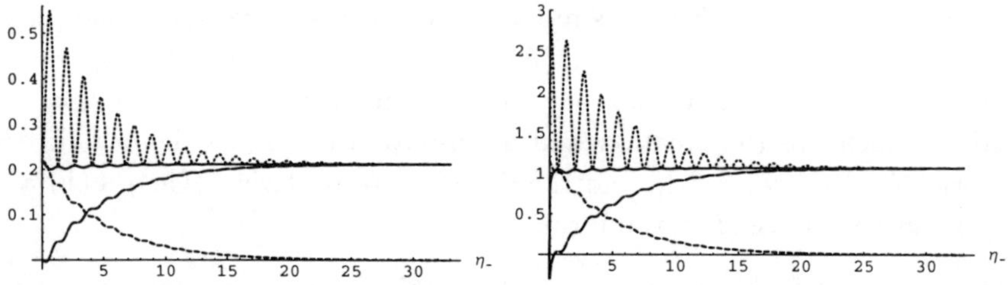

Fig. 1. The long-dashed lines are $\langle\, Q(\eta)^2 \,\rangle_{\rm v}$(left) and $\langle\, \dot{Q}(\eta)^2 \,\rangle_{\rm v}$(right), the short-dashed lines are $\langle\, Q(\eta)^2 \,\rangle_{\rm a}^{\rm qm}$(left) and $\langle\, \dot{Q}(\eta)^2 \,\rangle_{\rm a}^{\rm qm}$, and the solid lines are the sum of them ($\langle\, \Delta Q(\eta)^2 \,\rangle$(left) and $\langle\, \Delta \dot{Q}(\eta)^2 \,\rangle$(right)). The dotted lines are Λ_0-terms. Here we have taken $a = 1$, $\gamma = 0.1$, $\Omega = 2.3$, $m_0 = 1$ and $\Lambda_0 = 0.4$.

ing for the response to the vacuum fluctuations, while $\langle\, Q(\tau)Q(\tau') \,\rangle_{\rm a}$ corresponds to the intrinsic quantum fluctuations in the detector.

With the solution of $q^{(+)}$, the two-point functions of the detector $\langle\, Q(\eta)Q(\eta') \,\rangle_{\rm v} \sim \sum_{\mathbf{k}} q^{(+)}(\tau;\mathbf{k})q^{(-)}(\tau;\mathbf{k})$ can be explicitly obtained (here $\eta \equiv \tau - \tau_0$). The coincidence limit of it looks like

$$\langle\, Q(\eta)^2 \,\rangle_{\rm v} = \frac{\hbar\lambda_0^2\theta(\eta)}{(2\pi m_0\Omega)^2}\left[\Lambda_0 e^{-2\gamma\eta}\sin^2\Omega\eta + (\text{regular terms})\right], \qquad (5)$$

where $\Lambda_0 \sim -\ln|\tau_0' - \tau_0|$. Since $|\tau_0' - \tau_0|$ characterizes the time scale that the interaction is turned on, Λ_0 could be finite in real processes. In any case, for every finite value of Λ_0, the Λ_0-term vanishes as $\eta \to \infty$. In Figure 1(left), we show the evolution of the regular part of $\langle\, Q(\eta)^2 \,\rangle_{\rm v}$. Roughly speaking the curve saturates exponentially in the detector's proper time to a positive number.

The coincidence limit of the two-point function $\langle\, \dot{Q}(\eta)\dot{Q}(\eta') \,\rangle_{\rm v}$ looks similar:

$$\langle\, \dot{Q}(\eta)^2 \,\rangle_{\rm v} = \frac{\hbar\lambda_0^2\theta(\eta)}{(2\pi m_0\Omega)^2}\left[\Lambda_1\Omega^2 + \Lambda_0 e^{-2\gamma\eta}\left(\Omega\cos\Omega\eta - \gamma\sin\Omega\eta\right)^2 + \cdots\right] (6)$$

where $\Lambda_1 \sim -\ln|\tau - \tau'|$ corresponds to the resolution of this theory and can be subtracted safely. The regular part of $\langle\, \dot{Q}(\eta)^2 \,\rangle_{\rm v}$ is illustrated in Figure 1(right). Its behavior is quite similar to $\langle\, Q(\tau)^2 \,\rangle_{\rm v}$ in Figure 1(left).

For the expectation values of the detector two-point functions with respect to the coherent state, one has $\langle\, Q(\tau)Q(\tau') \,\rangle_{\rm a} = \langle\, Q(\tau)Q(\tau') \,\rangle_{\rm a}^{\rm qm} + \bar{Q}(\tau)\bar{Q}(\tau')$, where $\langle\, Q(\tau)Q(\tau') \,\rangle_{\rm a}^{\rm qm} \sim q^a(\tau)q^{a*}(\tau')$ is of purely quantum nature, and $\bar{Q} \sim {\rm Re}[q^a]$ does not involve $\hbar$. $\bar{Q}(\tau)\bar{Q}(\tau')$ is identified as the semiclassical part of the two-point functions. The coincidence limits of them are straightforward. $\langle\, Q(\eta)^2\,\rangle_{\rm a}^{\rm qm}$ and the variance (squared uncertainty) of Q, $\langle\, \Delta Q(\eta)^2\,\rangle \equiv \langle\, [Q(\eta)-\bar{Q}(\eta)]^2\,\rangle = \langle\, Q(\eta)^2\,\rangle_{\rm v} + \langle\, Q(\eta)^2\,\rangle_{\rm a}^{\rm qm}$, have been shown in Figure 1(left). One can see that $\langle\, Q(\eta)^2\,\rangle_{\rm a}^{\rm qm}$ decays exponentially due to the dissipation of the zero-point energy to the field. As $\langle\, Q(\eta)^2\,\rangle_{\rm a}^{\rm qm}$ decays, $\langle\, Q(\eta)^2\,\rangle_{\rm v}$ grows, then saturates asymptotically. Similar behavior can be found in Figure 1(right), in which $\langle\, \dot{Q}(\eta)^2\,\rangle_{\rm a}^{\rm qm}$ and $\langle\, \Delta\dot{Q}(\eta)^2\,\rangle$ are illustrated. Note that the two-point functions with respect to the coherent state are independent of the proper acceleration a.

4. Radiated Flux

Given the classical retarded field $\Phi_{\rm ret}$ induced by the uniformly accelerated UD detector, the stress-energy tensor $T^{tU}[\Phi_{\rm ret}]_{U\to 0}$ describes the radiated flux through the event horizon.[4,11] Now the quantum expectation value of the renormalized stress-energy tensor $\langle T_{\mu\nu}\rangle_{\rm ren}$ is obtained by calculating

$$\langle T_{\mu\nu}[\Phi(x)]\rangle_{\rm ren} = \lim_{x'\to x}\left[\frac{\partial}{\partial x^\mu}\frac{\partial}{\partial x'^\nu} - \frac{1}{2}g_{\mu\nu}g^{\rho\sigma}\frac{\partial}{\partial x^\rho}\frac{\partial}{\partial x'^\sigma}\right]G_{\rm ren}(x,x'), \qquad (7)$$

where $G_{\rm ren}$ is the renormalized two-point function of the field, defined by $G_{\rm ren}(x,x') \equiv \langle\, \hat{\Phi}(x)\hat{\Phi}(x')\,\rangle - G_{\rm v}^{00}(x,x')$ with the Green's function for free fields $G_{\rm v}^{00}$ subtracted. Similar to the two-point functions of the detector, for our factorized initial state, the two-point function of Φ could be split into $\langle\, \hat{\Phi}\hat{\Phi}\,\rangle = G_{\rm v}+G_{\rm a}$, where $G_{\rm a}(x,x') \sim f^a(x)f^{a*}(x') \propto \langle\, Q(\tau_-)Q(\tau'_-)\,\rangle_{\rm a}$ corresponds to the dissipation of the zero-point energy of the internal degree of freedom of the detector, while $G_{\rm v}(x,x') \sim \sum_{\bf k} f^{(+)}(x;{\bf k})f^{(-)}(x';{\bf k})$ accounts for the back reaction of the vacuum fluctuations of the scalar field on the field itself. Since the solution for $f^{(+)}$ is the sum of the free solution $f_0^{(+)}$ and the retarded solution $f_1^{(+)}$, $G_{\rm v}$ can be decomposed into four pieces, $G_{\rm v} = \sum_{i,j=0,1} G_{\rm v}^{ij}$ with $G_{\rm v}^{ij}(x,x') \sim \sum_{\bf k} f_i^{(+)}(x;{\bf k})f_j^{(-)}(x',{\bf k})$. One finds that $G_{\rm v}^{11}$ is proportional to $\langle\, Q(\tau_-)Q(\tau'_-)\,\rangle_{\rm v}$ and positive definite, and $G_{\rm v}^{10}$ accounts for the interference between $f_1^{(+)}$ and $f_0^{(+)}$.

After $G_{\rm ren}$ has been worked out, one obtains straightforwardly the quantum expectation value of T^{tU} near the event horizon $U\to 0$,

$$\langle\, T^{tU}\,\rangle_{\rm ren} \to \frac{2\lambda_0^2\theta(\eta_-)}{(2\pi)^2a^4}\left[\frac{\langle\, \dot{Q}(\tau_-)^2\,\rangle_{\rm tot}}{V^2(\rho^2+a^{-2})^2} + \frac{\rho^2\langle\, [\dot{Q}(\tau_-)+aQ(\tau_-)]^2\,\rangle_{\rm tot}}{(\rho^2+a^{-2})^4}\right]_{U\to 0} \qquad (8)$$

which has a similar form to the classical one. Here $\langle\, \dot{Q}^2\,\rangle_{\rm tot} \equiv \langle\, \dot{Q}^2\,\rangle + \varpi\Theta_{--}$, $\langle\, Q^2\,\rangle_{\rm tot} \equiv \langle\, Q^2\,\rangle + \varpi\Theta_{XX}$ and $\langle\, Q\dot{Q}\,\rangle_{\rm tot} \equiv \langle\, Q\dot{Q}\,\rangle + \varpi\Theta_{X-}$ with constant $\varpi \equiv \hbar/2\pi m_0$ and Θ_{ij} coming from the interfering term $G_{\rm v}^{10}+G_{\rm v}^{01}$.[8]

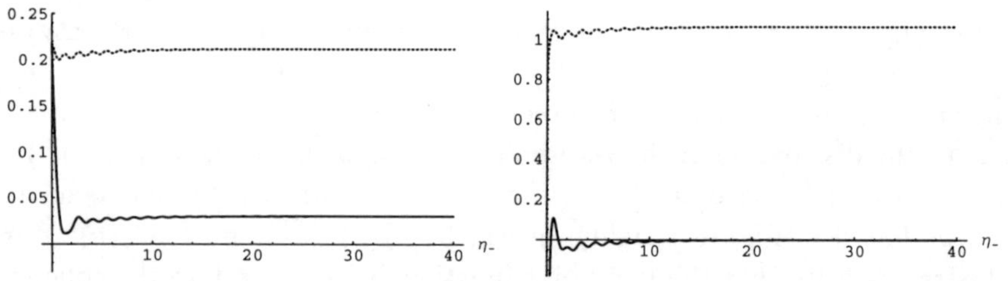

Fig. 2. The total variances $\langle\ \Delta Q^2\ \rangle_{\rm tot} \equiv \langle\ Q^2\ \rangle_{\rm tot} - \bar{Q}^2$ near the event horizon for the detector (left, solid line) and $\langle\ \Delta \dot{Q}^2\ \rangle_{\rm tot} \equiv \langle\ \dot{Q}^2\ \rangle_{\rm tot} - \bar{\dot{Q}}^2$ (right, solid line). The values of parameters are the same as those in Figure 1. $\langle\ \Delta Q^2\ \rangle_{\rm tot}$ in the left plot finally saturates to the value $\hbar a/2\pi m_0 \Omega_r^2$, while $\langle\ \Delta \dot{Q}^2\ \rangle_{\rm tot}$ in the right plot saturates to zero. One can compare $\langle\ \Delta Q^2\ \rangle_{\rm tot}$ and $\langle\ \Delta \dot{Q}^2\ \rangle_{\rm tot}$ with $\langle\ \Delta Q^2\ \rangle$ and $\langle\ \Delta \dot{Q}^2\ \rangle$ (dotted lines, which are the same as the solid lines in Figure 1) and see the suppression.

The $\langle\ \dot{Q}(\tau_-)^2\ \rangle_{\rm tot}$ term would be recognized as a monopole radiation by the Minkowski observer. To see the properties of quantum nature, we define the total variances by subtracting the semiclassical part from $\langle \cdots \rangle_{\rm tot}$. Their evolution against η_- are illustrated in Figure 2.

As shown in Figure 2(right), $\langle\ \Delta \dot{Q}(\tau_-)^2\ \rangle_{\rm tot}$ goes to zero at late times, so the corresponding monopole radiation ceases after the transient. This appears to agree with the claim that for a UAD in (1+1)D, emitted radiation is only associated with nonequilibrium process[12]. The interference between the quantum radiation induced by the vacuum fluctuations and the vacuum fluctuations themselves totally screens the information about the Unruh effect in this part of the radiation.

A natural definition of the energy of the "dressed" detector is $E(\eta) \equiv (m_0/2)[\langle\ \dot{Q}^2(\eta)\ \rangle + \Omega_r^2 \langle\ Q^2(\eta)\ \rangle]$. From our results, we find that

$$-\dot{E}(\eta) = \frac{\lambda_0^2}{4\pi} \langle\ \dot{Q}^2(\eta)\ \rangle_{\rm tot}\,, \tag{9}$$

for all $\eta > 0$. The left hand side of this equality is the energy-loss rate of the dressed detector, while the right hand side is the radiated power via the monopole radiation corresponding to $\langle\ \dot{Q}^2(\eta)\ \rangle_{\rm tot}$. Therefore (9) is simply a statement of energy conservation between the detector and the field, and the external agent which drives the detector along the trajectory $z^\mu(\tau)$ has no additional influence on this channel.

For the total variance $\langle\ \Delta Q^2\ \rangle_{\rm tot}$, Figure 2(left) shows that, right after the coupling is turned on, Θ_{XX} builds up and $\langle\ \Delta Q^2\ \rangle_{\rm tot}$ near the event horizon $U = 0$ is pulled down rapidly. Then $\langle\ \Delta Q^2\ \rangle_{\rm tot}$ turns into a tail, which exponentially approaches the saturated value $\hbar a/2\pi m_0 \Omega_r^2 \propto T_U$ with the time scale $1/2\gamma$. While all other terms ($\langle \Delta\dot{Q}^2 \rangle_{\rm tot}$ and $\langle \Delta Q \Delta\dot{Q} \rangle_{\rm tot}$) in (8) vanishes in steady state, there still exists a positive radiated power of quantum nature corresponding to $\langle \Delta Q(\infty)^2 \rangle_{\rm tot}$ and emitted by the uniformly accelerated UD detector. Note that $\langle \Delta Q^2(\eta_-(x)) \rangle_{\rm tot}$ is proportional to the quantum part of $G_{\rm ren}(x,x)$, whose counterpart in (1+1)D has been interpreted as the localized "vacuum polarization cloud".[7]

5. Conclusion

The uniformly accelerated UD detector in steady state does emit a positive radiated flux of quantum nature through the event horizon. This flux is proportional to the Unruh temperature T_U, so it could be interpreted as a hint of the Unruh effect. However, the total energy of the dressed detector is conserved only with the radiated energy of a monopole radiation, which ceases in steady state analogous to the radiation in (1+1)D detector theory. Hence the hint of the Unruh effect in the late-time radiated flux in (3+1)D is not directly from the energy flux experienced by the detector in Unruh effect.

References

1. S.W. Hawking, Commum. Math. Phys. **43**, 199 (1975).
2. W. G. Unruh, Phys. Rev. **D14**, 870 (1976).
3. J. D. Jackson, *Classical Electrodynamics* (Wiley, New York, 1983).
4. D. G. Boulware, Ann. Phys. (N.Y.) **124**, 169 (1980).
5. J. M. Leinaas, in *Quantum Aspects of Beam Physics*, 18th Advanced IFCA Beam Dynamics Workshop, edited by Pisin Chen (World Scientific, Singapore, 2001); P. Chen and T. Tajima, Phys. Rev. Lett. **83**, 256 (1999).
6. P. G. Grove, Class. Quan. Grav. **3**, 801 (1986); D. J. Raine, D. W. Sciama, and P. G. Grove, Proc. Roy. Soc. Lond. **A435**, 205 (1991); W. G. Unruh, Phys. Rev. **D46**, 3271 (1992); B. L. Hu and A. Raval, in [5], quant-ph/0012135.
7. S. Massar, R. Parentani and R. Brout, Class. Quantum Grav. **10**, 385 (1993).
8. S.-Y. Lin and B. L. Hu, [gr-qc/0507054].
9. B. S. DeWitt, in *General Relativity: an Einstein Centenary Survey*, edited by S. W. Hawking and W. Israel (Cambridge University Press, 1979).
10. C. R. Galley and B. L. Hu, Phys. Rev. **D72**, 084023 (2005); C. R. Galley, B. L. Hu and S.-Y. Lin, gr-qc/0603099.
11. S.-Y. Lin, Phys. Rev. **D68**, 104019 (2003).
12. B. L. Hu and P. R. Johnson, in [5], quant-ph/0012132.

Signatures of Spacetime Geometry Fluctuations

L.H. Ford and R. T. Thompson

Institute of Cosmology
Department of Physics and Astronomy
Tufts University, Medford, MA 02155 USA

The operational meaning of quantum fluctuations of spacetime geometry will be discussed. Three potential signatures of these fluctuations will be considered: luminosity fluctuations of a distant source, angular blurring of images, and broadening of spectral lines. To leading order, luminosity fluctuations arise only from passive geometry fluctuations, those driven by quantum fluctuations of the stress tensor. This effect can be described by a Langevin version of the Raychaudhuri equation. Angular blurring and line broadening can arise both from passive fluctuations and from the active fluctuations of the quantized gravitational field, and can be given a unified geometrical description using the Riemann tensor correlation function.

1. Introduction

Among the key phenomena which will distinguish a quantum theory of gravity from classical general relativity is the phenomenon of spacetime geometry fluctuations. These fluctuations can arise from the quantum fluctuations of stress tensors of matter fields (passive fluctuations)[1,2,3,4,5,6,7,8] as well as from the fluctuations of the dynamical degrees of freedom of the gravitational field (active fluctuations). In this talk, we will be concerned with the operational meaning and observability of spacetime geometry fluctuations.

The meaning of a classical spacetime geometry is ultimately encoded in the geodesics of test particles, both massive and massless, moving in the geometry. Similarly, the meaning of spacetime geometry fluctuations is encoded in the fluctuations in these geodesics, that is, their Brownian motion. The hope of finding observable consequences of various versions of quantum gravity theory has spawned a new field of "quantum gravity phenomenology" [9,10,11,12,13,14,15].

Here we will be concerned with three specific effects, luminosity fluctuations of a distant source [15], redshift fluctuations or line broadening, and angular blurring of the image of a distant source[16].

2. Luminosity Fluctuations

In a classical spacetime, the expansion, θ, of a congruence of geodesics satisfies the Raychaudhuri equation,

$$\frac{d\theta}{d\lambda} = -R_{\mu\nu}k^\mu k^\nu - a\,\theta^2 - \sigma_{\mu\nu}\sigma^{\mu\nu} + \omega_{\mu\nu}\omega^{\mu\nu}\,, \tag{1}$$

where $\sigma^{\mu\nu}$ is the shear and $\omega^{\mu\nu}$ is the vorticity of the congruence. The constant $a = 1/2$ for null geodesics, and $a = 1/3$ for timelike geodesics. Here we restrict our attention to the case where $\sigma^{\mu\nu} = \omega^{\mu\nu} = 0$ and the θ^2 term is small enough to neglect. Then we have

$$\frac{d\theta}{d\lambda} = -R_{\mu\nu}k^\mu k^\nu\,. \tag{2}$$

If the stress tensor of a matter field undergoes quantum fluctuations, passive fluctuations of the Ricci tensor will result, and we can interpret Eq. (2) as a Langevin equation[17]. The mean squared fluctuations of θ can be obtained by integrating this equation to find

$$\langle\theta^2\rangle - \langle\theta\rangle^2 = \langle(\Delta\theta)^2\rangle = \int_0^{\lambda_0} d\lambda \int_0^{\lambda_0} d\lambda'\, C_{\mu\nu\alpha\beta}(\lambda,\lambda')\, k^\mu(\lambda)k^\nu(\lambda)\, k^\alpha(\lambda')k^\beta(\lambda')\,, \tag{3}$$

where $C_{\mu\nu\alpha\beta}(x,x')$ is the Ricci tensor correlation function:

$$C_{\mu\nu\alpha\beta}(x,x') = \langle R_{\mu\nu}(x)R_{\alpha\beta}(x')\rangle - \langle R_{\mu\nu}(x)\rangle\langle R_{\alpha\beta}(x')\rangle\,. \tag{4}$$

Here we have assumed the initial condition of $\theta = 0$ at $\lambda = 0$. The Ricci tensor correlation function is simply related to the quantum stress tensor correlation function. However, the stress tensor correlation function is singular in the limit of coincident points, $x' \to x$, even if the expectation values of the stress tensor operators have been renormalized. This singularity can be removed by integrating over a spacetime region[18,19]. In Eq. (3), this means that we should really integrate over a worldtube describing a bundle of test particles, rather than along the path of a single test particle[15].

The physical significance of θ fluctuations is that they are linked to fluctuations in the apparent luminosity of a distant source. The expansion θ is the logarithmic derivative of the cross sectional area of a bundle of rays with respect to the affine parameter λ. Thus increases in θ lead to decreases in apparent luminosity, and *vice versa*. This is similar to the familiar "twinkling" of stars due to fluctuations in the density of the Earth's atmosphere, and was discussed in the context of stochastic classical gravity waves by Zipoy[20]. In general, the luminosity fluctuations require one to integrate a correlation function for the expansion:

$$\left\langle \left(\frac{\Delta L}{L}\right)^2 \right\rangle = \int_0^s \int_0^s dt'\, dt''\, \left[\langle \theta(t')\, \theta(t'') \rangle - \langle \theta(t') \rangle \langle \theta(t'') \rangle\right] . \tag{5}$$

Here we have used a time coordinate t as affine parameter and taken s to be the flight time in these coordinates. However, in many cases, where the flight time is long, $\Delta L/L$ can be related directly to the variance of the expansion[15]:

$$\left\langle \left(\frac{\Delta L}{L}\right)^2 \right\rangle = c\, s^2 \left\langle\, \theta^2(s)\, \right\rangle , \tag{6}$$

where c is a constant of order one. In the case of passive fluctuations caused by a massless scalar field in a thermal state in the high temperature limit

$$\left(\frac{\Delta L}{L}\right)_{rms} \approx \ell_P^2\, \frac{s^{\frac{3}{2}}}{b\, \beta^{\frac{5}{2}}} , \tag{7}$$

where ℓ_P is the Planck length, b is a typical cross sectional size of the bundle of test rays, and β is the thermal wavelength. The fact that the result depends inversely upon b reflects the singularity of stress tensor correlation functions noted above. Equation (7) applies in the high temperature limit where $\beta \ll b$. In the low temperature limit, $\beta \gg b$, the result is independent of b, and can be written as

$$\left(\frac{\Delta L}{L}\right)_{rms} \approx \ell_P^2\, \frac{s^{\frac{3}{2}}}{\beta^{\frac{7}{2}}} \approx 10^{-3} \left(\frac{s}{10^6 \text{km}}\right)^{\frac{3}{2}} \left(\frac{T}{1 \text{GeV}}\right)^{\frac{7}{2}} , \tag{8}$$

where T is the temperature of the bath. Clearly, this is a small effect unless either T or s are rather large.

3. Redshift Fluctuations and Angular Blurring

Two additional effects which can be caused by geometry fluctuations are the broadening of spectral lines and the blurring of images of distant sources. These two effects may be viewed as arising from variations in gravitational redshift and in angular deflection, respectively, and have been discussed by several previous authors[20,21,22,23,24]. These two effects were given a unified geometrical treatment in a recent paper[16], the results of which are summarized in this section. There it was shown that the variance of the fractional redshift and of the deflection angle can both be expressed as integrals of the Riemann tensor correlation function:

$$C_{\alpha\beta\mu\nu\,\gamma\delta\rho\sigma}(x,x') = \langle R_{\alpha\beta\mu\nu}(x) R_{\gamma\delta\rho\sigma}(x')\rangle - \langle R_{\alpha\beta\mu\nu}(x)\rangle\langle R_{\gamma\delta\rho\sigma}(x')\rangle, \quad (9)$$

where the indices $\alpha\beta\mu\nu$ refer to point x while the indices $\gamma\delta\rho\sigma$ refer to point x'.

3.1. *Redshift Fluctuations*

Let $\xi = \Delta\omega/\omega$ be the fractional redshift or blueshift of a photon emitted at angular frequency ω in the source's rest frame and detected at frequency $\omega + \Delta\omega$ in a detector's rest frame. In the case where a fluctuating gravitational field fills the intervening space between the source and the detector, there will be fluctuations in ξ given by an integral of the Riemann tensor correlation function $C_{\alpha\beta\mu\nu\,\gamma\delta\rho\sigma}(x,x')$. Consider the case where we have a source and a detector which are at rest with respect to each other in approximately flat regions of spacetime, with t^μ the four-velocity of the source, and $v^\mu = t^\mu$ that of the detector. Thus any observed variation in detected frequency is gravitational, rather than due to a Doppler effect. Suppose that the detector emits a photon at proper time τ_1 and a second photon at τ_2. Let $k^\mu = dx^\mu/d\lambda$ be the four-velocity of the photon, and λ is an affine parameter chosen so that $t^\mu\, k_\mu = -1$ at the detector. The effect of intervening spacetime curvature is to cause the photon four-velocity at the detector to change by Δk^μ between successive detections, as illustrated in Fig. 1. Fluctuations in the geometry lead to redshift fluctuations which are given by

$$\delta\xi^2 = \langle(\Delta\xi)^2\rangle - \langle\Delta\xi\rangle^2 = \int da \int da'\, C_{\alpha\beta\mu\nu\,\gamma\delta\rho\sigma}(x,x')\, t^\alpha k^\beta t^\mu k^\nu t^\gamma k^\delta t^\rho k^\sigma . \quad (10)$$

The integration over the spacetime region $\int da$ corresponds to $\int_{\tau_1}^{\tau_2} d\tau \int_0^{\lambda_0} d\lambda$ and similarly for da'. This is an integration over the spacetime region enclosed by the path ABCD in Fig. 1. The point x corresponds to the point (τ, λ) and similarly for x'.

3.2. *Angular Blurring*

We can also relate the degree of angular blurring of the source observed by the detector to the Riemann tensor correlation function. Let s^μ be a unit spacelike vector in a direction orthogonal to the direction of propagation of

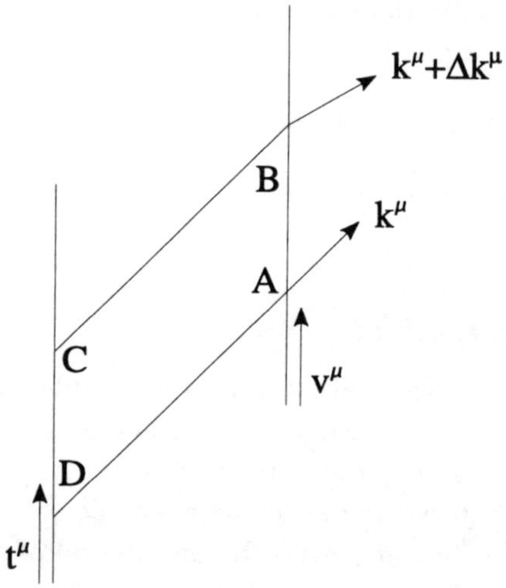

Fig. 1. A source moves along a worldline with tangent t^μ while a detector a proper distance s away moves along a worldline with tangent v^μ. The source emits a ray at point D which has tangent $k^\mu(\lambda = 0)$ at point D and tangent $k^\mu(\lambda_0)$ at A. Parallel propagation of k^μ around ABCD results in a slightly rotated vector $k^\mu + \Delta k^\mu$. The closed path ABCD encloses the spacetime region of interest.

the null rays; thus $s_\mu t^\mu = s_\mu k^\mu(\lambda = 0) = 0$. Then at the observation point

$$s_\mu k^\mu(\lambda_0) = \tan\Theta \approx \Theta, \tag{11}$$

where Θ is an angle in the plane defined by the pair of spacelike vectors s^μ and $n^\mu = k^\mu - t^\mu$ and is assumed to be small, $|\Theta| \ll 1$. The angle Θ is the angular deviation in the direction of s^μ of the image of the source from its classical flat space position. A treatment similar to that shown for fractional redshift allows us to express the change in angle between points A and B in Fig. 1 , $\Delta\Theta$, in terms of an integral of the Riemann tensor as

$$\Delta\Theta = s_\mu \Delta k^\mu = \int da\, R_{\alpha\beta\mu\nu} s^\alpha k^\beta t^\mu k^\nu . \tag{12}$$

A fluctuating spacetime results in an ensemble distribution of image positions about the classical flat space position. Analogously to the line broadening effects, the fluctuating angular position manifests itself as a blurring of the source's image. The variance of $\Delta\Theta$, $\delta\Theta^2$, due to fluctuations in the Riemann tensor is

$$\delta\Theta^2 = \langle(\Delta\Theta)^2\rangle - \langle\Delta\Theta\rangle^2 = \int da \int da'\, C_{\alpha\beta\mu\nu\,\gamma\delta\rho\sigma}(x,x') s^\alpha k^\beta t^\mu k^\nu s^\gamma k^\delta t^\rho k^\sigma , \tag{13}$$

is therefore a measure of the angular size of the blurred image.

3.3. *Examples: Gravitons in Squeezed States and in a Thermal State*

A bath of gravitons in a nonclassical state offers a simple example of a fluctuating spacetime geometry. Here we may use linearized gravity theory to describe the gravitons. One choice of nonclassical state is a single mode squeezed state. The squeezed states form a one-parameter family which include the coherent states (classical states) as one limit, and the squeezed vacuum states as another limit. The latter limit is of particular interest to us, as it describes a spacetime in which

$$\langle R_{\alpha\beta\mu\nu} \rangle = 0 \tag{14}$$

to linear order, that is, fluctuations around an average spacetime which is flat. Gravitons in a squeezed vacuum state are the natural result of quantum creation processes in a gravitational field[25].

Some explicit calculations have been made[16] of $(\Delta\xi)_{rms} = \sqrt{\delta\xi^2}$ and of $(\Delta\Theta)_{rms} = \sqrt{\delta\Theta^2}$ in squeezed vacuum states, and it was found that

$$(\Delta\xi)_{rms} \approx (\Delta\Theta)_{rms} \approx \ell_P \lambda_g \sqrt{: T_{00} :}\,, \tag{15}$$

where $: T_{00} :$ is the typical energy density of the gravitons and λ_g is their characteristic wavelength. This is only an order of magnitude result. More generally, $(\Delta\Theta)_{rms}$ depends both upon the direction in which the angular blurring is measured, and upon the polarization state of the gravitons. Suppose for example, a gravity wave with $\lambda_g = 1\,\text{ly} = 10^{18}\text{cm}$ and the closure energy density $: T_{00} : = 10^8 \text{cm}^{-4}$. Then with $\ell_{\text{Pl}} = 10^{-33}\text{cm}$,

$$(\Delta\xi)_{rms} \approx (\Delta\Theta)_{rms} \approx 10^{-11}\,. \tag{16}$$

In principle, the effect can be made as large as desired by increasing the squeezing parameter r, which increases the energy density of the wave. However, as shown from the estimate given, this is a very small effect and is likely to be unobservable in the present day universe.

4. Discussion

We have discussed several phenomenological effects that can in principle be produced by spacetime geometry fluctuations; namely, luminosity fluctuations, redshift fluctuations, and angular blurring of images. All of these effects tend to be small in the present day universe, as one would expect for effects in quantum gravity. Our interest has been primarily to illustrate some concrete effects which give an operational meaning to the concept of a fluctuating geometry.

However, it is useful to examine some of the general features of the three effects which we have discussed. All three are described by mean squared observables which are expressible in terms of integrals of the Riemann tensor correlation function, as is seen in Eqs. (3), (10), and (13). The luminosity fluctuations, to leading order, respond only to fluctuations of the Ricci part of the Riemann tensor, and hence to passive fluctuations driven by quantum stress tensor fluctuations. Although we chose to illustrate the redshift fluctuations and angular blurring with models involving active fluctuations, they can equally well arise from passive fluctuations. That is, in Eqs. (10) and (13), the Riemann tensor correlation function can include Ricci parts as well as Weyl parts. However, the effects of quantum stress tensor fluctuations require averaging over a spacetime volume and replacing the worldline integrations in Eqs. (10) and (13) by worldtube integrations. If we compare Eq. (7) with Eq. (15), we see that the former has two powers of the Planck length, ℓ_P, as compared to the single power in the latter. This is a general feature of passive versus active fluctuations. Passive fluctuations therefore tend to be suppressed compared to the active ones due to the extra power of ℓ_P.

We have seen that all of these efects tend to be very small in most astrophysical situations in the present universe. This is not surprising, as we are dealing with quantum gravity effects and observables proportional either to ℓ_P or to ℓ_P^2. However, it should be noted that we have selected only relatively conservative examples, such a thermal baths, based on established physics. This does not rule out the possibility that quantum gravity will produce new physics in which the effects discussed here are enhanced. An example might be "spacetime foam" at the Planck scale, which produces observable effects. This possibility has generated some discussion recently[11,12,13,14].

Even fluctuations due to well-established physics should be much larger in strong gravitational fields, such as in the early universe or near black holes. The possible consequences of spacetime fluctuations in these contexts remain to be thoroughly explored.

Acknowledgments

This work was supported in part by the National Science Foundation under Grant PHY-0244898.

References

1. L.H. Ford, Ann. Phys (NY) **144**, 238 (1982).
2. C.-I Kuo and L.H. Ford, Phys. Rev. D **47**, 4510 (1993), gr-qc/9304008.
3. B. L. Hu and K. Shiokawa, Phys. Rev. D **57**, 3474 (1998), gr-qc/9708023.
4. B. L. Hu, Int .J. Theor. Phys. **38**, 2987 (1999), gr-qc/9902064.
5. N.G. Phillips and B.L. Hu, Phys. Rev. D **63**, 104001 (2001), gr-qc/0010019.
6. B. L. Hu and E. Verdaguer, Class. Quant. Grav. **20**, R1 (2003), gr-qc/0211090.
7. R. Martin and E. Verdaguer, Phys. Lett. B **465**, 113 (1999), gr-qc/9811070; Phys. Rev. D **60**, 084008 (1999), gr-qc/9904021; Phys. Rev. D **61**, 124024 (2000), gr-qc/0001098.
8. B. L. Hu and E. Verdaguer, Living Rev. Rel. **7**, 3, (2004), gr-qc/0307032.
9. L.H. Ford, Phys. Rev. D **51**, 1692 (1995), gr-qc/9410047
10. J. Ellis, N.E. Mavromatos and D.V. Nanopoulos, Gen. Rel. Grav. **32**, 127 (2000), gr-qc/9904068.
11. Y. J. Ng and H. van Dam, Phys. Lett. B **477**, 429 (2000), gr-qc/9911054.
12. G. Amelino-Camelia, Lect. Notes Phys. **541**, 1 (2000), gr-qc/9910089.
13. Y. J. Ng, W. A. Christiansen and H. van Dam, Astrophys. J. **591** L87 (2003), astro-ph/0302372.
14. G. Amelino-Camelia, Gen. Rel. Grav. **36**, 539 (2004), astro-ph/0309174.
15. J. Borgman and L.H. Ford, Phys. Rev. D **70** 064032 (2004).
16. R. T. Thompson and L.H. Ford, gr-qc/0601137.
17. J.W. Moffat, Phys. Rev. D **56**, 6264 (1996), gr-qc/9610067.
18. C.-H. Wu and L. H. Ford, Phys. Rev. D **64,** 045010 (2001), quant-ph/0012144.
19. L. H. Ford and C.-H. Wu, Int. J. Theor. Phys. **42**, 15 (2003), gr-qc/0102063.
20. David M. Zipoy, Phys. Rev. **142**, 825 (1966).
21. R.K. Sachs, A.M. Wolfe, Astrophys. J. **147** 73 (1967).
22. W. Kaufmann, Nature **227** 158 (1970).
23. V.B. Braginsky, M.B. Menskii, JETP Lett. **13**, 417 (1971).
24. V.B. Braginsky, N.S. Kardashev, A.G. Polnarev, I.D. Novikov, Il Nuovo Cimento **105**, 1141 (1990).
25. L.P. Grishchuk and Y.V. Sidorov, Phys. Rev. **42** 3413 (1990).

DYADOSPHERES DON'T DEVELOP

DON N. PAGE

Institute for Theoretical Physics
Department of Physics, University of Alberta
Edmonton, Alberta, Canada T6G 2J1
and
Asia Pacific Center for Theoretical Physics (APCTP)
Hogil Kim Memorial Building #519
POSTECH, San 31
Hyoja-dong, Namgu, Pohang
Gyeongbuk 790-784, Korea

Pair production itself prevents the development of dyadospheres, hypothetical macroscopic regions where the electric field exceeds the critical Schwinger value. Pair production is a self-regulating process that would discharge a growing electric field, in the example of a hypothetical collapsing charged stellar core, before it reached 6% of the minimum dyadosphere value, keeping the pair production rate more than 26 orders of magnitude below the dyadosphere value.

1. Introduction

Ruffini and his group [1–16] have proposed a model for gamma ray bursts that invokes a *dyadosphere*, a macroscopic region of spacetime with rapid Schwinger pair production [17], where the electric field exceeds the critical electric field value

$$E_c \equiv \frac{m^2}{q} \equiv \frac{m^2 c^3}{\hbar q} \approx 1.32 \times 10^{16}\ \mathrm{V/cm}. \tag{1}$$

(Here m and $-q$ are the mass and charge of the electron, and I am using Planck units throughout.) The difficulty of producing these large electric fields is a problem with this model that has not been adequately addressed. Here I shall summarize calculations [18] showing that dyadospheres almost certainly don't develop astrophysically.

The simplest reason for excluding dyadospheres is that if one had an astrophysical object of mass M, radius $R > 2M$, and excess positive charge

Q in the form of protons of mass m_p and charge q at the surface, the electrostatic repulsion would overcome the gravitational attraction and eject the excess protons unless $qQ \leq m_p M$ or

$$\frac{E}{E_c} = \frac{qQ}{m^2R^2} \leq \frac{m_p M}{m^2 R^2} < \frac{m_p}{4m^2 M} < 1.2 \times 10^{-13} \left(\frac{M_\odot}{M}\right), \tag{2}$$

where $M_\odot$ is the solar mass. (If the excess charge were negative and in the form of electrons, the upper limit would be smaller by m/m_p.) Then the pair production would be totally negligible.

However, one might postulate the implausible scenario in which protons are bound to the object by nuclear forces, which in principle are strong enough to balance the electrostatic repulsion even for dyadosphere electric fields. Therefore, for the sake of argument, I did a calculation [18] of what would happen under the highly idealized scenario in which the surface of a positively charged stellar core with initial charge $Q_0 = M$ (the maximum allowed before the electrostatic repulsion would exceed the gravitational attraction on the entire core, not just on the excess protons on its surface) freely fell from rest at radial infinity along radial geodesics in the external Schwarzschild metric of mass M.

This idealization ignores the facts that a realistic charged surface would (a) not fall from infinity, (b) have one component of outward acceleration, relative to free fall, from the pressure gradient at the surface, (c) have another component of outward acceleration from the electrostatic repulsion, and (d) fall in slower in the Reissner-Nordstrom geometry if the gravitational effects of the electric field with $Q \sim M$ were included. Because of each one of these effects, the actual surface would fall in slower at each radius and hence have more time for greater discharge than in the idealized model. Hence the idealized model gives a conservative upper limit on the charge and electric field at each radius, even under the implausible assumption that the excess protons are somehow sufficiently strongly bound to the surface that they are not electrostatically ejected.

As we shall see, even in this highly idealized model, the self-regulation of the pair production process itself will discharge any growing electric field well before it reaches dyadosphere values. This occurs mainly because astrophysical length scales are much greater than the electron Compton wavelength, which is the scale at which the pair production becomes significant at the critical electric field value for a dyadosphere. Therefore, the electric field will discharge astrophysically even when the pair production rate is much lower than dyadosphere values.

These calculations lead to the conclusion that it is likely impossible astrophysically to achieve, over a macroscopic region, electric field values greater than a few percent of the minimum value for a dyadosphere, if that. The Schwinger pair production itself would then never exceed 10^{-26} times the minimum dyadosphere value.

2. Schwinger discharge of an electric field

In this section we shall analyze the pair production and discharge of an electric field produced by the collapse of the idealized hypothetical charged sphere or stellar core of mass M and initial positive charge Q_0, assuming that somehow the excess charge on its surface is not electrostatically ejected, and assuming that the surface falls in as rapidly as possible, which is free fall from rest at infinity in an assumed external Schwarzschild geometry.

As the surface radius R collapses, the electric field $E = Q/r$ outside $(r > R)$ produces pairs, with the positrons escaping and the electrons propagating in to the surface to reduce its charge Q. The pair production rate $\mathcal{N}$ per 4-volume [17] is given by

$$\mathcal{N} = \frac{q^2 E^2}{4\pi^3} \exp\left(-\frac{\pi m^2}{qE}\right) \equiv \frac{m^4}{4\pi} \frac{e^{-w}}{w^2}, \tag{3}$$

where

$$\begin{aligned} w &\equiv \frac{\pi m^2}{qE} \equiv \frac{\pi E_c}{E} = \frac{\pi m^2 r^2}{qQ} \\ &= \frac{4\pi m^2 M_\odot}{q} \left(\frac{M}{M_\odot}\right) \left(\frac{Q}{M}\right)^{-1} \left(\frac{r}{2M}\right)^2 \\ &\equiv \frac{1}{B} \left(\frac{M}{M_\odot}\right) \left(\frac{Q}{M}\right)^{-1} \left(\frac{r}{2M}\right)^2, \end{aligned} \tag{4}$$

$$B \equiv \frac{q}{4\pi m^2 M_\odot} \approx 42475. \tag{5}$$

A dyadosphere has $E \geq E_c \equiv m^2/q \Rightarrow w \leq \pi \Rightarrow$

$$\left(\frac{M}{M_\odot}\right) \left(\frac{Q}{M}\right)^{-1} \left(\frac{r}{2M}\right)^2 \leq \pi B \equiv \frac{q}{4m^2 M_\odot} \approx 1.33 \times 10^5. \tag{6}$$

For astrophysical electric fields anywhere near dyadosphere values, the electrons and positrons produced will quickly be accelerated to very near the speed of light, so one will effectively get a null number flux 4-vector $\mathbf{n}_+$ of highly relativistic positrons moving radially outward and another null number flux 4-vector $\mathbf{n}_-$ of highly relativistic electrons moving radially inward, with total current density 4-vector $\mathbf{j} = q\mathbf{n}_+ - q\mathbf{n}_-$.

It is most convenient to describe this current in terms of radial null coordinates, say U and V, so that the approximately Schwarzschild metric outside the collapsing core may be written as

$$ds^2 = -e^{2\sigma} dU dV + r^2(U,V)(d\theta^2 + \sin^2\theta d\phi^2). \tag{7}$$

Then Maxwell's equations (Gauss's law) gives

$$4\pi\mathbf{j} \equiv 4\pi q(n_+^V \partial_V - n_-^U \partial_U) = \nabla \cdot \mathbf{F} \equiv \frac{1}{r^2}\left(Q^{,V}\partial_V - Q^{,U}\partial_U\right). \tag{8}$$

The 4-divergence of each of the number flux 4-vectors $\mathbf{n}_+$ and $\mathbf{n}_-$ is equal to the pair production rate $\mathcal{N}$, which leads to the following relativistic partial differential equation for the pair production and discharge process:

$$Q_{,UV} = -2\pi q r^2 e^{2\sigma}\mathcal{N} = -\frac{q^3 Q^2 e^{2\sigma}}{2\pi^2 r^2}\exp\left(-\frac{\pi m^2 r^2}{qQ}\right), \tag{9}$$

or

$$8\pi q r^2 \mathcal{N} = \frac{2q^3 r^2 E^2}{\pi^2}\exp\left(-\frac{\pi m^2}{qE}\right) = {}^2\Box(r^2 E) = r\Box(rE) - r^3 E\Box\left(\frac{1}{r}\right), \tag{10}$$

where ${}^2\Box$ is the covariant Laplacian in the 2-dimensional metric ${}^2ds^2 = -e^{2\sigma}dUdV$ and where $\Box$ is the covariant Laplacian in the full 4-dimensional metric.

In my much more detailed paper [18], I have analyzed this partial differential equation for the charge distribution over the entire spacetime region exterior to the charged surface and have found an approximation that reduces it to a relativistic ordinary differential equation for the evolution of the charge at the surface itself. However, here I shall confine myself to a Newtonian approximation to this evolution equation, which turns out to give results very close to the relativistic approximation.

A convenient time parameter for describing the collapse of the surface is the velocity v the surface has in the frame of a static observer at fixed r when the surface radius R crosses that value of r. For free fall from rest at infinity in the external Schwarzschild metric of mass M, with τ being the proper time along the surface worldlines, one gets

$$v = -\frac{dR}{d\tau} = \sqrt{\frac{2M}{R}}, \tag{11}$$

so that the proper time remaining until the proper time τ_c at which the surface reaches the curvature singularity at $R = 0$ is $\tau_c - \tau = (4/3)M/v^3$

and $R = 2M/v^2 = (4.5M)^{1/3}(\tau_c - \tau)^{2/3}$. The Newtonian limit of this is when $R \gg 2M$, which gives $v \ll 1$ and $\tau \approx t$, the Schwarzschild coordinate time. In this limit, the surface moves negligibly during the time it takes for electrons to move inward from where they are created to the surface to reduce $Q(t)$.

If $w \equiv \pi E_c/E$ is defined at each point outside the surface, let

$$z(t) \equiv w(t, R(t)) \equiv \frac{\pi E_c}{E(R(t))} = \frac{\pi m^2 R(t)^2}{qQ(t)} = \frac{\pi m^2 M^2}{qQv^4} \tag{12}$$

be the value of w at the surface itself. Since we shall find that the electric field E always stays far below the critical dyadosphere value E_c, we have $z \gg 1$, which will be used for various approximations below.

Now the pair production rate (and assumed effectively instantaneous propagation of the electrons produced to the surface of the collapsing stellar core) gives

$$\frac{dQ}{dt} \approx -q \int_R^\infty \mathcal{N}(r) 4\pi r^2 dr \approx -q\frac{m^4 R^4}{z^2} \int_R^\infty \frac{dr}{r^2} e^{-z\frac{r^2}{R^2}} \approx -\frac{qm^4R^3}{2z^3e^z}. \tag{13}$$

Then using $R = 2M/v^2$ and $dv/dt \approx dr/d\tau = v^4/(4M)$ leads to the following ordinary nonlinear first-order differential equation for $z(v)$ in the Newtonian limit:

$$\frac{v}{z}\frac{dz}{dv} \approx -4\left[1 - \frac{(Mmq)^2}{\pi v^5 z^2 e^z}\right] = -4\left[1 - \frac{A\mu^2}{v^5 z^2 e^z}\right] = -4\left[1 - e^{-U}\right], \tag{14}$$

where

$$A \equiv \frac{(M_\odot mq)^2}{\pi} \approx 3.39643251 \times 10^{28}, \tag{15}$$

$$\mu \equiv \frac{M}{M_\odot}, \tag{16}$$

$$U \equiv z + \ln z - \ln A - 2\ln\mu + 5\ln v. \tag{17}$$

From Eq. (12), one can see that the boundary conditions for Eq. (14) are that initially ($\tau \approx t = -\infty \Leftrightarrow R = \infty \Leftrightarrow v = 0$) $v^4 z = \pi m^2 M^2/(qQ)$ with the surface charge Q having its asymptotically constant initial value Q_0, which will be taken to be its maximum allowed value, M, unless otherwise specified. One can see from this that both z and U start off initially at infinite values. The final value will be when the surface enters the event horizon of the black hole at $R = 2M$ or $v = 1$. This is beyond the applicability of the Newtonian approximation being used here, but it turned out that the relativistic analysis [18] gave very nearly the same answers.

One can now differentiate Eq. (17) for $U(v)$, using Eq. (14), to obtain

$$v\frac{dU}{dv} \approx -4(z+2)\left(1-e^{-U}\right)+5. \tag{18}$$

Since $z \gg 1$, this equation implies that U decreases to near zero (though it cannot reach zero, for if it could, the right hand side would be positive, contradicting the assumption that it dropped to zero and hence had a negative derivative on the left hand side). Then when $U \ll 1$, Eq. (17) may be solved approximately for $z(v)$ to give

$$\begin{aligned} z(v) &\approx \ln A + 2\ln\mu - 5\ln v - 2\ln\left(\ln A + 2\ln\mu - 5\ln v\right) \\ &\approx \ln A - 2\ln\ln A + \left(1 - \frac{2}{\ln A}\right)(2\ln\mu - 5\ln v) \\ &\approx 57.33 + 1.94\ln\mu + 4.85\ln\frac{1}{v}. \end{aligned} \tag{19}$$

This then gives the ratio of the electric field at the surface to the critical electric field of a dyadosphere as being

$$\frac{E}{E_c} = \frac{\pi}{z} \approx 0.0548 - 0.00185\ln\mu - 0.00463\ln\frac{1}{v} \ll 1. \tag{20}$$

One can improve this result by using a slightly improved formula (giving a roughly 2.5% correction for $z \sim 57$) for the radial integral in Eq. (13) for the pair production rate, by using a better explicit approximation for what U should approach at $v = 1$, and by using a numerical solution of the thus-corrected form of Eq. (17) for z with this expression for U, to estimate that when $\mu = 1$ and $v = 1$, $z \approx 57.5843$. Including the improved formula for the right hand side of Eq. (13) into the Newtonian differential equation (14) and then integrating it numerically from $v = 0$ to $v = 1$ for $\mu = 1$ gave the result at the horizon of the solar black hole of $z \approx 57.5845$, so only the 6th digit changed from the algebraic estimate obtained without numerically solving the Newtonian differential equation.

In [18] I used a relativistic ordinary differential equation approximation to the partial differential equation (9) and was able to deduce an explicit approximate relativistic result for $\mu = 1$ and $v = 1$ of $z \approx 57.60483$, whereas the numerical solution of the relativistic ordinary differential equation gave $z \approx 57.60480$, differing from the explicit formula (using as input the values of m, q, and $M_\odot$) by only about one part in two million. However, I would estimate that the relativistic ordinary differential equation approximation to the partial differential equation would itself introduce absolute errors of the order of 10^{-4}–10^{-3} in the value of z, so the numerical solution of that ordinary differential equation is not necessarily any better than the completely explicit approximate solution I also obtained.

The difference between the Newtonian and the relativistic approximations for z on the horizon ($v = 1$) of a solar mass collapsing core ($\mu = M/M_{\odot} = 1$) is about 0.02, which I would guess is considerably larger than the error of my relativistic approximation (not given here, but in [18]), but it is still a relative difference of only about one part in three thousand for the idealized upper limit of the value of the electric field of a hypothetical charged stellar core collapsing into a black hole after falling in freely (no nongravitational forces on the surface) from starting at rest at radial infinity with the external metric being Schwarzschild.

We can also give a heuristic derivation of the Newtonian result in the following way: We expect the self-regulation of the electric field to make $z = \pi E_c/E = \pi m^2 R^2/(qQ)$ change slowly, so

$$\frac{1}{Q}\frac{dQ}{dt} \sim \frac{2}{R}\frac{dR}{dt} \approx -\frac{1}{M}\left(\frac{2M}{R}\right)^{\frac{3}{2}} = -\frac{v^3}{M}. \tag{21}$$

Then the pair production rate per 4-volume, $\mathcal{N} = (m^4/4\pi)e^{-w}/w^2 \approx (m^4/4\pi z^2)e^{-zr^2/R^2}$ for $z \gg 1$ decreases roughly exponentially with r with e-folding length $\Delta r \sim R/(2z)$. Thus

$$\begin{aligned}\frac{1}{Q}\frac{dQ}{dt} &\approx -\frac{q}{Q}\int_R^\infty \mathcal{N}(r)4\pi r^2 dr \approx -\frac{q}{Q}\mathcal{N}(r=R)4\pi R^2 \Delta r \\ &\approx -q\frac{qz}{\pi m^2 R^2}\frac{m^4}{4\pi z^2}e^{-z}4\pi R^2\frac{R}{2z} = -\frac{1}{M}\frac{M^2 m^2 q^2}{\pi}\frac{e^{-z}}{z^2}\left(\frac{R}{2M}\right).\end{aligned} \tag{22}$$

Equating this to $-(1/M)(2M/R)^{3/2}$ gives

$$1 \approx \frac{M^2 m^2 q^2}{\pi}\frac{e^{-z}}{z^2}\left(\frac{R}{2M}\right)^{\frac{5}{2}} = \frac{A\mu^2}{v^5}\frac{e^{-z}}{z^2}, \tag{23}$$

which implies $z + 2\ln z \approx \ln A + 2\ln\mu - 5\ln v$, just as we got from the approximated solution of the ordinary differential equation in the Newtonian approximation.

A dyadosphere would have $E \geq E_c$, which implies $z = \pi E_c/E \leq z_c = \pi$ and $\mathcal{N} = (m^4/4\pi)e^{-z}/z^2 \geq \mathcal{N}_c = m^4 e^{-\pi}/(4\pi^3)$. But

$$\begin{aligned}\mathcal{N} &\approx \frac{m^4}{4\pi}\frac{v^5}{A\mu^2} = \frac{\pi^2 e^\pi v^5}{A\mu^2}\mathcal{N}_c = \frac{\pi^3 e^\pi v^5}{q^2 m^2 M}\mathcal{N}_c \\ &\approx 0.672\times 10^{-26}\,\frac{v^5}{\mu^2}\mathcal{N}_c < 10^{-26}\,\mathcal{N}_c,\end{aligned} \tag{24}$$

so the heuristic estimate gives the maximum pair production rate more than 26 orders of magnitude below that of a dyadosphere. By comparison, the numerical solution of the approximate relativistic ordinary differential equation [18] gave, at $v = 1$,

$$\frac{\mathcal{N}}{\mathcal{N}_c} \approx \frac{0.661168 \times 10^{-26}}{\mu^2}(1 + 0.0005545 \ln \mu - 0.00001759 \ln^2 \mu), \quad (25)$$

about 1.6% less than the heuristic estimate gives.

One can also calculate the maximum efficiency for converting the collapsing stellar core mass M into outgoing positron energy,

$$\begin{aligned}
\epsilon &\approx -\int \frac{QdQ}{MR} \approx \frac{2\mu^2}{B^2}\int_0^1 e^{-U}\frac{dv}{v^7 z^2} \\
&\approx \frac{1}{3}B^{-1/2}\left[\ln A + \frac{5}{4}\ln B - \frac{3}{4}\ln\ln(AB^2) + \frac{3}{4}\ln\mu\right]^{-1/2} \mu^{1/2}(Q_0/M)^{3/2} \\
&= \frac{m}{3M}\sqrt{\frac{4\pi Q_0^3}{q}}\left[\frac{1}{4}\ln\left(\frac{q^{13}M^3}{2^{10}\pi^9 m^3}\right) - \frac{3}{4}\ln\ln\left(\frac{q^4}{16\pi^3 m^2}\right)\right]^{-1/2} \\
&\sim 1.9 \times 10^{-4}\left[1 + \frac{\ln(M/M_\odot)}{300}\right]^{-1/2}\left(\frac{M}{M_\odot}\right)^{1/2}\left(\frac{Q_0}{M}\right)^{3/2} \\
&\sim 2 \times 10^{-4}\left(\frac{Q_0}{M}\right)^{3/2}\sqrt{\frac{M}{M_\odot}}. \qquad (26)
\end{aligned}$$

By comparison, the numerical solution of the approximate relativistic ordinary differential equation [18] gave the more precise result

$$\epsilon \approx 0.0001855\left(\frac{M}{M_\odot}\right)^{0.495}\left(\frac{Q_0}{M}\right)^{0.742}. \quad (27)$$

The dominant factor in an estimate of the coefficient 0.0001855 (the upper limit on the efficiency if $Q_0 = M = M_\odot$) is one-third the ratio of the electron mass to the proton mass, which is 0.0001815. This efficiency is too low for the pair production from these idealized collapsing charged cores to explain gamma ray bursts, even if it is admitted that this very implausible scenario (of the excess charge $Q \sim M$ not getting expelled from the collapsing core by the huge electrostatic forces on it) comes nowhere near being able to form a dyadosphere.

One can also calculate [18] that the probability of one of the particles annihilating with an antiparticle is less than 10^{-26}, so the direct interactions of individual particles is negligible, consistent with what was assumed above.

3. Conclusions

If protons are bound to a collapsing stellar core purely gravitationally, the maximum electric field is more than 13 orders of magnitude below dyadosphere values: $E_{\rm max} \leq 1.2 \times 10^{-13}(M_\odot/M)E_c$.

If protons are much more strongly bound, $E_{\rm max} \leq 0.055 E_c$ and $\mathcal{N}_{\rm max} \leq 10^{-26}\mathcal{N}_c$, where $\mathcal{N}_c$ is the minimal dyadosphere production rate.

The energy efficiency of the process for $M \sim M_\odot$ is very low, $\epsilon \lesssim 1.86 \times 10^{-4}(M/M_\odot)^{1/2}(Q_0/M)^{3/2} \approx (Q_0/M)^{3/2}\sqrt{M/(2.9 \times 10^7 M_\odot)}$.

If one relaxed the assumptions of this model, such as the spherical symmetry, one would expect to get similar results, perhaps changing the pair production rates by factors of order unity that depend upon the precise geometry. However, it seems very unlikely that any modification could increase the maximum possible pair production rate by any significant fraction of the 26 orders of magnitude that the model above fails to achieve dyadosphere values. Therefore, the example analyzed strongly suggests that dyadospheres do not form astrophysically.

Acknowledgments

I am grateful for being able to participate in the 9th Italian-Korean Symposium on Relativistic Astrophysics in South and North Korea and there learn about dyadospheres. I am thankful for the hospitality of Sang Pyo Kim at Kunsan University and for the hospitality of Beijing Normal University at the beginning of this work. Financial support was also provided in part by the Natural Sciences and Engineering Research Council of Canada

References

1. T. Damour and R. Ruffini, Phys. Rev. Lett. **35**, 463-466 (1975).
2. R. Ruffini, "Beyond the Critical Mass: The Dyadosphere of Black Holes," *Black Holes and High Energy Astrophysics, Proceedings of the Yamada Conference XLIX on Black Holes and High Energy Astrophysics held on 6-10 April, 1998 in Kyoto, Japan*, eds. H. Sato and N. Sugiyama (Frontiers Science Series No. 23, Universal Academic Press, Tokyo, 1998), p.167; "On the dyadosphere of black holes," astro-ph/9811232; Astronomy and Astrophysics Supplement **138**, 513-514 (1999).
3. G. Preparata, R. Ruffini, and S.-S. Xue, Astron. Astrophys. **338**, L87-L90 (1998), astro-ph/9810182; Nuovo Cim. **B115**, 915 (2000); J. Korean Phys. Soc. **42**, S99-S104 (2003), astro-ph/0204080.
4. R. Ruffini and S.-S. Xue, "Radiation and Time Scale Evolution of a P.E.M. Pulse from EMBH," *Abstracts of the 19th Texas Symposium on Relativistic*

Astrophysics and Cosmology, held in Paris, France, Dec. 14-18, 1998, eds. J. Paul, T. Montmerle, and E. Aubourg (CEA Saclay, 1998).

5. R. Ruffini, J. Salmonson, J. Wilson, and S.-S. Xue, Astronomy and Astrophysics Supplement **138**, 511-512 (1999), astro-ph/9905021; Astronomy and Astrophysics **350**, 334-343 (1999), astro-ph/9907030; Astronomy and Astrophysics **359**, 855-864 (2000), astro-ph/0004257.
6. C. L. Bianco, R. Ruffini, and S.-S. Xue, Astronomy and Astrophysics **368**, 377-390 (2001), astro-ph/0102060.
7. R. Ruffini, C. L. Bianco, F. Fraschetti, P. Chardonnet, and S.-S. Xue, Nuovo Cim. **B116**, 99 (2001), astro-ph/0106535.
8. R. Ruffini, C. L. Bianco, F. Fraschetti, S.-S. Xue, and P. Chardonnet, Ap. J. **555**, L107-L111 (2001), astro-ph/0106531; L113-L116 (2001), astro-ph/0106532; L117-L120 (2001), astro-ph/0106534.
9. R. Ruffini and L. Vitagliano, Phys. Lett. **B545**, 233-237 (2002), astro-ph/0209072.
10. R. Ruffini, S.-S. Xue, C. L. Bianco, F. Fraschetti, and P. Chardonnet, La Recherche **353**, 30-32 (2002).
11. R. Ruffini, C. L. Bianco, P. Chardonnet, F. Fraschetti, and S.-S. Xue, Ap. J. **581**, L19-L22 (2002), astro-ph/0210648; Int. J. Mod. Phys. **D12**, 173-270 (2003), astro-ph/0302141.
12. P. Chardonnet, A. Mattei, R. Ruffini, and S.-S. Xue, Nuovo Cim. **118B**, 1063-1070 (2003).
13. R. Ruffini, L. Vitagliano, and S.-S. Xue, Phys. Lett. **B559**, 12-19 (2003), astro-ph/0302549; Phys. Lett. **B573**, 33-38 (2003), astro-ph/0309022.
14. R. Ruffini, M. G. Bernardini, C. L. Bianco, P. Chardonnet, F. Fraschetti, and S.-S. Xue, Advances in Space Research **34**, 2715-2722 (2004), astro-ph/0503268.
15. R. Ruffini, C. L. Bianco, P. Chardonnet, F. Fraschetti, V. Gurzadyan, and S.-S. Xue, Int. J. Mod. Phys. **D13**, 843-852 (2004), astro-ph/0405284.
16. C. L. Bianco and R. Ruffini, Ap. J. **605**, L1-L4 (2004), astro-ph/0403379; **620**, L23-L26 (2005), astro-ph/0501390; **633**, L13-L16 (2005), astro-ph/0509621.
17. F. Sauter, Z. Phys. **69**, 742 (1931); W. Heisenberg and H. Euler, Z. Phys. **98**, 714 (1936); V. Weisskopf, K. Dan. Vidensk. Selsk. Mat. Fys. Medd. **14**, No. 6 (1936); J. S. Schwinger, Phys. Rev. **82**, 664 (1951); A. I. Nikishov, Nucl. Phys. B **21**, 346 (1970).
18. D. N. Page, "No Astrophysical Dyadospheres," astro-ph/0605432.

The quantized Schwarzschild Black Hole and the possible origin of Dark Matter

Liao Liu and Shouyong Pei

Department of Physics, Beijing Normal University, Beijing, 100875, PRC

If the situation of quantum gravity nowadays is nearly the same as that of quantum mechanics in its early time of Bohr and Sommerfeld, then a first-step study of the quantum gravity under Sommerfeld's quantum condition of action might be helpful. We present the spectra of quantum Schwarzschild black hole in non-relativistic quantum mechanics. It is found that the quantum of area is $\frac{8\pi}{3}l_p^2$, the quantum of entropy is $\frac{2\pi}{3}k_B$, and the Hawking evaporation will cease when the black hole reaches their ground state $m = \frac{1}{2\sqrt{3}}m_p$. So there are a lot of remnants of black hole that may be the origin of dark matter.

Though there are several roads try to quantize the gravity, no one can now establish a satisfied scheme of quantum gravity . It seems therefore that we may now be in a similar period as quantum mechanics in its early Bohr and Sommerfeld time. So a first step in a different road to treat the quantum schwarzschild black hole by using Sommerfeld's quantum condition of action might be a selectable road to the future of quantum gravity. We found in the year 2003,2004 that the Schwarzschild black hole (SBH) can be treated as a single periodic system in quantum mechanics, then all spectra of the quantum Schwarzschild black hole (QSBH) can easily be obtained after the Bohr-Sommerfeld's quantum condition of action is applied.[1]

Early in the year of 1916, Bohr and Sommerfeld successfully put forth the so-called Sommerfeld quantum condition of action variable[2]

$$\oint p dq = nh \tag{1}$$

in order to solve the spectra problem in atom physics. Landau and Lifshitz derived the Bohr-Sommerfeld quantum condition for the action variable of one-dimensional periodic motion of a particle by using the quasi-classical method and obtained[3]

$$\oint pdq = 2\pi\hbar\left(n + \frac{1}{2}\right) \tag{2}$$

that is, for any cyclic motion the action variable I_v of the system is quantized.

Now in this letter a brief introduction of Bohr-Sommerfeld action quantization principle is reminded. As is known, the Euclidean Kruskal section of SBH is a cyclic or single periodic system, whose metric reads

$$ds^2 = \frac{32}{r}m^3 \exp\left\{-\frac{r}{2m}\right\}(dT^2 + dR^2) + r^2 d\Omega_2^2, \quad (r > 2m) \tag{3}$$

where

$$iT = \left(\frac{r}{2m} - 1\right)^{\frac{1}{2}} \exp\{\frac{r}{4m}\} \sin\left(\frac{\tau}{4m}\right) \tag{4}$$

$$R = \left(\frac{r}{2m} - 1\right)^{\frac{1}{2}} \exp\{\frac{r}{4m}\} \cos\left(\frac{\tau}{4m}\right) \tag{5}$$

Obviously, both T and R in the parametric form Eqs.(4) and (5) are the periodic function of τ with period $8\pi m$. We would like to emphasize that this peculiar property of the Euclidean-Kruskal metric is very important to reveal the thermodynamics of the SBH. We shall see in the following that it is also of key importance for the recognition of its quantum property.

From classical general relativity, we know that the area A of the event horizon of the SBH and its action I are, respectively,

$$A = 16\pi m^2 \quad (G = C = 1) \tag{6}$$

$$I = 4\pi m^2 = I_v - \oint H'dt = \oint pdq - 8\pi m^2, \quad (G = C = \hbar = 1, \mathbf{T} = 8\pi m). \tag{7}$$

where $H' = H + \alpha$,

$$\alpha = \lim_{r \longrightarrow \infty} \oint_s \left(\frac{\partial h_{ij}}{\partial x^j} - \frac{\partial h_{ij}}{\partial x^i}\right) ds^i \tag{8}$$

is just the ADM mass m for asymptotic flat SBH[4]. We recall that h_{ij} is the induced metric of the three-dimensional space-like hypersurface and π_{ij} is the canonical conjugated momentum of h_{ij} . The important thing we would like to emphasize is that the Hamiltonian $H^{'}$ of any classical Einstein gravitational system with asymptotic flatness is always the ADM mass m of the system. Here we note that, for the vacuum Euclidean-Kruskal section, the volume integral of its action is zero, so the contribution to action comes only from the Gibbons-Hawking's surface term $4\pi m^2$. From Eqs. (6) and (7) the variation ΔA of A and ΔI of I have the relation

$$\Delta A = 4\Delta I, \quad (G = C = \hbar = 1) \tag{9}$$

Now noticing $H^{'} = m$ and $I = I_v - 8\pi m^2$ for the SBH, we can apply Sommerfeld's quantum condition of Eqs. (2) to (6) and (7) to obtain the spectrum of action I, the spectrum o fmass and the spectrum of area A of event horizon and the spectrum of entropy S of SBH as follows:

$$I = 4\pi m^2 = 2\pi\hbar(n + 1/2) - 8\pi m^2 \tag{10}$$

$$m^2 = \frac{1}{6}(n + \frac{1}{2})m_p^2 \tag{11}$$

$$A = 16\pi m^2 = \frac{8\pi}{3}(n + 1/2)l_p^2 \tag{12}$$

and

$$S = \frac{1}{4}A \tag{13}$$

where

$$I = S = \frac{1}{4}A \tag{14}$$

The minimum variation or quantum of the area of the event horizon δA , quantum of the entropy δS and quantum of the mass δm of SBH are, respectively,

$$\delta A = \frac{8\pi}{3}l_p^2 \tag{15}$$

$$\delta S = \frac{2\pi}{3}k_B \tag{16}$$

$$\delta m = \frac{1}{12} \frac{m_p^2}{m(T)} \tag{17}$$

where $l_p = (G\hbar C^{-3})^{\frac{1}{2}}$ is the Planck length, $l_p^2 = G\hbar C^{-3}$ is the Planck area, $m_p = (G^{-1}\hbar C)^{\frac{1}{2}}$ is the Planck mass and $m(T)$ is the mass of SBH at temperature T.

It seems that our area quantum (15) only refers to the event horizon. We have no reason to infer that this is a general result to all areas of any surface. It is appears that our result of the quantum area (15) is different from the recent value of $4 \ln 3 l_p^2$ obtained by Dreyer *et al*[5]. Equation (17) is the total quantum mass-energy loss (QML) of an SBH via Hawking evaporation in temperature T. If the black hole mass $m(T)$ has a lowest limit $\frac{1}{2\sqrt{3}} m_p$, then the QML has an upper limit $\Delta m = \frac{1}{2\sqrt{3}} m_p$.

From Eqs. (10), (11) and (12) we see $n = 0$ should correspond to the ground state of SBH. It is easy to show the ground state mass m_G of the SBH is

$$m_G = \frac{1}{2\sqrt{3}} m_p \tag{18}$$

It seems that there is no way to decrease the mass of an SBH under its ground state mass m_G. Therefore even Hawking evaporation will cease as a QSBH approaches its ground state; in other words, the QSBH will transit into something other than black hole[6]. The important thing is that there is a remnant left for any evaporating Schwarzschild black hole. The black hole cannot annihilate away totally due to Hawking evaporation.

Therefore, two important results are:

The first is an incoming pure vacuum state can never become an outgoing mixed state! No loss of quantum coherence will result in our quasi-classical scheme of quantum gravity. In other words, the long unsolved information puzzle in quasi-classical black hole physics may have a solution now! This is a very interesting result.

The second is that the ground state (or relics, remnants) of QSBH may be really called a dark star or a "dead black hole" without any kind of radiation the existence of them can only be detected by their gravitational action on other stars or the rotation curve of the galaxies..

We remember, early in the seventy years of last century, Zeldovich, Novikov [7], Carr and Hawking[8] denoted that a lot of primordial small black hole (PBH) may be created from the fluctuation of matter density of the very early universe. Hawking thought all the PBH that have their life-time shorter than that of the universe should evaporated whole away due to Hawking evaporation. Our work, however, denote that all the above mentioned PBH should existed finally as remnant of PBH or dark matter. So a very natural and bold conjecture is that the principal constituent if not totally of the dark matter may come from the above mentioned remnants of black hole or dark stars.

References

1. L. Liu and S. Y. Pei, Chin Phys Lett **21**, 1887(2004).
2. A. Sommerfeld *Ann. Physik.* **51**, 1(1916).
3. L. D. Landau and E. M. Lifshitz *Quantum Mechanics (Non-Relativity Theory)* (London: Pergamon) (1958).
4. R. Wald *General Relativity* 2nd ed. (Chicago: The University of Chicago Press) p469 (1984).
5. O. Dreyer, arXiv gr-qc/0211076.
6. E Livine Private communication (17 August 2003).
7. Ya.B. Zeldovich and I.D. Novikov, Sov. Astron. A.J. 10 602(1966).
8. B.J. Carr and S.W. Hawking, MNRAS 168 399(1974).

The Origin of the Immirzi Parameter in Loop Quantum Gravity

Hoi-Lai YU

Institute of Physics
Academia Sincia,
Nankang, Taipei, Taiwan
E-mail:hlyu@phys.sinica.edu.tw

Using quadratic spinor techniques we demonstrate that the Immirzi parameter can be expressed as ratio between scalar and pseudo-scalar contributions in the theory and can be interpreted as a measure of how Einstein gravity differs from a generally constructed covariant theory for gravity. This interpretation is independent of how gravity is quantized. One of the important advantage of deriving the Immirzi parameter using the quadratic spinor techniques is to allow the introduction of renormalization scale associated with the Immirzi parameter through the expectation value of the spinor field upon quantization.

1. Introduction

One of the most direct ways of approaching the quantization of Einstein's theory of gravity is to put it into a Hamiltonian form and then try to apply the procedures of canonical quantization. The fact that Einstein's theory is generally covariant makes the task one of most difficult problems in theoretical physics if not impossible. The complicated non-polynomial structure found for the standard Hamiltonian for general relativity raises another challenge to researchers. However, Ashtekar managed to make progresses using new canonical variables to reduce the constraints to polynomial form. The so called Ashtekar connection[1,2] $A = \Gamma + iK$ has a part Γ refers to intrinsic curvature on a spacelike 3-surface S, and another part K that refers to the extrinsic curvature to the spacetime M. Because of several technical difficulties, in order to make progress, a more general Barbero connection[3] was introduced, $A = \Gamma + \gamma K$ where γ is an arbitrary complex number. It is also known as the Immirzi parameter[4] (usually assumed to be a real number for $SU(2)$ Barbero connection).

The important achievement of quantizing the Ashtekar-Barbero connection variable is the construction of a kinematic Hilbert space using spin networks[2]. With spin networks, the area and volume spectra can be derived. If a spin network intersects a surface S transversely, then this surface has a definite area in this state, given as a sum over the spins j of the edges poking through S:

$$\text{Area}(S) = 8\pi\gamma \sum_j \sqrt{j(j+1)}$$

in units where the $\hbar = c = G = 1$, with a free Immirzi parameter γ. Due to the presence of the Immirzi parameter, the famous Bekenstein-Hawking entropy formula, $S_{BH} = \frac{A}{4}$ could not be uniquely determined. This has been viewed as the main unsatisfactory point of this approach for some years. Recently Dreyer[5] proposed a way to fix the Immirzi parameter using asymptotic behavior of the quasinormal modes of a Schwarzschild black hole. The result fixed the value $\gamma = \frac{\ln 3}{2\sqrt{2}\pi}$ with the lowest possible spin $j_{min} = 1$. Domagala, Lewandowski and Meissner[6] fixed an incorrect assumption that only the minimal value of the spin contributes. Their result involves the logarithm of a transcendental number instead of the logarithms of integers; $\gamma = 0.2375329...(> \frac{\ln 2}{\pi})$ with $j_{min} = \frac{1}{2}$. The calculation works for charged and rotating black holes and black holes coupled to a dilaton field, with the *same* value of γ. There appears to be no clear geometrical reason for a particular choice of the real number value γ and obscures its physical interpretation.

The appearance of the Immirzi parameter γ can be seen in the simplest tetrad-Palatini action of general relativity where one can add an additional term with coupling coefficient γ. This newly added term does not affect the equations of motion. In the case where torsion free connection that solves the equation of motion is employed to obtain the Einstein-Hilbert gravity action, the additional term in the action becomes identically zero. Arguing with this observation in mind , the effect of γ is therefore, not an physical observable in Einstein gravity. Recently Perez and Rovelli[7] had argued that one can observe physical effects of the Immirzi parameter γ by coupling Einstein gravity to fermionic degrees of freedom. The presence of matter field induces a torsion term in the connection and the additional term becomes non-vanishing. Freidel, Minic and Takeuchi[8] discussed parity violation and studied the coupling of fermionic degrees of freedom in the presence of torsion from the viewpoint of effective field theory. The importance of these works are to notice that physical effects arise from the Immirzi parameter γ is measurable and independent from how gravity is quantized. We believe however, if the Immirizi parameter γ is a physical property of the gravity sector, then it should be observable without the introduction of other matter field. In the following, we shall introduce a Quadratic Spinor Representation of General Relativity[9,10,11] formalism where the physical meanings and effects of the Immirzi parameter γ become transparent in general relativity. In this formalism, the Immirzi parameter becomes a ratio between scalar and pseudo-scalar contributions in the theory and measures how a generally formulated general theory of relativity differs from Einstein gravity. More importantly, one can acquire this ratio a renormalization scale upon quantization.

2. Quadratic spinor representation of General Relativity

The canonical formulation of Loop Quantum Gravity can be derived by the Holst action[12],

$$S[\vartheta,\omega] = \alpha \int *(\vartheta^a \wedge \vartheta^b) \wedge R_{ab}(\omega) + \beta \int \vartheta^a \wedge \vartheta^b \wedge R_{ab}(\omega), \qquad (1)$$

where the Immirzi parameter is,

$$\gamma = \frac{\alpha}{\beta}. \qquad (2)$$

In the above, $a, b ... = 0, 1, 2, 3$ being the internal indices of the internal orthonormal frame. The field ϑ^a being the tetrad field; ω is the $SU(2)$ connection; R_{ab} being the curvature of ω and $*R$ being its dual. This is comparable with the Quadratic Spinor Lagrangian[9,10],

$$L_\psi = 2D(\overline{\psi}\vartheta)\gamma_5 D(\vartheta\psi), \qquad (3)$$

where $\vartheta = \vartheta^a \gamma_a$ and γ_a being the Dirac gamma matrices. The auxiliary spinor field ψ in the Quadratic Spinor Lagrangian was first introduced by Witten[13] as a convenient tool used in the proof of positive energy theorem in Einstein gravity. In a more general context, this auxiliary spinor field provides a nice gauge condition to pick up the relevant variables in the theory. The key to these successes is a "spinor-curvature identity":

$$\begin{aligned} &2D(\overline{\psi}\vartheta)\gamma_5 D(\vartheta\psi) \\ &= \overline{\psi}\psi R_{ab} \wedge *(\vartheta^a \wedge \vartheta^b) + \overline{\psi}\gamma_5\psi R_{ab} \wedge \vartheta^a \wedge \vartheta^b \\ &\quad + d[D(\overline{\psi}\vartheta)\gamma_5\vartheta\psi + \overline{\psi}\vartheta\gamma_5 D(\vartheta\psi)] \end{aligned} \qquad (4)$$

Note that in the above expression the boundary terms provide an important condition in obtaining a finite action at spatial infinity and consequently a well defined Hamiltonian. Here, the spinor field ψ plays a key rule which allows one to pick up the correct gauges in obtaining a well defined Hamiltonian of the theory. The equation of motion for the connection $\omega[\vartheta]$ is:

$$D[\overline{\psi}\psi * (\vartheta^a \wedge \vartheta^b) + \overline{\psi}\gamma_5\psi(\vartheta^a \wedge \vartheta^b)] = 0, \tag{5}$$

where D is the covariant derivative defined by the connection variable $\omega[\vartheta]$. For $\overline{\psi}\psi = 1$ and $\overline{\psi}\gamma_5\psi = 0$ the torsion free spin connection $\omega[\vartheta]$ of the tetrad field ϑ solves the above field equation. Therefore, if we set

$$\gamma = \frac{\overline{\psi}\psi}{\overline{\psi}\gamma_5\psi} \tag{6}$$

then the choice of $\gamma = \infty$ ($\overline{\psi}\psi = 1$ and $\overline{\psi}\gamma_5\psi = 0$), is the Einstein-Hilbert action in equation(1); and $\gamma = i$ is the self-dual action in the Ashtekar canonical gravity framework, and $\gamma = 1$ corresponds to the action for the Hamiltonian considered by Barbero[3]. The Immirzi parameter γ in this setting becomes a measure of how Einstein gravity differs from a most generally formulated gravitation theory which satisfies general coordinate covariance. It is also the ratio between scalar and pseudo-scalar contributions in the theory as can be seen from the explicit expression of γ. Another important feature revealing in this derivation is the possibility of introducing a renormalization scale μ associated with the Immirzi parameter γ upon quantization where expectation of $< \overline{\psi}\psi >_\mu$ and $< \overline{\psi}\gamma_5\psi >_\mu$ at some scale μ should be employed. Thus the "Quadratic Spinor Representation of General Relativity" provides a transparent interpretation of the Immirzi parameter. A technical drawback of the above derivation is that $\overline{\psi}\gamma_5\psi$ is not in general a real function. This can be easily seen from using a particular representation of the Dirac algebra. In order for $\overline{\psi}\gamma_5\psi$ to be always real to render a corresponding real Ashtekar-Barbero variable, one has to use anti-commuting spinor.

3. Concluding remarks

In summary, we have demonstrated how Quadratic Spinor Representation of general theory of relativity with auxiliary spinor field ψ can provide a systematic way to derive the physical properties of the Immirzi parameter γ. From the explicit expression for γ, one can see that it is the ratio between scalar and pseudo-scalar contributions in the theory. An important feature of this derivation is the possibility of introducing a renormalization scale associated with γ upon quantization.

References

1. A. Ashtekar, *Phys. Rev. Lett.* ***57*** 2244 (1986); *Phys. Rev. D* ***36*** 1587 (1987).
2. For a reveiw see: C Rovelli, Quantum Gravity *(Cambridge University Press, 2004)*; T. Thiemann, Modern Canonical General Relativity. *(Cambridge University Press, 2005)*.
3. F. Barbero, *Phys. Rev. D51* 5507 (1995); *Phys. Rev. D51* (1995) 5498.
4. G. Immirzi, *Class. Quantum Grav. 14*, L177 (1997).
5. O. Dreyer, *Phys. Rev Lett.* ***90*** 081301 (2003); gr-qc/0211076.
6. M. Domagala, J. Lewandowski, *Class. Quantum Grav.* ***21*** (2004) 5233; gr-qc/0407051; K. A. Meissner, *Class. Quantum Grav.* ***21*** (2004) 5245; gr-qc/0407052.
7. A. Perez and C. Rovelli, "Physical effects of the Immirzi parameter", gr-qc/0505081.
8. L. Freidel, D. Minic and T. Takeuchi, "Quantum gravity, torsion, parity violation and all that", hep-th/0507253.
9. J. M. Nester and R. S. Tung, *Gen. Rel. Grav.* ***27*** , 115 (1995).
10. R. S. Tung and T. Jacobson, *Class. Quantum Grav.* ***12***, L51 (1995).
11. R. S. Tung and J. M. Nester, *Phys. Rev. D* ***60***, 021501 (1999).
12. S. Holst, *Phys. Rev. D* ***53***, 5966 (1996).
13. E. Witten, *Commun. Math. Phys.* ***80***,, 381(1981).

Quantum Entanglement in Non-inertial frames

R.B. Mann*† ‡ and I. Fuentes-Schuller†◇ ‡,

‡*Perimeter Institute, 31 Caroline Street North Waterloo, Ontario Canada N2L 2Y5*

◇ *Instituto de Ciencias Nucleares, UNAM, A-postal 70-543 04510, Mexico D.F.*

† *Department of Physics, University of Waterloo, Waterloo, Ontario Canada N2L 3G1*

We discuss the phenomenon of quantum entanglement between relatively accelerating observers, illustrating that it is an observer-dependent concept due to its degradation from the Rindler vacuum.

1. Introduction

Of all of the phenomena connected with quantum mechanics, entanglement is perhaps the most intriguing. It yields correlations between physical quantities that are spacelike separated, yet in such a way that observers cannot make use of these correlations to superluminally send signals to one another. Some amount of classical communication is always required when transmitting information using quantum correlations. Furthermore entanglement plays a central role in quantum information theory and provides a resource for quantum communication and teleportation [1].

Most of quantum information theory presupposes an non-relativistic setting for the physical systems under its consideration. While this is a reasonable requirement for many situations, it is ultimately unsatisfactory. From a fundamental viewpoint relativity is indispensable in any complete theoretical consideration of the world, and so a full theory of quantum information must take this into account. Inclusion of relativity can also be important in practical situations, as when considering the implementation of quantum information tasks between observers in arbitrary relative motion.

Only recently has the importance of understanding quantum information – particularly entanglement – in a relativistic setting received considerable attention [2,3,4,5]. For inertial observers, entanglement was shown

*presenter of seminar at ICAG7

†Published before with maiden name Fuentes-Guridi

to be an invariant quantity [2], one that can redistributed among the different degrees of freedom for different inertial observers, but nevertheless conserved over-all. However the situation is quite different in non-inertial frames. For uniformly accelerated observers, horizons appear and regions of spacetime become observationally inaccessible. This implies that accelerating observers perceive themselves to be in a heat bath, whose temperature is proportional to the acceleration[6]. It also has a profound effect on the nature of entanglement and its physical effects. In the case of teleportation between observers in relative uniform acceleration [3], there is a loss of information yielding degradation of the entanglement. The acceleration of the observer effectively produces a kind of "enviromental decoherence", limiting the fidelity of the process. This is in strong contrast to the situation for inertial observers, for whom entanglement is invariant.

More generally, entanglement has recently been shown to be an observer-dependent phenomenon. This was first explicitly shown by considering the entanglement between two modes of a free scalar field seen by two relatively accelerated observers [4]. These results have recently been extended to the spin-1/2 case [7], which is qualitatively similar, but has a number of distinct quantitative features. Further investigation into expanding curved spacetimes has indicated that entanglement can encode information concerning the underlying spacetime structure [5].

Here we will review the situation concerning entanglement and its quantification in non-inertial frames. Even two inertial observers will not agree on the degree of entanglement of a given bipartite quantum state of some quantum field if they are in a curved spacetime, since they will be relatively accelerated in a manner described by the geodesic deviation equation. An understanding of this phenomenon is therefore important for both fundamental physics and for the implementation of quantum information tasks between accelerated partners [4].

2. Entanglement and Its Measures

The concept of entanglement is fairly straightforward to present. The description of a physical system in quantum mechanics is that of a state $|\Psi\rangle$. This state time-evolves under unitary transformations. Now suppose the system can be broken into two subsystems A and B, with each subsystem being in either of two possible states, either $|\phi\rangle$ or $|\psi\rangle$. Then if

$$|\Psi\rangle_{AB} = |\phi\rangle_A \, |\psi\rangle_B$$

we say the state is separable. Such states can be prepared through a series of Local Operations and Classical Communications (LOCC). However the linearity of quantum mechanics admits another possibility, namely that

$$|\Psi\rangle_{AB} = \frac{1}{\sqrt{2}}\left(|\phi\rangle_A |\phi\rangle_B + |\psi\rangle_A |\psi\rangle_B\right)$$

in which case we say that the state of AB is entangled. The physical system can no longer be cleanly separated – for the above example, whatever state we measure subsystem A to be in, subsystem B must also be in that state. Entangled states cannot be prepared via LOCC.

Intuitively then, quantum entanglement should be a measureable quantity (at least indirectly), varying from a minimum of zero to some maximal value for any physical system. Quantifying entanglement for pure bipartite systems is well understood.

A pure (entangled) state can be written as

$$|\Psi\rangle_{AB} = \sum_{ij} \omega_{ij} |i\rangle_A |j\rangle_B$$

or, through a change of basis, as

$$|\Psi\rangle_{AB} = \sum_{n} \omega_{n} |n\rangle_A |n\rangle_B$$

a process called Schmidt decomposition. From this the density matrix

$$\rho_{AB} = |\Psi\rangle\langle\Psi|$$

can be formed. Entanglement is quantified in this case by the von Neumann entropy

$$S(\rho_A) \equiv -\mathrm{Tr_A}[\rho_\mathrm{A} \ln(\rho_\mathrm{A})] > 0$$

of the reduced density matrix

$$\rho_A = \mathrm{Tr_B}[\rho_\mathrm{AB}]$$

Note that the pure density matrix ρ_{AB} has a single eigenvalue of unity, and so

$$S(\rho_{AB}) = 0.$$

However quantifying entanglement for mixed states is not a straightforward task. Unfortunately making use of the von Neumann entropy as a measure of entanglement is unsatisfactory for mixed states, whose density matrices are of the form

$$\rho_{AB} = \sum_{ijkl} \sigma_{ijkl} |i\rangle_A |j\rangle_{BA} \langle k|_B \langle l|$$

Such entangled states have no Schmidt decomposition, and the von Neumann entropy no longer quantifies entanglement.

Given an arbitrary quantum state, we are presented with two problems: that of determining separability, and that of measuring entanglement. The first problem is relatively easy to address. A necessary condition for non-separability is that the partial transpose of the density matrix has at least one negative eigenvalue [8]. However the second problem is quite difficult to deal with and in general there is no unique measure of entanglement. However there have been several proposals [9]. The most popular are formation and distillation of entangled states.

Consider the number m of maximally entangled pairs that can be created using only LOCC on a number n of non-maximally entangled states. We can take the asymptotic conversion ratio, $\frac{m}{n}$ in the limit of infinitely many copies [10],

$$E_F(\rho_{ab}) = \min \sum_i p_i S(\rho_a^i),$$

where the minimum is taken over all the possible realizations of the state $\rho_{ab} = \sum_i p_i |\Psi^i_{ab}\rangle\langle\Psi^i_{ab}|$. The quantity $E(F)$ is called the entanglement of formation. For two qubits it can be explicitly calculated via

$$\begin{aligned} E_F = & \frac{1+\sqrt{1-C^2}}{2}\log_2\frac{1+\sqrt{1-C^2}}{2} \\ & + \frac{1-\sqrt{1-C^2}}{2}\log_2\frac{1-\sqrt{1-C^2}}{2}. \end{aligned}$$

where C is the concurrence. For two qubits, the concurrence of a pure state $|\psi\rangle$ is

$$C(\psi) \equiv |\langle\psi|\sigma_y \otimes \sigma_y|\psi^*\rangle|,$$

where the star denotes complex conjugation in the standard basis and $\sigma_y \otimes \sigma_y |\psi^*\rangle$ represents the 'spin-flip' of $|\psi\rangle$.

Entanglement of distillation $E(D)$ is essentially the opposite process. It is the asymptotic rate of converting non-maximally entangled states into maximally entangled states by means of a purification procedure. In general it is smaller than that the entanglement of formation, and shows that the entanglement conversion is irreversible. This is because there is a loss of classical information about the decomposition of the state. One consequence of this is the existence of bound entangled states: no entanglement can be distilled from them even though they are inseparable.

Unfortunately these (and other measures) are in general very difficult to calculate, particularly for the problem we wish to address. Fortunately there is a quantity called the the logarithmic negativity [11], which is by comparison much easier to compute. It is not an entanglement measure because it does not satisfy the requirement of being equal to the von-Neuman entropy in the pure case. However it is an entanglement monotone that is an upper bound to the entanglement of distillation. It is defined as

$$N(\rho) = \log_2 |\rho^T||_1$$

where $||\rho^T||_1$ is the trace-norm of the partial transpose density matrix ρ^T. In fact we have

$$E(D) \leq N, \quad E(D) \leq E(F).$$

Another quantity of use is that of mutual information [12], defined as

$$I(\rho_{AR}) = S(\rho_A) + S(\rho_R) - S(\rho_{AR})$$

where $S(\rho) = -\mathrm{Tr}(\rho \log_2(\rho))$ is the entropy of the density matrix ρ. This will allow us to estimate the total amount of correlations (classical plus quantum) in a mixed state.

3. Relativistic Entanglement

We wish to address the problem of quantum entanglement between two states shared by two observers Alice and Rob, where Rob is uniformly accelerating relative to Alice.

To formalize this, consider two modes, k and s, of a free massless scalar field in Minkowski spacetime. From an inertial perspective the maximally entangled quantum field is in a state

$$|\Psi_{ks}\rangle^{\mathcal{M}} = \frac{1}{\sqrt{2}}\left(|0_s\rangle^{\mathcal{M}}|0_k\rangle^{\mathcal{M}} + |1_s\rangle^{\mathcal{M}}|1_k\rangle^{\mathcal{M}}\right). \tag{1}$$

where $|0_j\rangle^{\mathcal{M}}$ and $|1_j\rangle^{\mathcal{M}}$ are the vacuum and single particle excitation states of the mode j in Minkowski space. Suppose that Alice has a detector which only detects mode s and Rob has a detector sensitive only to mode k. The states corresponding to mode k must be specified in Rindler coordinates in order to describe what Rob sees, since Rob undergoes uniform acceleration a. These Rindler coordinates are defined by

$$\begin{aligned} t &= a^{-1}e^{a\xi}\sinh a\tau, \quad z = a^{-1}e^{a\xi}\cosh a\tau, \quad |z| < t, \\ t &= -a^{-1}e^{a\xi}\sinh a\tau, \quad z = a^{-1}e^{a\xi}\cosh a\tau, \quad |z| > t, \end{aligned} \tag{2}$$

where the hyperbolae correspond to the space-like coordinates ξ and τ is the proper time, i.e., the length of the hyperbolic world line measured by the Minkowski metric.

World lines of uniformly accelerated observers in Minkowski coordinates correspond to hyperbolae, to the left (region I) and right (region II) of the origin, bounded by light-like asymptotes constituting the Rindler horizon. The Minkowski vacuum state, defined as the absence of any particle excitation in any of the modes

$$|0\rangle^{\mathcal{M}} = \prod_j |0_j\rangle^{\mathcal{M}}, \tag{3}$$

can be expressed in terms of a product of two-mode squeezed states of the Rindler vacuum, [13]

$$|0_k\rangle^{\mathcal{M}} \sim \frac{1}{\cosh r} \sum_{n=0}^{\infty} \tanh^n r \; |n_k\rangle_I \, |n_k\rangle_{II} \,, \tag{4}$$

$$\cosh r = (1 - e^{-2\pi\Omega})^{-1/2}, \quad \Omega = |k|c/a. \tag{5}$$

where $|n_k\rangle_I$ and $|n_k\rangle_{II}$ refer to the mode decomposition in region I and II, respectively, of Rindler space. Each Minkowski mode j has a Rindler mode expansion given by Eq. (4). For simplicity we shall take all modes of the field to be in the vacuum, except for the single Minkowski mode s for Alice and k for Rob. We can trace over all other modes and obtain a pure state because different modes j and j' do not mix.

Using Eq. (4) and

$$|1_k\rangle^{\mathcal{M}} = \frac{1}{\cosh^2 r} \sum_{n=0}^{\infty} \tanh^n r \sqrt{n+1} \, |(n+1)_k\rangle_I \, |n_k\rangle_{II} \,,$$

we rewrite the state $|\Psi_{ks}\rangle^{\mathcal{M}}$ and obtain

$$\rho_{AR} = \frac{1}{2\cosh^2 r} \sum_n (\tanh r)^{2n} \rho_n \,, \tag{6}$$

$$\begin{aligned} \rho_n &= |0n\rangle\langle 0n| + \frac{\sqrt{n+1}}{\cosh r} |0n\rangle\langle 1n+1| \\ &+ \frac{\sqrt{n+1}}{\cosh r} |1n+1\rangle\langle 0n| + \frac{(n+1)}{\cosh^2 r} |1n+1\rangle\langle 1n+1| \end{aligned}$$

by tracing over the states in region II, since Rob is causally disconnected from this region. Here $|nm\rangle = |n_s\rangle^{\mathcal{M}} |m_k\rangle_I$.

Checking the eigenvalues of the partial transpose of the density matrix (obtained by interchanging Alice's qubits), we find

$$\lambda_{\pm}^n = \frac{\tanh^{2n} r}{(4\cosh^2 r)} \left[\left(\frac{n}{\sinh^2 r} + \tanh^2 r \right) \pm \sqrt{Z_n} \right],$$

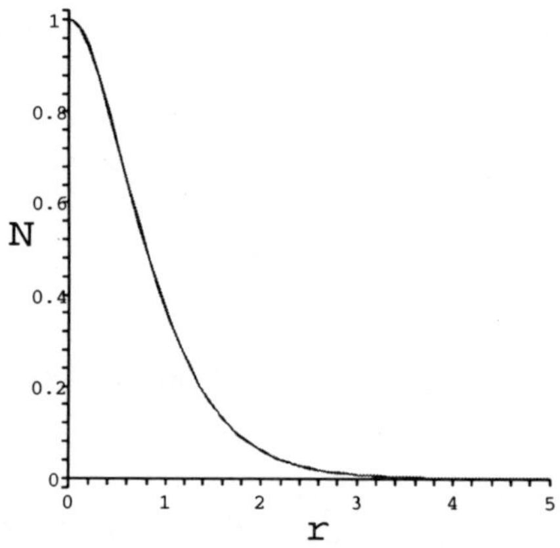

Fig. 1. The negativity as a function of the acceleration r.

for the eigenvalues in the $(n, n+1)$ block, where

$$Z_n = \left(\frac{n}{\sinh^2 r} + \tanh^2 r\right)^2 + \frac{4}{\cosh^2 r}.$$

Clearly one eigenvalue is always negative for finite acceleration ($r < \infty$) and so the state is always entangled. Only in the limit $r \to \infty$ could the negative eigenvalue possibly go to zero.

Turning next to the logarithmic negativity, we find

$$\Sigma = \sum_{n=0}^{\infty} \frac{\tanh^{2n} r}{2\cosh^2 r} \sqrt{\left(\frac{n}{\sinh^2 r} + \tanh^2 r\right)^2 + \frac{4}{\cosh^2 r}}.$$

which is plotted in fig. 1.

We see that when $r = 0$ the negativity is unity, as expected. For any finite value of the acceleration the entanglement is degraded, and it is not too difficult to show that in the infinite acceleration limit the negativity is exactly 0, indicating that the state no longer has any distillable entanglement.

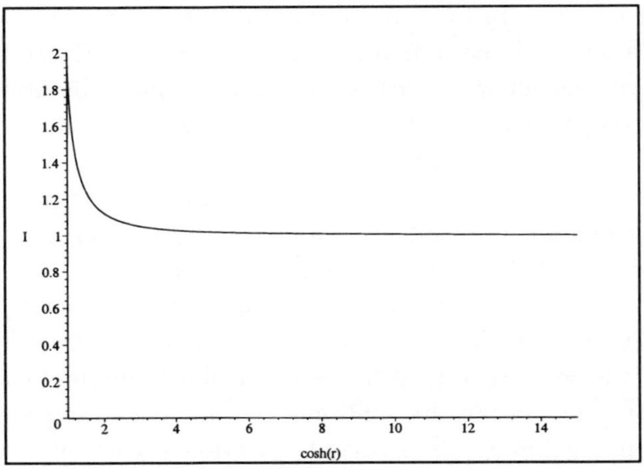

Fig. 2. Mutual information as a function of cosh(r).

We find the mutual information to be

$$I(N) = 1 - \frac{1}{2}\log_2\left(\tanh^2 r\right) - \frac{1}{2\cosh^2 r}\sum_{n=0}^{N}\tanh^{2n} r\mathcal{D}_n,$$

$$\mathcal{D}_n = (1+\frac{n}{\sinh^2 r})\log_2\left(1+\frac{n}{\sinh^2 r}\right)$$
$$- (1+\frac{n+1}{\cosh^2 r})\log_2\left(1+\frac{n+1}{\cosh^2 r}\right),$$

which is plotted in Fig. (2). From a maximal value of 2 in the zero acceleration case, it monotonically decreases to unity in the infinite acceleration limit. Since the distillable entanglement in this limit is zero, the total correlations are classical correlations plus bound entanglement.

4. Observer-dependent Entanglement

Where did the entanglement go? First note that $S(\rho_{ARI})$, the entropy of the density matrix for Rob and Alice in region I is equal to $S(\rho_{RII})$, the density matrix for Rob in region II. This must be so because the entropies of the reduced density matrices of any bipartite division of a physical system are equal. When the bosons are maximally entangled, for vanishing acceleration, there is no distillable entanglement with region II. For finite acceleration, the entanglement between the bosons is degraded as the entanglement with region II grows. In the limit of infinite acceleration $S(\rho_{ARI}) = 1$ and the modes in region II become maximally entangled to the state in region I. Since entanglement in tripartite pure states cannot be arbitrarily distributed amongst the subsystems [14], the entanglement between the bosons is degraded as acceleration grows.

Suppose Alice were also accelerated. Then the density matrix would be mixed to a higher degree, resulting in a higher degradation of entanglement. It is only for inertial observers that the state under observation is maximally entangled. Consequently entanglement is an observer-dependent quantity in non-inertial frames. Since inertial observers have a preferred role in flat spacetime, it is possible to prescribe a well-defined notion of entanglement in that setting by stating that only inertial observers are good observers of entanglement. However in a curved spacetime even two nearby inertial observers are relatively accelerated, and so even they will not agree on the degree of entanglement of a given bipartite quantum state of some quantum field.

We close by making some brief remarks concerning black holes. The infinite acceleration limit can be interpreted as Alice falling into a black hole while Rob hovers outside it. To escape falling in he must be (approximately) uniformly accelerated, since close to the black hole horizon spacetime is flat. Rob therefore loses information about the state in the whole of spacetime. There will thus be a reduced fidelity in any information task performed by Alice and Rob using this state since the entanglement resource is degraded.

We are grateful for discussions and correspondence with a number of people throughout the course of this work, including Frederic P. Schuller, Don Marolf, Martin Plenio, Jonathan Ball, Daniel Terno, and Valeri Frolov. This work was supported in part by the Natural Sciences and Engineering Research Council of Canada. Robert Mann would like to thank the organizers of the 7th International Conference on Gravitation and Astrophysics in the Asia Pacific Area for their invitation to present this work.

References

1. "The Physics of Quantum Information", Springer-Verlag 2000, D. Bouwmeester, A. Ekert, A. Zeilinger (Eds.).
2. . Czachor, Phys. Rev. A 55, 72 (1997); D.R. Terno and A. Peres, Rev. Mod. Phys. **76**, 93 (2004); A. Peres, P.F. Scudo and D.R. Terno, Phys. Rev. Lett. **88**, 230402 (2002); Czachor, PRL 94, 078901 (2005); P.M. Alsing and G.J. Milburn, Quant. Inf. Comp. **2**, 487 (2002); R.M. Gingrich and C. Adami,

Phys. Rev. Lett. **89**, 270402 (2002); J. Pachos and E. Solano, *QIC Vol. 3, No. 2, pp.115* (2003); W.T. Kim and E.J. Son, quant-ph/0408127; D. Ahn, H.J. Lee, Y.H. Moon, and S.W. Hwang, Phys. Rev. A 67, 012103 (2003).[6] D. Ahn, H.J. Lee, and S.W. Hwang, e-print quant-ph:/0207018 (2002); D. Ahn, H.J. Lee, S.W. Hwang, and M.S. Kim, e-print quant-ph:/0207018 (2003). [[8] H. Terashima and M. Ueda, Int. J. Quant. Info. 1, 93 (2003). [9] A.J. Bergou, R.M. Gingrich, and C. Adami, Phys. Rev. A 68, 042102 (2003). [10] C. Soo and C.C.Y. Lin, Int. J. Quant. Info. 2, 183 (2003).

3. P.M. Alsing and G.J. Milburn, Phys. Rev. Lett. **91**, 180404 (2003); P.M. Alsing, D. McMahon and G.J. Milburn ,quant-ph/0311096.
4. I. Fuentes-Schuller and R.B. Mann, Phys.Rev.Lett. 95 (2005) 120404
5. J. Ball, I. Fuentes-Schuller and F.P. Schuller, quant-ph/0506113
6. P.C.W. Davies, J. of Phys. A8, 609 (1975); W.G. Unruh, Phys. Rev. D14, 870 (1976);
7. P.M. Alsing, I. Fuentes-Schuller, R.B. Mann, T.E. Tessier, quant-ph/0603269.
8. A. Peres , Phys. Rev. Lett. 77, 1413, (1996).
9. For a review on entanglement see Dagmar Bruss J. Math. Phys., (43): 4237, 2002.
10. W. K. Wootters, Phys. Rev. Lett. 80 2245 (1998).
11. G. Vidal and R. F. Werner Phys. Rev.A (65):032314, 2002; Plenio quant-ph/0505071, 2005.
12. R. S. Ingarden, A. Kossakowski and M. Ohya," Informa- tion Dynamics and Open Systems - Classical and Quan- tum Approach", (Kluwer Academic Publishers, Dor- drecht, 1997). M.B.
13. D.F.Walls and G.J. Milburn, Quantum Optics, Springer- Verlag, New York (1994). S.A. Fulling, Phys. Rev. D7, 2850 (1973).
14. V. Coffman, J. Kundu, and W. Wootters, Phys. Rev. A **61**, 052306 (2000). T. Tessier, A. Delgado, I. Deutsch and I. Fuentes-Guridi, Phys. Rev. A 68, 062316 (2003).

Black hole fluctuations and dynamics from back-reaction of Hawking radiation: Current work and further studies based on stochastic gravity

B. L. Hu

Department of Physics, University of Maryland,
College Park, Maryland 20742-4111

E-mail: hub@physics.umd.edu

Albert Roura

Theoretical Division, T-8, Los Alamos National Laboratory,
M.S. B285, Los Alamos, NM 87545

E-mail: roura@lanl.gov

We give a progress report of our research on spacetime fluctuations induced by quantum fields in an evaporating black hole and a black hole in quasi-equilibrium with its Hawking radiation. We note the main issues involved in these two classes of problems and outline the key steps for a systematic quantitative investigation. This report contains unpublished new ideas for further studies.

1. Black hole back-reaction and fluctuations: Our current work

Back-reaction here refers to the influence of particle creation as in the Hawking radiation and other quantum effects (such as trace anomaly) on the structure and dynamics of the background spacetime. It is believed to grow in importance when the energy reaches the Planck scale, as in the very early universe and at the final stages of black hole evolution. In semiclassical black hole physics this is perhaps one of the more difficult but important unsolved problems. It is important because its solution is necessary for us to understand better the black hole end-state and information loss puzzles. It also provides a check on the range of validity for Hawking's derivation of black hole radiance in the framework of semiclassical gravity. As we found out from the *stochastic gravity* perspective [1,2], the back-reaction problem is tied to dissipation and fluctuation phenomena. Thus it provides a natural generalization to *non-equilibrium black hole thermodynamics* and can reveal deeper connections between gravity, quantum field theory and statistical mechanics.

Difficulties in the black hole back-reaction problems start with finding a renormalized energy momentum tensor[a]. The stochastic gravity program [2] introduces fresh insight and new methodology into the back-reaction problem by a) stressing the importance of an in-in (closed-time-path, or CTP) formulation [3,4] which gives real and causal equations of motion, b) imparting a statistical mechanical meaning to the back-reaction effects via quantum open system concepts and techniques [5,6], c) noting the necessity of including fluctuations in conjunction with dissipation in the system dynamics and showing how noise arises with the help of the influence functional formalism [7,8,9]. The quantities of importance are the dissipation kernel, which enters into the Einstein-Langevin (E-L) equation [10], and the noise kernel, which characterizes the correlations of its stochastic source.

The back-reaction problems of interest to us fall into two classes: 1) A black hole in a box in quasi-equilibrium with its Hawking radiation. 2) An evaporating black hole emitting Hawking radiation under fully non-equilibrium conditions. These problems have been treated before in varying degrees of completeness. The former is easier to understand but has some remaining technical challenges. We approach this problem at two levels: i) At the *quantum field theory level* is the derivation of the influence action, which demands a computation of the noise kernel for quantum fields near the Schwarzschild horizon. ii) At the *statistical thermodynamic level* we approach this problem by viewing the back-reaction as an embodiment of the fluctuation-dissipation relation (FDR). We have demonstrated the usefulness of this way of thinking in cosmological back-reaction problems [8]. The problems for evaporating black holes have only been treated sparsely and qualitatively, even with contradictory claims. Our current effort has focused on clarifying some existing conceptual confusion and building a unifying framework capable of producing more quantitative results.

[a]For a list of papers dedicated to this task, see footnote 1 of [36]. Some notable work under the framework of semiclassical gravity (in addition to those few discussed in more detail below) includes Anderson *et al.* [32] for a quantum scalar field in a spherically symmetric spacetime, the back-reaction in the interior of a black hole [33] (easier technically since it is a cosmological Kantowski-Sachs spacetime) and for two-dimensional dilaton gravity theory [34].

2. Quasi-equilibrium conditions: Black hole in a box

A black hole can be in quasi-equilibrium with its Hawking radiation if placed inside a box of the right size [17,20b], or in an anti-de Sitter universe [18]. We divide our consideration into the far-field case and the near-horizon case. The far field case has been studied before [35]. In [36] we consider the near-horizon case. Using the model of a black hole described by a radially-perturbed quasi-static metric and Hawking radiation by a conformally coupled massless quantum scalar field, we showed that the closed-time-path (CTP) effective action yields a non-local dissipation term as well as a stochastic noise term in the equation of motion. We have presented the overall structure of the theory and the strategy of our approach in [36], but due to the lack of an analytic form of the Green function for a scalar field in the strong field region of the Schwarzschild metric, numerical calculations may be the only way to go. Being based on a quasilocal expansion, Page's approximation [19], though reasonably accurate for the stress tensor, is insufficient for the noise kernel since one needs to consider arbitrarily separated points in that case. In [36] we also presented an alternative derivation of the CTP effective action in terms of the Bogoliubov coefficients, thus connecting with the interpretation of the noise term as measuring the difference in particle production in alternative histories. [This will be useful for a grand-canonical ensemble description of black hole fluctuations. (See below.)]

2.1. *Connecting different approaches*

On the black-hole-in-a-box back-reaction problem we take a two-prong approach: via quantum field theory and statistical thermodynamics. We are currently performing a calculation of the noise kernel near the Schwarzschild horizon, making use of the results of [14] but with points kept separate [15]. We will use the result of this calculation to show the existence of a fluctuation-dissipation relation (described below). In addition we want to integrate the master equation approach of Zurek [27] and the transition probability approach of Bekenstein, Meisel [24] and Schiffer [25] (see also [26]) under the stochastic gravity framework. Within the thermodynamic descriptions we want to compare the results from the canonical with the microcanonical [21]

[b]Note that one needs to introduce the appropriate redshift factors that account for the finite size of the box. Moreover, for a sufficiently small box, as required in general to have (meta)stable equilibrium, curvature corrections, which were not included in Ref. [35] may not be negligible.

and the grand canonical ensembles [22c]. For the last task we will invoke results obtained earlier [23] for the number fluctuations in particle creation and the relation we obtained recently in [36] expressing the CTP effective action in terms of the Bogoliubov coefficients for the black hole particle creation, to derive the susceptibility and isothermal compressibility functions of the black hole. This would move us a step closer to establishing a linear response theory (LRT) of non-equilibrium black hole thermodynamics.

2.2. *Back-reaction manifested through a fluctuation-dissipation relation*

Historically Candelas and Sciama [28] first suggested a fluctuation-dissipation relation for the depiction of dynamic black hole evolution. They proposed a classical formula relating the dissipation in area linearly to the squared absolute value of the shear amplitude. The quantized gravitational perturbations (they choose the quantum state to be the Unruh vacuum) approximate a flux of radiation emanating from the hole at large radii. Thus they claim that the dissipation in area due to the Hawking flux of gravitational radiation is related to the quantum fluctuations of the metric perturbations. However, as was pointed out in [45], it is not clear that their relation corresponds to a FDR in the correct statistical mechanical sense and does not include the effect of matter fields. The FDR for the contribution of the matter fields should involve the fluctuations of the stress tensor (the "generalized force" acting on the spacetime), which are characterized by the noise kernel. With an explicit calculation of the noise kernel in a similar context one could obtain the correct FDR.

For the quasi-static case Mottola [29] introduced a FDR based on linear response theory for the description of a black hole in quasi-equilibrium with its Hawking radiation. He showed that in some generalized Hartle-Hawking state a fluctuation-dissipation relation (FDR) exists between the expectation values of the commutator and anti-commutator of the energy-momentum tensor. This is the standard form of FDR. However, in his case the dynamical equation for the linear metric perturbations describes just their mean evolution, which corresponds to taking the stochastic average of the linearized E-L equation. The E-L equation from stochastic gravity does more in providing the dynamics of the fluctuations.

[c]Note, however, that for massless particles with vanishing chemical potential the canonical and grand canonical ensembles coincide

In a recent paper [36] we laid out the road map for treating the quasi-static case in the stochastic gravity framework and point out that a non-local dissipation and a fluctuation term will arise which should match with the analytic results in the far field limit derived earlier [35]. These terms are absent in York's work [30] because the approximate form he uses for the semiclassical Einstein equation corresponds to the variation of terms in the CTP effective action which are linear in the metric perturbations around the Schwarzschild background geometry, whereas the non-local dissipation and noise kernels appear at quadratic order. The noise kernel, which is connected to the fluctuations of the stress tensor, would give no contribution to his equation for the mean evolution, even if higher order corrections were included.

3. Evaporating black hole: Non-equilibrium conditions

This time-dependent problem has a very different physics from the quasi-equilibrium case, and is in general more difficult to treat. Starting in the early 80's, it has been approached by Bardeen [37], Israel [38] and Massar [40], who considered the mean evolution. The role of fluctuations was initially studied by Bekenstein [41] and has received further attention in recent years by Ford [46,47], Frolov [49], Sorkin [43], Marolf [44], and their collaborators (see also Ref. [42]), largely based on qualitative arguments. On some issues, such as the size of black hole horizon fluctuations, there are contradictory claims.

To make progress one needs to introduce a theoretical framework where all prior claims can be scrutinized and compared. Because of its non-equilibrium nature we expect the stochastic gravity program [1] to provide some useful quantitative results. In [45] we wrote down the Einstein-Langevin equation for the fluctuations in the mass of an evaporating black hole and found that the fluctuations compared to the mean is small unless the mean solution is unstable with respect to small perturbations. Our 1998 paper emphasized the first part of this statement, which is in apparent contradiction to what Bekenstein claimed in his 1984 paper [41]. Recently we revisited this question with a closer analysis [51]. Since for an evaporating black hole there exist unstable perturbations around the solution of the semiclassical Einstein equation for the mean evolution, the second part of the statement applies.

A key assumption in these studies is that the fluctuations of the incoming energy flux near the horizon are directly related to those of the outgoing one (a condition that does hold for the mean flux). If this condition were true, one can indeed show that the fluctuations of a black hole horizon become important (growing slowly, but accumulating over long times), as Bekenstein claimed. However, we have serious reservations on its validity for energy flux fluctuations (see below). Using the E-L equation we also point out how the different claims of Bekenstein and Ford-Wu can be reconciled by recognizing the different physical assumptions they used in their arguments. The stochastic gravity theory which our present work is based on should provide a platform for further investigations into this important issue.

3.1. *Bardeen's evaporating black hole and Bekenstein's fluctuation theory*

In 1981 Bardeen [37] considered the back-reaction of Hawking radiation on an evaporating black hole by invoking a Vaidya-type metric. (This model was later used by Hiscock [39] for similar inquiries.) Bardeen's calculation with this model affirmed the validity of the semiclassical picture assumed in Hawking's original derivation of thermal radiance. His results including back-reaction indicate that the black hole follows an evolution which is largely determined by the semiclassical Einstein equations down to where the black hole mass drops to near the Planck mass ($\sim 10^{-5}$ g), the point where most practitioners of semiclassical gravity would agree that the theory will break down. Bardeen's result was developed further by Massar [40].

For black hole mass fluctuations, in 1984 Bekenstein [41] considered the mass fluctuations of an *isolated* black hole due to the energy fluctuations in the radiation emitted by the hole, and asks when such fluctuations become large. According to his calculation, depending on the initial mass, mass fluctuations can grow large well before the mass of the black hole reaches the Planck scale. On the other hand, of the few studies in this setting, the result of Wu and Ford [47] supports a scenario in which fluctuations do not become important before reaching the Planckian regime. In view of such contradictory claims in the literature, it is highly desirable to have a more solid and complete theoretical framework where all prior claims can be scrutinized and compared. We expect the stochastic gravity program to be useful for this purpose. At the most rudimentary level, Bekenstein's approach shares similar conceptual emphasis as does the stochastic gravity program in that both attribute importance to the fluctuations of stress tensor and the black hole mass, characterized by their correlation functions.

3.2. *Non-equilibrium conditions*

Investigations in this case may assist in answering two important questions: *1) Are the fluctuations near the horizon large or small? 2) How reliable are earlier results from test quantum fields in fixed curved spacetime?*

As far as the first question is concerned, one should distinguish between fluctuations with short and long characteristic time scales. For time scales comparable to the evaporation time, one expects spherically-symmetric modes ($l = 0$) to be dominant. Moreover, if one assumes a direct correlation between the energy flux fluctuations on the horizon and those far from it, as done in earlier work, an explicit result supporting Bekenstein's conclusion can be obtained and the origin of the discrepancy with Ford and Wu's result can be understood. However, a more careful analysis reveals that such an assumption is not correct (see below) and it is necessary to compute the noise kernel near the horizon to get an accurate answer. One can present arguments to the effect that this kind of fluctuations may not modify in a drastic way the result obtained by Hawking for test fields evolving on a fixed black hole spacetime, and later extended by Bardeen and Massar to include their back-reaction effect on the mean evolution of the spacetime geometry. On the other hand, in principle, fluctuations with much shorter correlation times, which also involve higher multipoles ($l \neq 0$), could alter substantially Hawking's result.

Detailed calculations within the framework of stochastic gravity can address those issues in a natural way, at least for fluctuations with typical scales much larger than the Planck length. Nevertheless, one needs to pay attention to the subtleties. Preliminary results, briefly described below, signal a possible breakdown of the geometric optics approximation for the propagation of test fields when fluctuations are included. This would require finding alternative ways of probing satisfactorily those metric fluctuations and extracting physically meaningful information.

3.2.1. *Spherically symmetric sector ($l = 0$)*

Consideration of this problem with restriction to spherically symmetric modes was done in [27,41,47,49,50] as well as two-dimensional dilaton-gravity models [34]. All those studies restricted from the outset the contribution to the classical action to s-wave modes for both the metric and the matter fields, which only allows $l = 0$ modes for each matter field to contribute to the noise kernel. Our approach goes well beyond that approximation since modes with all possible values of l for the matter fields can contribute to the $l = 0$ sector of the noise kernel.

We focus on the physics in the *adiabatic regime*, which is the time when the black hole mass M remains much larger than the Planck mass m_{p}. This allows one to use $\langle \hat{T}_{ab} \rangle_{\mathrm{ren}}$ in the Schwarzschild spacetime for each instant of time but with a mass slowly evolving in time. Technically this saves one the trouble of solving the integro-differential (semiclassical Einstein) equation to obtain the mean evolution.

In the adiabatic regime, one can show from energy-momentum conservation that the outgoing (for $r \gg 2M$) and ingoing (for $r \approx 2M$) *mean fluxes* are related. If a similar argument is employed to relate the outgoing and ingoing energy fluxes for the *stochastic source* characterizing the stress tensor fluctuations, one can proceed in the same way as was done for the mean fluxes to provide a justification for Bekenstein's approach.

This energy conservation argument has been assumed to be valid for energy flux fluctuations by Bekenstein [41] as well as Wu and Ford [47] for an evaporating black hole. With this assumption one can also clarify the apparent discrepancy with Wu and Ford as follows. The growth in time of the fluctuations can be understood in terms of the following "instability" exhibited by the mean evolution: if the initial mass of a macroscopic black hole with $M \gg m_{\mathrm{p}}$ is slightly perturbed by a small amount of the order of the Planck mass, the difference between the masses of the perturbed and unperturbed black holes becomes of the same order as the mass of the unperturbed black hole when the latter becomes of order $(m_{\mathrm{p}}^2 M)^{1/3}$, *i.e.*, still much larger than the Planck mass. This growth of the fluctuations, first found by Bekenstein, is a consequence of the secular effect of the renormalized stress energy tensor of the perturbations, whose effect builds up in time and gives a contribution to the mass evolution of the same order as the mean evolution for times of the order of the black hole evaporation time even when the mass of the black hole is still much larger than the Planck mass. This term was not taken into account by Wu and Ford, which explains why they found much smaller fluctuations than Bekenstein for times of the order of the evaporation time.[d]

Moreover, due to the nonlinear dependence of the flux of radiated energy on the mass of the black hole, terms of higher order in the perturbations become relevant when the fluctuations become of the same order as the

[d] An interesting related question is whether decoherence effects render those fluctuations effectively classical, so that they can be regarded as fluctuations within an incoherent statistical ensemble rather than coherent quantum fluctuations.

mean value of the mass. As pointed out by Bekenstein, this implies a deviation from the usual semiclassical Einstein equation for the evolution of the ensemble/stochastic average of the mass.[e]

3.2.2. *No correlation between the fluctuations of the energy flux crossing the horizon and far from it*

At the time of this meeting (Nov. 2005) we had serious doubts that the same argument that connects the outgoing flux and the flux crossing the horizon, valid for the expectation value of the stress tensor, could also hold for the stochastic source accounting for the stress tensor fluctuations. The reason we gave was the following: while the time derivative of the expectation value, being of higher order in $(m_{\rm p}/M)$, is negligible in the adiabatic regime, that is not always the case for the stochastic source. Therefore, when integrating the conservation equation and computing the correlation function for the flux crossing the horizon, one gets additional terms besides the correlation function for the outgoing flux. Soon after we came up with a proof that this relation does not hold. This can be found in [52], where the assumption of a simple correlation between the fluctuations of the energy flux crossing the horizon and far from it, which was made in earlier work on spherically-symmetric induced fluctuations, was carefully analyzed and found to be invalid (see also Ref. [50] for a related result in an effectively two-dimensional model). This recent finding would invalidate the working assumption of prior results on black hole event horizon fluctuations based on semiclassical gravity, and points to the necessity of doing the calculation of the noise kernel near the horizon in all seriousness, an effort barely gotten started a few years ago [13,14].

3.2.3. *Non-spherically symmetric sector ($l \neq 0$)*

Sorkin [43], Casher *et al.* [42] and Marolf [44] have provided qualitative arguments for the existence of large quantum fluctuations of the event horizon involving time scales much shorter than the evaporation time[f] which would

[e]This effect can be interpreted as follows: due to the growth of the mass fluctuations, the higher order radiative corrections to the semiclassical equation, which involve Feynman diagrams with internal lines corresponding to correlation functions of the metric perturbations (mass perturbations in this case), can no longer be neglected.

[f]In all cases a Schwarzschild spacetime rather than that of an evaporating black hole was considered. However, this is a good approximation if one is interested in analyzing fluctuations with correlation times much shorter than the evaporation time. Moreover,

give rise to an effective width of order $(R_S l_p^2)^{1/3}$, much larger than the Planck length (in all cases induced rather than intrinsic fluctuations are implicitly considered). However, Ford and Svaiter [46] pointed out that Casher *et al.*'s result was probably an artifact from invoking a wrong vacuum to evaluate the fluctuations.[g] Sorkin's result is based on Newtonian gravity, but Marolf's work is intended to be a generalization of Sorkin's to the general relativistic case. We intend to clarify these apparently contradictory claims and treat it with the E-L equation and the noise kernel calculations for this case.

Additional insight into this problem can be gained by studying induced metric fluctuations in de Sitter spacetime. A static observer in de Sitter spacetime perceives the quantum fluctuations of the Bunch-Davies vacuum as a thermal equilibrium distribution in the same way a static observer outside a black hole event horizon would perceive the quantum fluctuations of the Hartle-Hawking vacuum. The high degree of symmetry of de Sitter spacetime makes it easier to obtain exact analytical results. We can check the validity of claims of large black hole horizon fluctuations by studying the corresponding problems in the de Sitter spacetime.

4. Probing metric fluctuations near the horizon

Our recent finding that there exists no simple connection between the outgoing flux and the flux crossing the horizon implies that one needs to compute the noise kernel near the horizon. Ideally one would compute the noise kernel everywhere in a Schwarzschild background, but the difficulties mentioned above make it very hard to obtain an analytical result. On the other hand, having a good approximation for the noise kernel near the horizon might be enough to get the main features because the emission of Hawking radiation is mostly dominated by what is happening near the horizon.[h]

one expects for similar reasons that if those large fluctuations did actually exist, they would also be present in the equilibrium case.

[g] One expects that a small region near the event horizon for a very large black hole should be very similar to a Rindler horizon. The arguments of Casher *et al.* would lead to large fluctuations (actually infinite in the limit of infinite radius), but one knows that the fluctuations in Minkowski spacetime are small.

[h] The effect from the fluctuations of the potential barrier for the equation of motion of the radial component could also be important, but it is not taken into account here.

4.1. *Computing the noise kernel near the horizon*

The key ingredient to compute the noise kernel in a given background spacetime and for a given (vacuum) state of the quantum matter fields is the Wightman function $G^+(x,y)$. Page developed an approximation to obtain two-point Green functions for spacetimes which are a vacuum solution of the Einstein equation and are conformally related to an ultrastatic spacetime. Thus, it can be applied in particular to the Schwarzschild spacetime. Page's approximation involves (among other things) an expansion in terms of the geodetic interval σ between the two points in the Green function starting at order $1/\sigma$ and valid up to order σ^2. Page used it to obtain the renormalized expectation value of the stress tensor operator. The expansion up to order σ^2 was enough for his purpose because, after applying the appropriate differential operators and subtracting the divergent terms in the renormalization process, the contribution from terms of order σ^2 or higher in the expansion of the Green function vanishes when the coincidence limit is finally taken.

On the other hand, when computing the noise kernel (which involves a product of two Wightman functions) using Page's approximation for the Wightman function one obtains an expansion in powers of σ starting at order $1/\sigma^4$, which coincides with the flat space result, and valid through order $1/\sigma$. Furthermore, since the only additional scale that appears in the problem is the Schwarzschild radius of the black hole ($= 2M$, the mass in geometrical/Planckian units), one can conclude by dimensional analysis that the dimensionless expansion parameter is proportional to σ/M. In contrast to the expectation value of the stress tensor, one does not need to take the coincidence limit $\sigma \to 0$ when computing the noise kernel. In fact, when projecting onto the subset of spherically symmetric multipoles, one needs to integrate over the whole solid angle for the two points appearing in the noise kernel. Unfortunately, that involves considering pairs of points with $\sigma \sim M$, for which Page's expansion in terms of σ/M would break down. However, since the first few terms contain inverse powers of σ/M, the integral of the noise kernel over the whole solid angle is dominated by the contribution from small angle separations, *i.e.*, pairs of points with $\sigma/M \ll 1$. Therefore, one expects that using Page's approximation for the Wightman function when computing the integral of the noise kernel over the whole solid angle would provide a fairly good approximation to the actual result. [i]

[i]Note, however, that one needs to choose the appropriate analytical continuation to

4.2. *Probing metric fluctuations*

Actually the contribution from small separation angles to the integral discussed in the previous paragraph not only dominates the integral but also gives a divergent result. This implies that the event horizon is no longer well defined as a three-dimensional null hypersurface when the effect of fluctuations is included since the amplitude of its fluctuations is infinite. In order to get a finite result, it is necessary to introduce some additional smearing along the transverse null direction. The final result will then depend on the characteristic size of the smearing functions employed. This means that in addition to seeking workable prescriptions one should also understand the physical meaning of the smearing introduced for this task. This can be achieved by analyzing how the propagation of test fields probes the metric fluctuations of the underlying geometry.

For this purpose a natural first step is to study the effect of those metric fluctuations on a bundle of null geodesics near the horizon. This is expected to provide a good characterization of the propagation of a test field whenever the geometric optics approximation is valid [49], which is certainly the case when studying Hawking radiation in the absence of fluctuations. In Ref. [49], where the particular form of the metric fluctuations was simply assumed rather than derived from first principles, the authors found no dramatic effect due to the fluctuations on the Hawking radiation associated with the test field propagating in the black hole spacetime. In contrast, following the stochastic gravity program, an estimate of the metric fluctuations exhibits a much more singular correlation function. Our preliminary analysis suggests a larger effect on the propagation of a bundle of null geodesics near the horizon, which would imply a substantial modification of the Hawking effect derived under a test field approximation.

However, it is likely that the large fluctuations found this way for a bundle of null geodesics sufficiently close to the horizon is signaling a breakdown of the geometric optics approximation rather than an actual drastic modification of Hawking's result. This assessment stems from a similar situation in flat space: Near a Rindler horizon large fluctuations arise from using the geometric optics approximation, but they are artifacts when examined in a more detailed quantum field theory calculation. If confirmed, this would mean that the propagation of a bundle of geodesics does not constitute an

go from the Euclidean Green function obtained by Page to the Lorentzian Wightman function. Page's scheme provides the Green function for Hartle-Hawking vacuum, but the Unruh vacuum should be considered for an evaporating black hole.

adequate probe of spacetime fluctuations in this context. Although computationally more involved, a more reliable way to extract physically meaningful information about the effect of metric fluctuations on the Hawking radiation is to consider the Wightman two-point function for the test field, which characterizes the response of a particle detector for that field, and analyze the radiative corrections when including the interaction with the quantized metric perturbations.

We thank Paul Anderson, Larry Ford, Valeri Frolov, Ted Jacobson, Don Marolf, Emil Mottola, Don Page, Renaud Parentani and Rafael Sorkin for their interest in and useful discussions on various aspects of this subject. This work is supported by NSF under Grant PHY-0601550. A. R. is also supported by LDRD funds from Los Alamos National Laboratory.

References

1. B. L. Hu, "Stochastic Gravity" Int. J. Theor. Phys. **38**, 2987 (1999) [gr-qc/9902064]
2. B. L. Hu and E. Verdaguer, "Stochastic gravity: Theory and Applications" in Living Reviews in Relativity 7 (2004) 3. [gr-qc/0307032] lrr-2004-3. "Stochastic gravity: A primer with applications", Class. Quant. Grav. 20 (2003) R1-R42 [gr-qc/0211090]. "Recent Advances in Stochastic Gravity: Theory and Issues" Erice Lectures May 2001, in *Advances in the Interplay between Quantum and Gravity Physics* edited by P. Bergmann and V. De Sabbata, (Kluwer, Dortrecht, 2002). gr-qc/0110092
3. E. Calzetta and B. L. Hu, Phys. Rev. D **35**, 495 (1987)
4. A. Campos and E. Verdaguer, Phys. Rev. D **49**, 1861 (1994)
5. B.-L. Hu. "Dissipation in Quantum-Fields and Semiclassical Gravity" *Physica* A **158**, 399 (1989).
6. E. Calzetta and B.-L. Hu, "Noise and fluctuations in semiclassical gravity" *Phys. Rev.* D **49**, 6636 (1994).
7. B.-L. Hu and A. Matacz, "Back reaction in semiclassical gravity: The Einstein-Langevin equation" *Phys. Rev.* D **51**, 1577 (1995).
8. B.-L. Hu and S. Sinha, "Fluctuation-dissipation relation for semiclassical cosmology" *Phys. Rev.* D **51**, 1587 (1995).
9. A. Campos and E. Verdaguer, Phys. Rev. D **53**, 1927 (1996)
10. B. L. Hu and A. Matacz, Phys. Rev. D **51**, 1577 (1995); B. L. Hu and S. Sinha, Phys. Rev. D **51**, 1587 (1995); A. Campos and E. Verdaguer, Phys. Rev. D **53**, 1927 (1996); F. C. Lombardo and F. D. Mazzitelli, Phys. Rev. D **55**, 3889 (1997).
11. R. Martin and E. Verdaguer, "On the semiclassical Einstein-Langevin equation" *Phys. Lett.* B **465**, 113 (1999).
12. R. Martin and E. Verdaguer, "Stochastic semiclassical gravity" *Phys. Rev.*

D **61**, 084008 (1999).

NOISE KERNEL

13. N. G. Phillips and B. L. Hu, "Noise Kernel in Stochastic Gravity and Stress Energy Bi-Tensor of Quantum Fields in Curved Spacetimes" Phys. Rev. D **63**, 104001 (2001).
14. N. G. Phillips and B. L. Hu, "Noise Kernel and Stress Energy Bi-Tensor of Quantum Fields in Hot Flat Space and the Schwarzschild Metric: Failure of the Gaussian Approximation" Phys. Rev. D67 (2003) 104002 [gr-qc/0209056]
15. Ardeshir Eftekharzadeh, B. L. Hu and A. Roura, "Noise Kernel of Quantum Fields near the Schwarzschild horizon" (in preparation)

BLACK HOLE IN THERMAL RADIANCE

16. G. Gibbons and M. J. Perry, Proc. Roy. Soc. Lon. A**358**, 467 (1978).
17. See work of Page, Penrose, Smolin and Sorkin on the thermodynamics of the black hole in a box, as quoted in references below.
18. S. W. Hawking and D. N. Page, Commun. Math. Phys. **87**, 577 (1983).
19. D. N. Page, Phys. Rev. **D25**, 1499 (1982).
20. B. F. Whiting and J. W. York, Jr. Phys. Rev. Lett. 61, 1336 (1988)
21. J. D. Brown and J. W. York Jr., "Microcanonical functional integral for the gravitational field" Phys. Rev. D 47, 1420-1431 (1993)
22. H. W. Braden, B. F. Whiting, and J. W. York, Jr., "Density of states for the gravitational field in black-hole topologies", Phys. Rev. D 36, 3614-3625 (1987); H. W. Braden, J. D. Brown, B. F. Whiting, and J. W. York, Jr. "Charged black hole in a grand canonical ensemble", Phys. Rev. D 42, 3376-3385 (1990).
23. B. L. Hu, G. Kang and A. Matacz, Int. J. Mod. Phys. A9, 991 (1994)
24. J. D. Bekenstein and A. Meisels, Phys. Rev. D 15, 2775 (1977)
25. M. Schiffer, Phys. Rev. D 48, 1652 (1993)
26. P. Panangaden and R. M. Wald, "Probability distribution for radiation from a black hole in the presence of incoming radiation" Phys. Rev. D 16, 929 (1977)
27. W. H. Zurek, Phys. Lett. 77A (1980) 399

BLACK HOLE BACK-REACTION FROM RADIANCE

28. P. Candelas and D. W. Sciama, Phys. Rev. Lett. **38**, 1372 (1977).
29. E. Mottola, Phys. Rev. D **33**, 2136 (1986).
30. J. W. York, Jr., Phys. Rev. D **28**, 2929 (1983); D **31**, 775 (1985); D **33**, 2092 (1986).
31. J. W. York, Jr. "Black hole in thermal equilibrium with a scalar field: The back-reaction" *Phys. Rev.* D **31**, 775 (1985).
32. P. R. Anderson, W. A. Hiscock and D. A. Samuel, Phys. Rev. D **51**, 4337 (1995).
33. W. A. Hiscock, S. L. Larson and P. A. Anderson, "Semiclassical effects in black hole interiors" gr-qc/9701004 Phys. Rev. D **56**, 3571 (1997).
34. S. Bose, L. Parker and Y. Peleg, Phys. Rev. Lett. 76, 861 (1996) Phys. Rev.

D53, 7089 (1996) and erarlier references therein. F. C. Lombardo, F. D. Mazzitelli, Phys. Rev. D58 (1998) 024009; F. C. Lombardo, F. D. Mazzitelli and J. G. Russo, Phys. Rev. D59 (1999) 064007.

35. A. Campos and B. L. Hu, Phys. Rev. D **58** (1998) 125021; Int. J. Theor. Phys. **38** (1999) 1253-1271.
36. S. Sinha, A. Raval and B. L. Hu, "Black Hole Fluctuations and Back-reaction in Stochastic Gravity", in Foundations of Physics 33 (2003) 37-64 [gr-qc/0210013]
37. J. M. Bardeen, Phys. Rev. Lett. 46, 382 (1981).
38. P. Hajicek and W. Israel, Phys. Lett. 80A, 9 (1980).
39. W. A. Hiscock, "Models of evaporating black holes. I" Phys. Rev. D 23, 2813-2822 (1981) "II. Effects of the outgoing created radiation" Phys. Rev. D 23, 2823-2827 (1981)
40. S. Massar, Phys. Rev. **D 52**, 5857 (1995).

BLACK HOLE HORIZON FLUCTUATIONS

41. J. D. Bekenstein, in *Quantum Theory of Gravity*, edited by S. M. Christensen (Adam Hilger, Bristol, 1984).
42. A. Casher, F. Englert, N. Itzhaki, S. Massar, R. Parentani, "Black Hole Fluctuations" Nucl. Phys. B484,419(1997) [hep-th/9606016]
43. R. D. Sorkin, "How Wrinkled is the Surface of the Black Hole?" in *Proceedings of the First Australasian Conference on General Relativity and Gravitation* February 1996, Adelaide, Australia, edited by D. Wiltshire, pp. 163-174 (University of Adelaide, 1996) [gr-qc/9701056]; R. D. Sorkin and D. Sudarsky, Class. Quantum Grav. **16**, 3835 (1999). [gr-qc/9701056]
44. D. Marolf, "On the quantum width of a black hole horizon"[hep-th/0312059] Adriatic 2003 conference.
45. B. L. Hu, A. Raval and S. Sinha, "Notes on Black Hole Fluctuations and Back-reaction" in *Black Holes, Gravitational Radiation and the Universe: Essays in honor of C. V. Vishveshwara* eds. B. Iyer and B. Bhawal (Kluwer Academic Publishers, Dordrecht, 1998); gr-qc/9901010.
46. L. H. Ford and N. F. Svaiter, Phys. Rev. D56, 2226(1997).
47. C. H. Wu and L.H. Ford, Phys. Rev. D60, 104013(1999)
48. S. Massar and R. Parentani, "How the change in horizon area drives black hole evaporation," Nucl. Phys. B575, 353(2000)
49. C. Barrabes, V. Frolov and R. Parentani, Phys. Rev. D **59**, 124010 (1999); **62**, 044020 (2000).
50. R. Parentani, Int. J. Theor. Phys. **41**, 2175 (2002).
51. B. L. Hu and A. Roura, "Fluctuations of an evaporating black hole from back reaction of its Hawking radiation: Questioning a premise in earlier work" Int. J. Theor. Phys. (2006) [gr-qc/0601088];
52. B. L. Hu and A. Roura, (in preparation for Phys. Rev. D)

Classical Gravity

Growth of primordial black holes *

Tomohiro Harada
Department of Physics, Kyoto University,
Kyoto 606-8502, Japan
E-mail: harada@scphys.kyoto-u.ac.jp

Primordial black holes have important observational implications through Hawking evaporation and gravitational radiation as well as being a candidate for cold dark matter. Those black holes are assumed to have formed in the early universe typically with the mass scale contained within the Hubble horizon at the formation epoch and subsequently accreted mass surrounding them. Numerical relativity simulation shows that primordial black holes of different masses do not accrete much, which contrasts with a simplistic Newtonian argument. We see that primordial black holes larger than the 'super-horizon' primordial black holes have decreasing energy and worm-hole like struture, suggesting the formation through quamtum processes.

1. Introduction

Primordial black holes (PBHs) may have formed in the early universe [1]. Those black holes may contribute to current gamma-ray and cosmic ray backgrounds through Hawking evaporation. They may also contribute to the cosmic density and behave as cold dark matter. They could be a promising target for ground-based interferometric gravitational wave detectors [2,3]. Thus we can obtain information of the early universe. In particular, we can constrain the probability of PBH formation in the early universe [4]. See [5] for a recent review of theoretical and observational background and development in the studies of PBHs.

PBHs are usually assumed to have formed with the mass scale $M_{\rm h,f} \simeq G^{-1}c^3 t_{\rm f}$ which was contained within the Hubble horizon at the formation epoch, where G, c and $t_{\rm f}$ are the gravitational constant, the light speed

*This article is based on the collaboration with B. J. Carr.

and the formation time from big bang, respectively. This is based on the argument on the Jeans scale, gravitational radius and separate universe condition [4,6]. The typical mass scale of the formed PBHs is still the horizon mass scale at the formation epoch, i.e.,

$$M_{\rm PBH,f} \simeq M_{\rm h,f} \simeq \frac{c^3 t_{\rm f}}{G} \simeq 1 M_{\odot} \left(\frac{T_{\rm f}}{100 \text{ MeV}} \right)^{-2} . \tag{1}$$

2. Newtonian argument of PBH growth

The accretion onto a PBH could change the mass scale of PBHs in principle. If we assume spherically symmetric and quasi-stationary flow onto a PBH, we can estimate the mass accretion rate of the black hole as

$$\frac{dM}{dt} = 4\pi \alpha r_{\rm A}^2 v_{\rm s} \rho_{\infty} , \tag{2}$$

where $r_{\rm A} = GM/v_{\rm s}^2$ is the accretion radius, $v_{\rm s}$ is the sound speed, ρ_{∞} is the density at infinity and α is a constant of order unity. To apply the above equation for the growth of PBHs in the early universe, we assume that ρ_{∞} is given by the density of the background Friedmann universe and $v_{\rm s}$ is the order of the speed of light. Then the above equation is integrated to give [6,7]

$$M = \frac{At}{1 + \frac{t}{t_{\rm f}} \left(\frac{A t_{\rm f}}{M_{\rm f}} - 1 \right)} , \tag{3}$$

where $A \simeq c^3/G$ is a constant and $M_{\rm f}$ is the PBH mass at the time $t_{\rm f}$ of formation. Figure 1 shows three categories of solutions expressed by Eq. (3). In this argument, the effects of cosmological expansion are neglected. Therefore, the above analysis for PBHs much smaller than the cosmological horizon scale is expected to be valid. On the other hand, for PBHs as large as or larger than the cosmological horizon, which would be naturally realised just after the formation, the above analysis is suspect.

3. Numerical relativity of a scalar field

Since several independent observations strongly suggest the accelerated expansion of the presenr Universe, the so-called dark energy which enables the acceleration have attracted much attention. One of the simplest and most natural models for dark energy is a scalar field slowly rolling down on the potential, which is called quintessence. Recently, Bean and Magueijo [8] applied a quasi-Newtonian argument for the accretion of a quintessence field onto a PBH, which leads to solutions given by Eq. (3), and claimed that

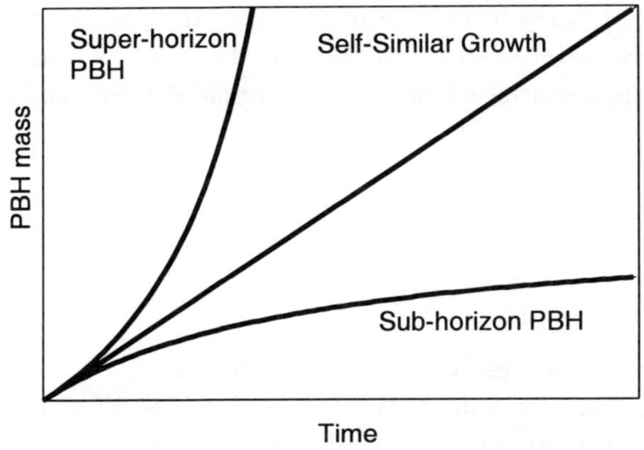

Fig. 1. PBH mass growth based on the Newtonian argument. Three categories of solutions, sub-horizon, self-similar and super-horizon, are shown.

PBHs of inflation origin could be the seeds for supermassive black holes, based on a quantitative analysis. Actually, if we assume stationary flow and neglect cosmic expansion, we necessarily reach the solutions (3) whether we take relativistic effects of accretion into account.

To get insight into the growth of horizon-scale PBHs in the context of the quintessence/scalar field cosmology, we implement fully general relativistic numerical simulation of the growth of PBHs in a universe containing a massless scalar field. The line element in spherically symmetric spacetimes is given by

$$ds^2 = -a^2(u,v)dudv + r^2(u,v)(d\theta^2 + \sin^2\theta d\phi^2), \tag{4}$$

in the double-null coordinates and henceforth we use the geometrised units $G = c = 1$. Governing equations in this coordinate system are given in [9] explicitly. For the present problem, this scheme is advantageous because it has no apparent coordinate singularity and it also fits the characteristics of the propagation of the scalar field. The details of the numerical scheme and implementation are described in [9,10].

In the double-null formulation, it is most natural to provide initial data on the null hypersurfaces $u = u_0$ and $v = v_0$ and solve a diamond region $u_0 < u < u_1$ and $v_0 < v < v_1$. As for initial data we adopt the simplest model, in which the Schwarzschild region is surrounded by the flat Friedmann background. To avoid the discontinuity at the matching surface, we set the smoothing region to retain the numerical accuracy.

Since we have null infinity in the flat Friedmann spacetime, we can define an event horizon. On the other hand, the notion of trapping horizons [11], which is very similar to apparent horizons, is also useful. This is defined as a hypersurface foliated by marginal surfaces and the definition is local. Although the event horizon and the trapping horizon coincide for the Schwarzschild black hole, they are different from each other for general dynamical spacetimes.

4. Numerical results

Here we present the results for three models. See [10,12] for the details of the chosen parameter values. We define the mass of the black hole using the Misner-Sharp mass [13] on the event horizon. Figure 2 shows the time evolution of the PBH mass and the mass contained within the cosmological apparent horizon denoted as 'POTH' as the abbreviation of the past outer trapping horizon. We can see that sub-horizon PBHs do not accrete much, the accretion onto horizon-scale PBHs is suppressed and super-horizon PBHs even decrease their masses.

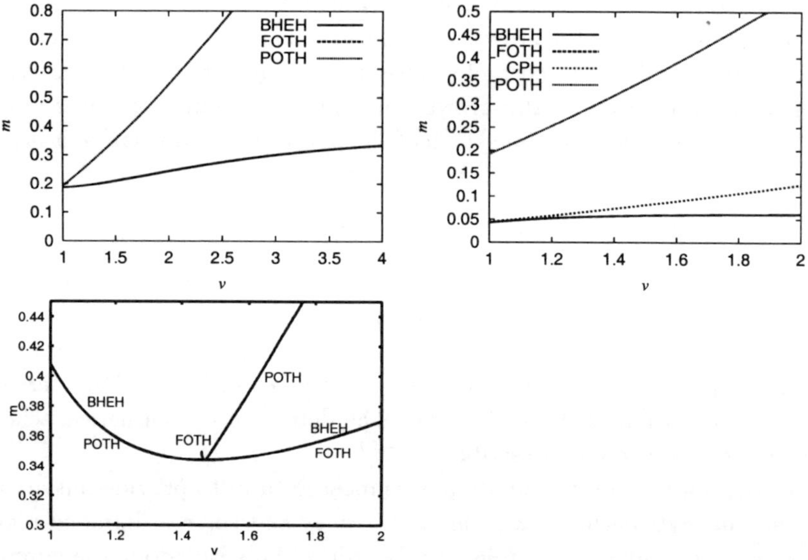

Fig. 2. The masses of the black hole event horizon and the cosmological apparent horizon as functions of v. The upper left, upper right and lower left panels show the results for the initial mass ratio of the PBH to the cosmological apparent horizon to be 0.22, 0.97 and 2.12, respectively.

The evolution of PBHs of different masses is understood in terms of the mass accretion equation [10,12]:

$$m_{\mathrm{BHEH},v} = \frac{8\pi r^2}{a^2}(T_{uv}r_{,v} - T_{vv}r_{,u}), \tag{5}$$

where $T_{\mu\nu}$ is the stress-energy tensor and the time derivative is taken with respect to v because the event horizon is given by $u = \text{const}$. Since $T_{uv} = 0$ and $T_{vv} \geq 0$ for a massless scalar field, the sign of the mass growth rate is governed by $r_{,u}$ on the event horizon. $r_{,u} = -1/2$ for the Minkowski spacetime with $a = 1$. This corresponds to the expansion of ingoing null geodesic congruence on the event horizon. This is negative inside the cosmological horizon, but zero on and positive outside it. Figure 3 schematically summarises the results of general relativistic numerical simulations in contrast to Fig. 1.

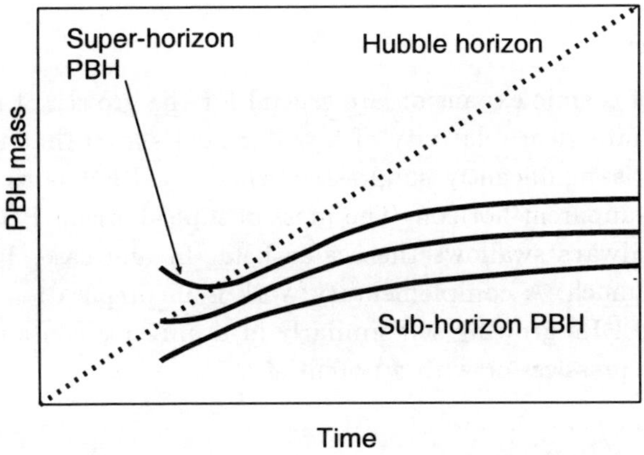

Fig. 3. Schematic figure showing the mass growth of PBHs in a universe containing a massless scalar field, based on general relativistic numerical simulations.

For the models where the initial event horizon is past trapped, we can show that both first outgoing null ray $u = u_0$ and ingoing null ray $v = v_0$ reach infinity. This means we have two distinct null infinities but no regular centre and this results cannot be embedded into the standard diagram of the PBH. Figure 4 shows a possible causal struture, which is a very similar structure to the cosmological wormhole as a result of first-order vacuum phase transition [14]. This implies the quantum origin of super-horizon PBHs. On the ther hand, we should note that our matter model is without a potential term and all dominant, strong and weak energy conditions are always satisfied.

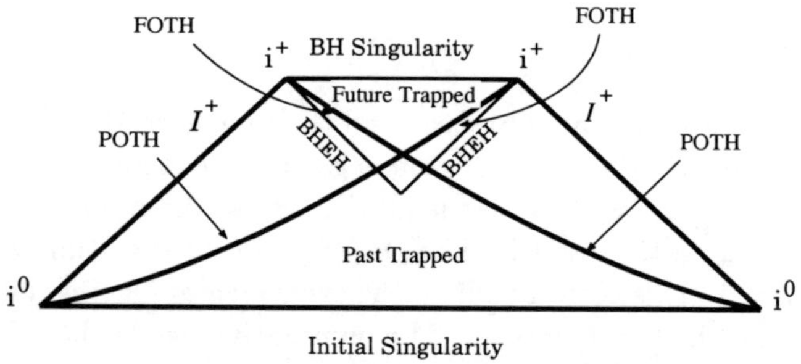

Fig. 4. A possible causal struture of a super-horizon PBH in a scalar field universe. The spatial section has a wormhole structure, although it is not traversable.

5. Summary

The effects of cosmic expansion are crucial for the growth of horizon-scale PBHs. The numerical relativity of a scalar field shows that the accretion onto a PBH is significantly suppressed when the PBH is as large as the cosmological apparent horizon. The mass of super-horizon PBHs decreases although it always swallows the scalar field. In any case, PBHs do not accrete very much. A complementary work is in preparation on the non-existence of PBHs growing self-similarly in a universe containing a scalar field whether massless or with a potential.

References

1. S. W. Hawking, *Phys. Rev. Lett.* **26**, 1344 (1971).
2. T. Nakamura, M. Sasaki, T. Tanaka and K. S. Thorne, *Astrophys. J.* **487**, L139 (1997).
3. B. Abbott, *Phys. Rev.* D**72**, 082002 (2005).
4. B. J. Carr, *Astrophys. J.* **201**, 1 (1975).
5. B. J. Carr, to appear in *Proceedings of Inflating Horizon of Particle Astrophysics and Cosmology* (*Preprint* astro-ph/0511743) (2005).
6. T. Harada and B. J. Carr *Phys. Rev.* D**71**, 104009 (2005).
7. Y. B. Zel'dovich and I. D. Novikov, *Sov. Astron.* **10**, 602 (1967).
8. R. Bean and J. Magueijo, *Phys. Rev.* D**66**, 063505 (2002).
9. R. S. Hamadé and J. M. Stewart, *Class. Quantum Grav.* **13**, 497 (1996).
10. T. Harada and B. J. Carr, *Phys. Rev.* D**71**, 104010 (2005).
11. S. A. Hayward, *Preprint* gr-qc/9303006 (1993).
12. T. Harada and B. J. Carr, *Phys. Rev.* D**72**, 044021 (2005).

13. C. W. Misner and D. H. Sharp, *Phys. Rev.* **136**, B571 (1964).
14. K. Sato, S. Sasaki, H. Kodama and K. Maeda, *Prog. Theor. Phys.* **65**, 1443 (1981).

De Sitter invariant special relativity

C.-G. Huang[1] and H.-Y. Guo[2]
[1] *Institute of High Energy Physics, Chinese Academy of Sciences*
P.O. Box 918-4, Beijing 100049
E-mail: huangcg@ihep.ac.cn
[2] *Institute of Theoretical Physics, Chinese Academy of Sciences*
P.O.Box 2735, Beijing 100080
E-mail: hyguo@itp.ac.cn

In de Sitter space, de Sitter invariant special relativity can be established more or less parallel to the Einstein special relativity, based on the principle of special relativity and the postulate of two-universal constants. There are two kinds of simultaneity. One is the Beltrami-time simultaneity. By using the simultaneity, a new kind of inertial motion and a series of classical observables can be defined. In addition, the temperature of the horizon of de Sitter space for a set of inertial observers should be zero! Therefore, there is no need to explain the statistical origin of the entropy for the horizon of de Sitter space. Another is the proper-time simultaneity. With the proper-time simultaneity, the metric takes the Robertson-Walker-like form, which shows that the space has positive spatial curvature of order Λ. This has already been shown by the CMB power spectrum from WMAP and should be further confirmed in future. The existence of the two kinds of simultaneity also makes possible to explain the cosmic background or origin of inertial motion. Further, in de Sitter invariant special relativity, dynamics for a particle, including the pseudo-Hamiltonian mechanics, can be established. The non-relativistic limit of de Sitter invariant special relativity gives rise to the Newton-Hooke mechanics. The possibility of a test of de Sitter invariant special relativity is also studied tentatively. Finally, a kind of the doubly special relativity may be viewed as the de Sitter invariant special relativity in energy-momentum space.

1. Introduction

The astronomical observations show that the component of dark energy in our universe is positive (i.e. $\Omega_\Lambda > 0$)[1,2], that the ratio of the total energy to the critical energy is slightly greater than 1 ($\Omega_{\rm tot} = 1.02 \pm 0.02$)[2], and that the equation of state for dark energy is consistent with a cosmological constant at 95% confidence[3]. These imply that our universe is probably asymptotically de Sitter (dS).

It is well known that dS-space is empty and of constant curvature without intrinsic singularity but there is a cosmological horizon surrounding an observer at the spatial origin. In ordinary approach, one can define, on the horizon, the surface gravity $\kappa_s = 1/R = \sqrt{\Lambda/3}$, Hawking temperature $T = (2\pi R)^{-1}$, and entropy $S = \pi R^2$ and so on. These, however, lead to many puzzles. For example, why is dS like a black hole[4]? How to explain the statistical origin of the entropy of this horizon[4]? How to define an observable in dS space[4]? Energy is not definitely positive[5]. The finite entropy implies the finite dimension of the dS Hilbert space which, in turn, requires finite degree of freedom in dS-space[5]. One cannot in dS-space make sense in a precise way of what we usually regard as local particle physics quantities[5]. Whether is the method of imaginary-time Green functions always valid? $\cdots$

In fact, there is another viewpoint to understand dS-space in which the above puzzles may be eliminated more or less. It is the viewpoint of de Sitter-invariant special relativity ($SR_{c,R}$). In this talk, we shall review its development briefly.

2. De Sitter-invariant special relativity

In dS-space, there is a special coordinate system, called Beltrami coordinate system. In the system, the metric of dS-space takes the form

$$ds^2 = \left(\sigma^{-1}(x)\eta_{\mu\nu} + R^{-2}\sigma^{-2}(x)\eta_{\mu\lambda}\eta_{\nu\sigma}x^\lambda x^\sigma\right) dx^\mu dx^\nu, \tag{1}$$

$$\eta_{\mu\nu} = \mathrm{diag}(1,-1,-1,-1),\ \sigma(x) = \sigma(x,x) = 1 - R^{-2}\eta_{\mu\nu}x^\mu x^\nu > 0. \tag{2}$$

It is the Beltrami metric, which was first introduced in studying Lobachevski geometry in 19$^{\rm th}$ century[6], and then used by de Sitter in the debate with Einstein on "relative inertia"[7] and mentioned in Euclidean signature by Pauli in his book[8]. To construct a quantized space-time model in 1947, Snyder used the projective geometry approach to dS-space of momenta[9], which is also closely related to Beltrami metric.

In the system, the timelike and null geodesics can be integrated out

$$x^\mu = v^\mu w + b^\mu \tag{3}$$

so that

$$dx^i/dt = v^i, \qquad d^2x^i/dt^2 = 0, \qquad (x^0 = ct \text{ and } i = 1,2,3). \tag{4}$$

The metric (1) and the linearity of rectilinear uniform-velocity motion (3) or (4) are *invariant* under fractional linear transformations of group $SO(1,4)$

$$\begin{aligned} x^\mu \to \tilde{x}^\mu &= \pm\sigma^{1/2}(b)\sigma^{-1}(b,x)(x^\nu - b^\nu)D^\mu_\nu, \\ D^\mu_\nu &= L^\mu_\nu + R^{-2}\eta_{\nu\lambda}b^\lambda b^\sigma(\sigma(b)+\sigma^{1/2}(b))^{-1}L^\mu_\sigma, \quad L \in SO(1,3). \end{aligned} \tag{5}$$

The facts were first noted by Lu in 1970[10].

The motion which can be written in the above linear form is referred to as an *inertial motion*. An inertial motion moves along a straight line and at constant-component-coordinate velocity ($dx^i/dt = v^i = const.$), and has vanishing 3-coordinate-acceleration ($d^2x^i/dt^2 = 0$) and 4-acceleration ($D^2x^\mu/ds^2 = 0$).

A set of inertial observers at rest with respect to each other constitute an *inertial reference frame*. A Beltrami coordinate system is referred to as an *inertial coordinate system*. It plays the role of Minkowski coordinate system in Special Relativity (SR_c). In the same inertial coordinate system the two static ($dx^i/dt = 0$) observers satisfy, in addition to $d(\delta x^i)/dt = 0$,

$$d^2(\delta x^i)/dt^2 = 0. \tag{6}$$

It is remarkable that (6) can be obtained from geodesic deviation equation

$$D^2(\delta x^\mu)/ds^2 + R^\mu{}_{\lambda\nu\sigma}\delta x^\nu(dx^\lambda/ds)(dx^\sigma/ds) = 0. \tag{7}$$

In view of the invariance of metric (1) and of the linearity of rectilinear uniform-velocity motion (3) or (4) the *Principle of Relativity* (PoR) may be reiterated as: there exist a set of inertial reference frames, in which free particles and light signals move with uniform velocities along straight lines, the laws of nature without gravity are invariant under the transformations among them. Further, the postulate of constancy of speed of light may be generalized as: there exist two invariant universal constants, one being with dimension of velocity (c), and another being of length (R). It is known as *Postulate of Invariant speed and length* (PoI). Based on PoR and PoI, one may set up $SR_{c,R}$[11–13,15]. It can be shown that SR_c is the limit of $SR_{c,R}$ at $R \to \infty$.

3. Two kinds of simultaneity

It is well known that in order to make physical measurements, one needs first define simultaneity. In $SR_{c,R}$, there are two kinds of simultaneity.

The first one is the Beltrami-time simultaneity. By the Beltrami-time simultaneity, two events P and Q being simultaneous or not is determined by their Beltrami time equal to each other or not. Only with respect to the definition of Beltrami-time simultaneity, inertial motion and inertial frame in dS-space make sense. The Beltrami-time simultaneity is the generalization of the simultaneity of Minkowski *coordinate* time in SR_c. Note that in SR_c, Minkowski coordinates have physical meaning: the difference in time coordinate stands for the time interval.

In dS-space with a Beltrami metric, 4-momentum pseudo-vector

$$p^\mu := m_\Lambda \sigma^{-1}(x) dx^\mu/ds = m_\Lambda \sigma^{-1}(x)(cdt/ds, dx^i/ds) = (E/c, P^i) \tag{8}$$

is conserved along straight line, i.e.

$$dp^\mu/ds = 0 \Longleftrightarrow dE/dt = 0, dP^i/dt = 0, \text{ for } d(\sigma^{-1}(x)dx^\mu/ds)/dt = 0, \tag{9}$$

where E is energy, P^i 3-momentum, and m_Λ rest mass of a particle in dS-space. In addition, 4-d angular momentum

$$L^{\mu\nu} := x^\mu p^\nu - x^\nu p^\mu = \begin{pmatrix} 0 & cK^1 & cK^2 & cK^3 \\ -cK^1 & 0 & J^3 & -J^2 \\ -cK^2 & -J^3 & 0 & J^1 \\ -cK^3 & J^2 & -J^1 & 0 \end{pmatrix} \tag{10}$$

is also conserved along straight line, i.e.

$$dL^{\mu\nu}/ds = 0 \Longleftrightarrow dJ^i/dt = 0, \ dK^i/dt = 0, \tag{11}$$

where J^i is 3-d angular momentum and K^i boost. They constitute a set of classical "basic observables" and satisfy the generalized Einstein's relation

$$E^2 = m_\Lambda^2 c^4 + P^2 c^2 + c^2 R^{-2} J^2 - c^4 R^{-2} K^2. \tag{12}$$

Under the Wick rotation, the Beltrami metric for dS-space becomes the one for 4-d Riemann sphere,

$$\begin{aligned} ds_E^2 &= \left(\sigma_E^{-1}(x)\delta_{\mu\nu} - R^{-2}\sigma_E^{-2}(x)\delta_{\mu\lambda}\delta_{\nu\sigma}x^\lambda x^\sigma\right) dx^\mu dx^\nu \\ \sigma_E(x) &= 1 + R^{-2}\delta_{\mu\nu}x^\mu x^\nu. \end{aligned} \tag{13}$$

It should be noted that the imaginary Beltrami coordinate time has no period. Also the relation between imaginary Beltrami time and the imaginary cosmic time in the static metric of dS-space,

$$\tau_B = R\tan(c\tau_s/R), \tag{14}$$

shows that the imaginary cosmic time has a period $2\pi R/c$ while the imaginary Beltrami time has no period. Further, *the 'surface gravity' at horizon is zero* in the viewpoint of $SR_{c,R}$. This is because for static observers in an inertial frame,

$$dx^i/dt = 0, \quad D^2x^\mu/ds^2 = 0, \quad d^2x^i/dt^2 = 0, \quad d^2(\delta x^i)/dt^2 = 0. \quad (15)$$

Thus, no force is needed for an inertial observer to hold a test mass in the place where $x^i = const.$. Namely,

$$\kappa_B = 0. \quad (16)$$

All these facts show that the temperature at horizon should be zero in the viewpoint of $SR_{c,R}$. Therefore, there is no need to explain the statistical origin of entropy of the dS horizon. These facts also show that the T, S in other dS-spaces do *not* arise from gravity but non-inertial motions.

The second kind of simultaneity is the *proper-time simultaneity*. It is the one with respect to the proper time $t_{\rm p}$ of the clock rest at the spatial origin of a Beltrami coordinate system. Two events $P(p^\mu)$ and $Q(q^\mu)$ are said to be proper-time simultaneous if and only if

$$p^0\sigma^{-1/2}(p) = q^0\sigma^{-1/2}(q). \quad (17)$$

The proper time scale is given by

$$t_{\rm p} = R\sinh^{-1}(R^{-1}\sigma^{-1/2}(x)x^0). \quad (18)$$

The proper-time simultaneity is closely linked with the cosmological principle (CP). If $t_{\rm p}$ is taken as a "cosmic time", the Beltrami metric becomes a RW-like metric with a positive spatial curvature

$$ds^2 = dt_{\rm p}^2 - \cosh^2(R^{-1}t_{\rm p})dl^2_{\Sigma_{t_{\rm p}=0}},$$

where $dl^2_{\Sigma_{t_{\rm p}=0}}$ is the Beltrami metric of 3-d Riemann sphere at $t_{\rm p} = 0$. This shows that the 3-d cosmic space is S^3 rather than flat. The deviation from the flatness is of order Λ. As our universe is probably asymptotically dS, the 3-d real cosmos should also be so. The property seems more or less already indicated by the CMB power spectrum from WMAP[2] and should be further checked in future.

The two definitions of simultaneity make sense in two different kinds of measurements. The first one is concerned with the measurements in a laboratory and is related to the PoR of $SR_{c,R}$, while the second one is concerned with cosmological effects and is related to the CP. The relation between x^0 and $t_{\rm p}$ links the PoR and CP.

The origin of the law of inertia and the origin of inertial mass in both Newtonian Mechanics and SR_c are *two* related but *different* long-standing problems. The former means the origin of the inertial motions and inertial frames, which is referred to as origin of inertia in brief. Based on the two kinds of simultaneity, we may proposed the *postulate on the origin of inertia* as follows. In dS-space there should exist the inertial motions and inertial systems as the maximal symmetry indicated, and their origin should be displayed by the cosmic background with the cosmological constant[15,16].

Finally, we point out that it should be distinguished the origin of inertia from the origin of *local* inertia in General Relativity, — the local inertial systems in local Minkowski space and local inertial motions along geodesics, respectively.

4. Dynamics of a massive particle

The 2nd law of a relativistic massive particle has the form[15]

$$dp^\mu/ds = f^\mu, \qquad \text{and} \qquad dL^{\mu\nu}/ds = M^{\mu\nu}. \tag{19}$$

with $M^{\mu\nu} = x^\mu f^\nu - x^\nu f^\mu$. They are invariant under fractional linear transformation between Beltrami coordinate systems. Note that f^μ is not a 4-vector but a 4-d pseudo-vector.

In Ref. 17, the Hamiltonian formalism for a free particle in dS-space has been analyzed in a standard way, in which the obtained Hamiltonian is not the conserved energy defined in Eq.(8). Here, we show that it is possible to construct a pseudo-Hamiltonian formalism so that the pseudo-Hamiltonian is the conserved energy defined in Eq.(8)[18]. To do so, we begin with the pseudo-Lagrangian for a free massive particle

$$\tilde{L}_t = -mc(c^2 - \delta_{ij}\acute{x}^i\acute{x}^j - (1/2)R^2\eta_{\mu\nu}\eta_{\lambda\sigma}\omega^{\mu\lambda}\omega^{\nu\sigma})^{1/2}, \tag{20}$$

where $\acute{x}^\mu := dx^\mu/dt$, $\omega^{\mu\lambda}$ is the 4-d angular velocity with respect to the origin of a Beltrami coordinate system, which is defined by

$$\omega^{\mu\lambda} := R^{-2}(x^\mu\acute{x}^\lambda - x^\lambda\acute{x}^\mu) =: \acute{\varphi}^{\mu\lambda}. \tag{21}$$

In the following, $\acute{x}^i$ and $\omega^{\mu\lambda}$ are treated as independent variables! The pseudo-canonical momenta are

$$\begin{aligned} p_i &:= \partial(\tilde{L}_t)/\partial\acute{x}^i = P_i, \;\; J^{ij} := \partial(\tilde{L}_t)/\partial\omega^{ij} = (x^iP^j - x^jP^i), \\ K_i &:= \partial(\tilde{L}_t)/\delta\omega^{0i} = \delta_{ij}x^j - tP_i. \end{aligned} \tag{22}$$

The pseudo-Hamiltonian is defined by

$$\begin{aligned} H &= \acute{x}^ip_i + \omega^{ij}J_{ij}/2 + \omega^{0i}K_i - \tilde{L}_t \\ &= \left(m^2 + P^2 + R^{-2}J^2 - R^{-2}K^2\right)^{1/2} = E. \end{aligned} \tag{23}$$

Under the Poisson bracket

$$\{f,g\} := \frac{\partial f}{\partial x^i}\frac{\partial g}{\partial p_i} - \frac{\partial f}{\partial p_i}\frac{\partial g}{\partial x^i} + \frac{\partial f}{\partial \varphi^{\mu\lambda}}\frac{\partial g}{\partial J_{\mu\lambda}} - \frac{\partial f}{\partial J_{\mu\lambda}}\frac{\partial g}{\partial \varphi^{\mu\lambda}}. \tag{24}$$

the Hamiltonian equation of motion now becomes

$$\begin{aligned} \acute{x}^i &= \{x^i, H\} = P^i/E, & \acute{p}_i &= \{p_i, H\} = 0, \\ \acute{\varphi}^{ij} &= \{\varphi_{ij}, H\} = J^{ij}/(R^2E), & \acute{J}_{ij} &= \{J_{ij}, H\} = 0, \\ \acute{\varphi}^{0i} &= \{\varphi_{0i}, H\} = K^i/(R^2E), & \acute{K}_i &= \{K_i, H\} = 0. \end{aligned} \tag{25}$$

Thus,

$$\acute{H} = \{H, H\} = 0. \tag{26}$$

Namely, the energy for free massive particle, defined in $SR_{c,R}$, is conserved.

In the Newton-Hooke limit, $c \to \infty$, $R \to \infty$ but $c/R \to \nu$ is finite. Eqs.(20)–(26) reduces to their correspondences in the Newton-Hooke mechanics. In particular, when $F^i \neq 0$, the equation of motion becomes

$$dP^i/dt = F^i. \tag{27}$$

For the force which is independent of $\acute{x}^i$, ϕ^{0i}, ω^{0i}, we have

$$F^i = (1 - \nu^2 t^2)^{-1}\partial_i V. \tag{28}$$

This is consistent with the results in Ref.19.

At last, a few words about the experimental test on $SR_{c,R}$ should be mentioned. Since $c^2R^{-2} \sim 10^{-35}s^{-2}$ is extremely small, it seems impossible to prove or disprove $SR_{c,R}$ in local experiments. Therefore, to test $SR_{c,R}$, one should consider large-scale proceesses. Ref.20 gives a tentative study in this direction. It shows that GZK cutoff[21] cannot be used to prove or disprove $SR_{c,R}$.

5. Concluding remarks

$SR_{c,R}$ can be set up in dS-space based on PoR and PoI. The relativistic kinematics and dynamics in the framework of $SR_{c,R}$ can be established. In addition, we may establish the field theory in the framework of $SR_{c,R}$. The formulas in $SR_{c,R}$ can be written in the form in 5-d embedding space. In that case, an inertial motion becomes a uniform "great-circular" motion. The 4-d momentum and 4-d angular momentum are incorporated in 5-d angular momentum. The 4-d force and the moment of a 4-d force are incorporated in the moment of a 5-d force. All above results reduce to the ones in SR_c when $R \to \infty$ and to the ones in Newton-Hooke mechanics in the non-relativistic limit. Of course, to verify the theory experiments and/or observations are needed.

Finally, a kind of the doubly special relativity may be viewed as the $SR_{c,R}$ in energy-momentum space, in which the two fundamental constants are group velocity c and κ near the Planck energy $E_{\rm pl}$[22].

Acknowledgments

The author would like to thank Professors Z. Chang, Q.-K. Lu, Z. Xu and M.-L. Yan as well as Drs. Y. Tian and B. Zhou for successful collaborations and/or valuable discussions. This work is partly supported by NSFC under Grant Nos. 90403023, 90503002 and 10375087.

References

1. A. G. Riess, *et al*, *Astro. J.* **116** (1998) 1009; S. Perlmutter, *et al*, *Astrophys. J.* **517** (1999) 565.
2. C. L. Bennett, *et al*, *Astrophys. J. (Suppl.)* **148** (2003) 1; D. N. Spergel, *et al*, *Astrophys. J. (Suppl.)* **148** (2003) 175; M. Tegmark , *et al*, *Phys. Rev. D* **69** (2004) 103501.
3. Y. Wang and M. Tegmark, *Phys. Rev. Lett.* **92** (2004) 241302; Y. Wang and P. Mukherjee, *Astrophys. J.* **606** (2004) 654.
4. A. Strominger, Talk at the String Satellite Conference to ICM 2002, August, 2002, Beijing.
5. E. Witten, hep-th/0106109.
6. E. Beltrami, Opere Matematiche Vol. **I** 1868 pp.374-405.
7. W. de Sitter, *Month. Not. Roy. Astr. Soc.* **78** (1917) 3.
8. W. Pauli, Theory of Relativity, Pergamon Press, London, 1958.
9. H.S. Snyder, *Phys. Rev.* **71** (1947) 38.
10. K. H. Look (Qi-Keng Lu), Why the Minkowski metric must be used? (1970) unpublished; K. H. Look, C. L. Tsou (Z. L. Zou) & H. Y. Kuo (H.-Y. Guo), *Acta Physica Sinica* **23** (1974) 225; *Nature* (Shanghai, Suppl.) **1** (1980) 97 (in Chinese); H.-Y. Guo, in *Proc. 2nd Marcel Grossmann Meet. on GR*, ed. by R. Ruffini, (North-Holland Pub., 1982) 801; *Nucl. Phys. B* (Proc. Suppl.) **6** (1989) 381.
11. H.-Y. Guo, *et al*, *Mod. Phys. Lett.* **A19** (2004) 1701.
12. H.-Y. Guo, *et al*, *Phys. Lett.* **A331** (2004) 1.
13. H.-Y. Guo, *et al*, *Chin. Phys. Lett.* **22** (2005) 2477.
14. H.-Y. Guo, C.-G. Huang and B. Zhou, *Europhys. Lett.* **72** (2005), 1045.
15. H.-Y. Guo, *et al*, *Acta Phys. Sin.* **54** (2005) 2494 (in Chinese).
16. H.-Y. Guo, C.-G. Huang, Y. Tian, Z. Xu and B. Zhou, hep-th/0405137.
17. M.-L. Yan, *et al*, hep-th/0512319.
18. C.-G. Huang *et. al.*, In preparation.
19. C.-G. Huang, *et al*, hep-th/0403013; Y. Tian, *et al*, *Phys. Rev.* **D**71 (2005) 044030.

20. Z. Chang, S.-X. Chen and C.-G. Huang, *Chin. Phys. Lett.* **22** (2005) 791.
21. K. Greisen, *Phys. Rev. Lett.* **16** (1966) 748; G. T. Zatsepin and V. A. Kuz'min, *JETP Lett.* **4** (1966) 78.
22. See, e.g. J. Kowalski-Glikman, *Phys. Lett.* **B547** (2002) 291; J. Kowalski-Glikman and S. Nowak, *Class. Quant. Grav.* **20** (2003) 4799.

Scalar Field Contribution to Rotating Black Hole Entropy – Semiclassical Method –

M. KENMOKU

Department of Physics, Nara Women's University,
Nara 630-8506, Japan

E-mail: kenmoku@asuka.phys.nara-wu.ac.jp

Scalar field contribution to entropy is studied in arbitrary D-dimensional one parameter rotating spacetime by semiclassical method. By introducing the zenithal angle dependent cutoff parameter, the generalized area law is derived. The non-rotating limit can be taken smoothly and it yields known results. The derived area law is applied to the Kerr-Newman black hole in (3+1) dimension. This work is the collaboration with K. Ishimoto, K.K. Nandi and K. Shigemoto of the paper [1].

1. Introduction

Extensive investigations in black hole physics have thrown up riches of information that are significant from the theoretical and/or observational point of view [2]. Black holes are assumed to attract many matter fields due to their strong gravitational force but they scatter out almost all the matter information in the form of gravitational or scalar wave radiation so that the final stage is characterized by only three hairs: their mass, angular momentum and charge [3].

Black holes may interact strongly with matter fields and are thought of as thermal objects. The area law of the black hole entropy has played a pivotal role in the understanding of black hole physics in general [4,5]. A remarkable consequence of the area law is its connection with the holographic principle which produces Einstein's equations and other thermodynamical relations [6]. The matter field contribution to the black hole entropy has also been studied extensively using some of the methods [7,8,9]. Among them, the semiclassical method with the brick wall regularization scheme seems to be more transparent from mathematical and physical points of view [9]. These features become evident when the method is applied to cases of static Schwarzschild and Reissner-Nordström black holes[10].

However, the statistical mechanics for the matter field contribution in the background of a rotating black hole is somewhat problematic. The entropy of the Kerr-Newman black hole in (3+1) dimension has been studied by several authors [11,12,13] and the extra divergent structures were pointed out to appear due to the superradiant mode. The non-rotating limit of the entropy cannot be taken in these calculations. The problems outlined here revealed themselves in the analyses dealing with specific solutions. It is therefore desirable that the calculations be carried out in a sufficiently general framework such that the above difficulties are either removed or at least minimized as far as possible. A solution independent generalized area law is expected to provide a better physical insight in the understanding of rotating black hole thermodynamics.

In this paper, we calculate the scalar field contribution to the black hole entropy in an arbitrary D dimensional rotating black hole spacetime without assuming any particular exact solution.

2. Scalar Field Contribution to Rotating Black Hole Entropy by Semiclassical Method

In this section, we study the statistical mechanics for the scalar field in D dimensional one parameter rotating spacetime by the semiclassical method. Our analysis is general in the sense that the metric does not depend on the explicit black hole solution. We adopt units such that $G = c = \hbar = k_{\rm B} = 1$.

2.1. *Stephan-Boltzmann's law in rotating black hole spacetime*

We set the D-dimensional polar coordinate as

$$x^\rho = (x^0, x^1, x^2, \cdots, x^{D-1}) = (t, r, \phi, \theta^3, \cdots, \theta^{D-1}) \ , \tag{1}$$

where the ranges are $t \in (-\infty, \infty)$, $r \in [0, \infty)$, $\phi \in [0, 2\pi]$ and $\theta^3, \cdots, \theta^{D-1} \in [0, \pi]$. The invariant line element is assumed to be of the form

$$ds^2 = g_{tt}dt^2 + g_{rr}dr^2 + g_{\phi\phi}d\phi^2 + 2g_{t\phi}dtd\phi + \sum_{i=3}^{D-1} g_{ii}(d\theta^i)^2 \ , \tag{2}$$

where the off diagonal metric $g_{t\phi}$ induces the rotation of the system with the angular velocity $-g_{t\phi}/g_{\phi\phi}$. The metric components in Eq.(2) do not depend on t, ϕ. Consequently, two Killing vectors exist: $\xi_t = \partial_t$ and $\xi_\phi = \partial_\phi$ which imply the conservations of the total energy E and the azimuthal angular momentum m of a scalar field.

The matter action for the scalar field Φ of mass μ in D dimension is

$$I_{\text{matter}}(\Phi) = \int d^D x\sqrt{-g}\left(-\frac{1}{2}g^{\rho\sigma}\ \partial_\rho\Phi\partial_\sigma\Phi - \frac{1}{2}\mu^2\Phi^2\right) . \quad (3)$$

From this action, the field equation for the scalar field is obtained:

$$\frac{1}{\sqrt{-g}}\partial_\rho\left(\sqrt{-g}g^{\rho\sigma}\partial_\sigma\Phi\right) - \mu^2\Phi = 0 . \quad (4)$$

We take the ansatz for the scalar field Φ in the semiclassical method as

$$\Phi(t, r, \phi, \theta^3, \cdots, \theta^{D-1}) \simeq \exp\left(\pm i\sum_{\rho=0}^{D-1}\int^x p_\rho dx^\rho\right) , \quad (5)$$

where the end point of inegral is $x^\rho = (t, \phi, r, \theta^3, ..., \theta^{D-1}, r)$ and the integrand is momentum: $p_\rho = (-E, p_r, m, p_3, \cdots, p_{D-1})$. Putting the scalar field function Eq.(5) into the field equation Eq.(4), the on-shell energy-momentum relation is obtained

$$-\mu^2 = g^{\rho\sigma}p_\rho p_\sigma = g^{tt}E^2 + g^{rr}p_r^2 + g^{\phi\phi}p_\phi^2 - 2g^{t\phi}Em + \sum_{i=3}^{D-1} g^{ii}p_i^2 , \quad (6)$$

where the contravariant components of the metric are obtained from the original metric Eq.(2) as

$$\begin{aligned} g^{tt} &= g_{\phi\phi}/\Gamma \ , \quad g^{rr} = 1/g_{rr} \ , \quad g^{\phi\phi} = g_{tt}/\Gamma \ , \\ g^{t\phi} &= -g_{t\phi}/\Gamma \ , \quad g^{ii} = 1/g_{ii} \ \ (i = 3, \cdots, D-1) \ , \end{aligned} \quad (7)$$

with $\Gamma := g_{tt}g_{\phi\phi} - g_{t\phi}^2$.

Since the black hole is rotating, its energy can be transferred to scalar particles in the ergo region by the Penrose process [?]. However, not all the black hole energy can be mined out. We can obtain the restriction on energy E and angular momentum m of the scalar particle in the following way [?]. Consider a new Killing vector combining two Killing vectors linearly:

$$\eta := \xi_t + \xi_\phi\Omega_H , \quad (8)$$

where the angular velocity on the horizon r_H is defined as

$$\Omega_H := \left.\frac{g^{t\phi}}{g^{tt}}\right|_{r_H} = -\left.\frac{g_{t\phi}}{g_{\phi\phi}}\right|_{r_H} . \quad (9)$$

This vector is light like in future direction on the horizon and the inner product of it with momentum becomes non-positive, which provides the restriction on the energy:

$$p \cdot \eta := \sum_{\rho=0}^{D-1} p_\rho \eta^\rho \le 0 \quad \Rightarrow \quad m\Omega_H \le E \ . \tag{10}$$

This means that the angular momentum of scalar particle is inverse in sign to the angular velocity of the black hole if the negative energy particle is absorbed into the black hole.

The number of the quantum state with energy not exceed E is the sum of phase space K divided by the unit quantum volume such that

$$\sum_K \simeq \frac{1}{2\pi^{D-1}} \int dr dp_r d\phi dm d\theta^3 dp_3 \cdots d\theta^{D-1} dp_{D-1} \ , \tag{11}$$

where the integration range is the range shown below Eq.(1) for angle variables and the range satisfying the energy-momentum condition Eq.(6) with the restriction Eq.(10) for the momentum variables near the horizon region.

Next we consider the partition function of the scalar field in the rotating black hole geometry of the temperature $T = 1/\beta$, viz.,

$$\begin{aligned} Z &= \sum_{\{n^{(K)}\}} \exp\left(-\beta \sum_{\{K\}} n^{(K)} (E_{(K)} - m\Omega_H) \right) \\ &= \prod_{\{K\}} \left[1 - \exp\left(-\beta(E_{(K)} - m\Omega_H)\right)\right]^{-1} \ . \end{aligned} \tag{12}$$

The exponent of the Boltzmann factor $E - m\Omega_H$ is understood taking account of the rotation effect according to the Hartle-Hawking argument [7] and is positive as shown in Eq.(10) ensuring that the partition function is well defined. The summation with respect to $n^{(K)}$ and K are the occupation number sum and the phase space sum in Eq.(12) respectively.

The free energy F is obtained through the partition function :

$$\beta F = -\log Z = \sum_{\{K\}} \log\left[1 - \exp\left(-\beta(E_{(K)} - m\Omega_H)\right)\right] \ . \tag{13}$$

Using the semiclassical phase space sum in Eq.(11) and changing the integration variable from p_r to E, we obtain the expression of the free energy after the integration by parts in the form

$$F = -\frac{1}{2\pi^{D-1}} \int_{m\Omega_H}^{\infty} dE \int d\phi dm d\theta^3 dp_3 \cdots d\theta^{D-1} dp_{D-1}$$
$$\times \int dr \frac{p_r}{e^{\beta(E-m\Omega_H)} - 1} , \tag{14}$$

where the radial momentum p_r is determined by the mass-shell energy-momentum condition Eq.(6) as

$$p_r = \frac{1}{(g^{rr})^{1/2}} \left(-g^{tt} E^2 - g^{\phi\phi} m^2 + 2g^{t\phi} Em - \sum_{i=3}^{D-1} g^{ii} p_i^2 - \mu^2 \right)^{1/2}$$
$$= (g_{rr})^{1/2} \left(-g^{tt} (E + \frac{g_{t\phi}}{g_{\phi\phi}} m)^2 - \frac{m^2}{g_{\phi\phi}} - \sum_{i=3}^{D-1} \frac{p_i^2}{g_{ii}} - \mu^2 \right)^{1/2} . \tag{15}$$

In view of the situation of rotating geometry, we introduce a new energy variable as

$$E' := -p \cdot \eta = E - m\Omega_H . \tag{16}$$

We make the near horizon approximation: $-g_{t\phi}/g_{\phi\phi} \simeq \Omega_H$ for $r \simeq r_H$, so that the radial momentum becomes

$$p_r \simeq (g_{rr})^{1/2} \left(-g^{tt} E'^2 - \frac{m^2}{g_{\phi\phi}} - \sum_{i=1}^{D-1} \frac{p_i^2}{g_{ii}} - \mu^2 \right)^{1/2} . \tag{17}$$

Then the free energy is expressed as

$$F = -\frac{1}{2\pi^{D-1}\beta^D} \int_0^{\infty} dx \frac{x^{D-1}}{e^x - 1} \int dr \int d\phi d\theta^3 \cdots d\theta^{D-1} (-g^{tt})^{D/2-1/2}$$
$$\times (g_{rr} g_{\phi\phi} g_{33} \cdots g_{D-1D-1})^{1/2} \left(1 + \frac{\mu^2}{g^{tt} E'^2} \right)^{D/2-1/2} \frac{v_{\rm unit}}{2^{D-2}} , \tag{18}$$

where $(D-1)$-dimensional unit sphere is $v_{\rm unit} = 2^{D-1}\pi^{D/2-1}\Gamma(D/2)/\Gamma(D)$.

The free energy for the massless case ($\mu = 0$) is obtained as

$$F = -\frac{\zeta(D)\Gamma(D)}{(2\pi)^{D-1}\beta^D} v_{\rm unit} V_{\rm opt} , \tag{19}$$

where the optical volume is

$$V_{\rm opt} := \int dr d\phi d\theta^3 ... d\theta^{D-1} (-g^{tt})^{D/2-1/2} (g_{rr} g_{\phi\phi} g_{33} \cdots g_{D-1D-1})^{1/2} . \tag{20}$$

The entropy is obtained:

$$S := \beta^2 \frac{\partial F}{\partial \beta} = \frac{\zeta(D)\Gamma(D+1)}{(2\pi\beta)^{D-1}} v_{\rm unit} V_{\rm opt} \tag{21}$$

The rotation effects are included in $V_{\rm opt}$ and may be in β.

Note also that the non-rotating limit can be taken smoothly and the resultant free energy and entropy Eq.(19) and Eq.(21) agree with the known non-rotating results[10].

2.2. *Area law in rotating black hole spacetime*

In this section, we perform the real space integration of $V_{\rm opt}$ in Eq.(21). For this the horizon and temperature conditions should be imposed and the brick wall regurlarization scheme [9] will be adopted.

1. Horizon Condition
   We require simple zeros for the inverse metric components $1/g^{tt}$ and $1/g_{rr}$ at the black hole horizon r_H, which is the radius corresponding to the outer zero of these inverse metric components:

$$\frac{1}{g^{tt}} \simeq C_t(\theta)(r - r_H) \ , \quad \frac{1}{g_{rr}} \simeq C_r(\theta)(r - r_H) \ , \tag{22}$$

   where the coefficient functions $C_t(\theta), C_r(\theta)$ are defined by

$$C_t(\theta) := \partial_r \frac{1}{g^{tt}}\bigg|_{r_H} \ , \quad C_r(\theta) := \partial_r \frac{1}{g_{rr}}\bigg|_{r_H} \ . \tag{23}$$

2. Temperature Condition
   The temperature is defined by the condition that no conical singularity is required in the Rindler space not to depend on zenithal angle θ:

$$\frac{2\pi}{\beta_H} = \frac{-\partial_r(1/g^{tt})}{2\sqrt{-g_{rr}/g^{tt}}}\Bigg|_{r_H} = \frac{1}{2}\left(C_t(\theta)C_r(\theta)\right)^{1/2} = \text{indep. on } \theta. \tag{24}$$

Under these two conditions, we estimate the radial integration part in the optical volume Eq.(20) near the horizon and obtain

$$\int_{r_H+\epsilon}^{L} dr\,(-g^{tt})^{D/2-1/2}(g_{rr})^{1/2} \simeq C_t(\theta)^{-D/2+1/2}C_r(\theta)^{-1/2}\frac{\epsilon^{-D/2+1}}{D/2-1} \ , \tag{25}$$

where ϵ and L are the short distance and large distance regularization parameter respectively in the brick wall regularization scheme with their magnitudes restricted by the relation $0 < \epsilon \ll r_H \ll L < \infty$. Here we change the idea from the θ independent cutoff parameter ϵ to the θ dependent cutoff parameter $\epsilon(\theta)$ to define the θ independent invariant cutoff parameter $\epsilon_{\rm inv}$:

$$\epsilon_{\rm inv} := \int_{r_H}^{r_H+\epsilon(\theta)} dr\sqrt{g_{rr}} \ , \tag{26}$$

which is inversely solved near the horizon as $\epsilon(\theta) = C_r(\theta)\epsilon_{\rm inv}^2/4$. Using $\epsilon_{\rm inv}$, the entropy can be expressed as

$$S = \frac{\zeta(D)D\Gamma(D/2-1)}{2^D\pi^{3D/2-1}}\frac{A}{(\epsilon_{\rm inv})^{D-2}} \ , \tag{27}$$

where A is the surface area of the rotating black holes on the horizon

$$A := \int d\phi d\theta^3 \cdots d\theta^{D-1}(g_{\phi\phi}g_{33}\cdots g_{D-1D-1})^{1/2}\Big|_{r_H} \ . \tag{28}$$

Eq.(27) exhibits the desired area law of the entropy in D-dimensional rotating black hole spacetime. Note that we have derived the result without using any explicit expression for the metric solutions but using only two conditions; Horizon Condition Eq.(22) and Temperature Condition Eq.(24). Note also that we can take the smooth non-rotating limit of the generalized area law, which then reproduces the known expressions.

3. Application to Kerr-Newman Black Hole

Next we treat the Kerr-Newman black hole as an application to the (3+1) dimensional spacetime. This black hole solution is interesting because the metric components depend on zenithal angles θ as well as on the radial coordinate r.

The line element of the Kerr-Newman black hole is given by

$$ds^2 = g_{tt}dt^2 + g_{rr}dr^2 + g_{\phi\phi}d\phi^2 + 2g_{t\phi}dtd\phi + g_{\theta\theta}d\theta^2 \ , \tag{29}$$

with the metric components

$$\begin{aligned} g_{tt} &= -\frac{\Delta - a^2\sin^2\theta}{\Sigma} \ , \quad g_{t\phi} = -\frac{a\sin^2\theta(r^2+a^2-\Delta)}{\Sigma} \ , \\ g_{\phi\phi} &= \frac{(r^2+a^2)^2 - \Delta a^2\sin^2\theta}{\Sigma}\sin^2\theta \ , \quad g_{\theta\theta} = \Sigma \ , \\ g_{rr} &= \frac{\Sigma}{\Delta} \ , \end{aligned} \tag{30}$$

where

$$\Sigma(r,\theta) := r^2 + a^2\cos^2\theta \ , \quad \Delta(r) := r^2 - 2Mr + a^2 + e^2 \ . \tag{31}$$

Here M, a and e denote the three hairs of the black holes: mass, angular momentum per unit mass and charge respectively. The time component of the inverse metric is $g^{tt} = -g_{\phi\phi}/(\Delta \sin^2\theta)$. The outer zero of $1/g^{tt}$ and $1/g_{rr}$, and thus $\Delta = 0$, define the horizon radius r_H:

$$r_H = M + (M^2 - a^2 - e^2)^{1/2} \quad , \quad (a^2 + e^2 \leq M^2) \ . \tag{32}$$

The angular velocity and the inverse temperature on the horizon are given:

$$\Omega_H = \frac{a}{r_H^2 + a^2} \quad , \quad \frac{2\pi}{\beta_H} = \frac{r_H - r_-}{2(r_H^2 + a^2)} \ . \tag{33}$$

Instead of using the constant cutoff parameter, the θ dependent cutoff parameter is used:

$$\epsilon(\theta) = \frac{(r_H - r_-)^2}{4(r_H^2 + a^2\cos^2\theta)} \epsilon_{\rm inv}^2 \ , \tag{34}$$

and the generalized area law in the Kerr-Newman case is obtained

$$S = \frac{1}{360\pi} \frac{A}{\epsilon_{\rm inv}{}^2} \ , \tag{35}$$

where the area of the Kerr-Newman black hole is $A = 4\pi(r_H^2 + a^2)$.

The final form Eq.(35) is the same form as the non-rotating black hole cases, but the rotating effects are included in A and implicitly in $\epsilon_{\rm inv}$ through θ dependent $\epsilon(\theta)$ paremeter.

4. Summary

We have studied the statistical mechanics of the scalar field in D-dimensional rotating spacetime. We have imposed the physically admissible minimal set of conditions on the one parameter black hole metric. Under the general metric conditions, we derived the Stefan-Boltzmann's law and the area law of the rotating black holes in arbitrary D-dimensional spacetime. One of the key point is the introduction of the zenithal angle dependent cutoff parameter $\epsilon(\theta)$, which leads to the constant invariant cutoff parameter $\epsilon_{\rm inv}$ and the generalized area law in a compact form for rotating black holes. The generalized area law is applied to the Kerr-Newman black hole in (3+1) dimension. Non-rotating limit of these thermal quantities can be taken smoothly and they straightforwardly reproduce the known results.

There are other methods to study the statistical mechanics of the scalar field in black hole spacetime. Among them, the quasi-normal mode approach [15] and the Green's function approach are interesting because rigorous treatments can be possible in those approaches, especially in the (2+1) dimension [14,16].

References

1. M. Kenmoku, K. Ishimoto and K. Shigemoto, *Phys. Rev.* **D73**,12006 (2006).
2. For a textbook of general relativity, see for example, C.W. Misner, K.S. Thorne and J.A. Wheeter, *Gravitation*, W.H. Freeman and Company (1972).
3. R. Ruffine and J. Wheeler, Physics Today, 24(1), 30 (1971); B. Carter, "Properties of the Kerr metric", in DeWitt and DeWitt eds., *Black Holes*, Proceedings of 1972 session of Ecole d'éle de physique theorique, Gordon and Breach, New York (1973).
4. J.D. Bekenstein, *Phys. Rev.* **D7**, 2333 (1973); *ibid.* **D9**, 3292 (1974).
5. S.W. Hawking, *Commun. Math. Phys.* **43**, 199 (1975).
6. T. Jacobson, *Phys. Rev. Lett.* **75**, 1260 (1995).
7. J.B. Hartle and S.W.Hawking, *Phys. Rev.* **D13**, 2188 (1976).
8. S.P. de Alwis and N. Ohta, *Phys. Rev.* **D52**, 3529 (1995).
9. G. 't Hooft, *Nucl. Phys.* **B256**, 727 (1985).
10. M. Kenmoku, K.K. Nandi and K. Shigemoto, *Class. Quant. Grav.* **22**, 3923 (2005).
11. M-H. Lee and J.K. Kim, *Phys. Rev.* **D54**, 3904 (1996).
12. R.B. Mann and S.N. Solodukhin, *Phys. Rev.* **D54**, 3932 (1996).
13. V. Frolov and V. Fursaev, *Phys. Rev.* **D61**, 024007 (2000); S. Hod, *Phys. Rev.* **D61**, 024018 (2000); B. Wang and E. Abdalla, *Phys. Rev.* **D62**, 044030 (2000); J. Jing and M-L. Yan, *Phys. Rev.* **D62**, 104013 (2000).
14. I. Ichinose and Y. Satoh, *Nucl. Phys.* **B447**, 340 (1995).
15. O. Dreyer, *Phys. Rev. Lett.* **90**, 081301 (2003); A. Corichi, *Phys. Rev.* **D67**, 087502 (2003).
16. D. Birmingham, *Phys. Rev.* **D64**, 064024 (2001).

Further simplification of the constraints of four-dimensional gravity

Chopin Soo
Department of Physics,
National Cheng Kung University,
Tainan 701, Taiwan
E-mail: cpsoo@mail.ncku.edu.tw

The super-Hamiltonian of 4-dimensional gravity as simplified by Ashtekar through the use of gauge potential and densitized triad variables can furthermore be succinctly expressed as a Poisson bracket between the volume element and other fundamental gauge-invariant elements of 3-geometry. This observation naturally suggests a reformulation of non-perturbative quantum gravity wherein the Wheeler-DeWitt Equation is identical to the requirement of the vanishing of the corresponding commutator. Moreover, this reformulation singles out spin network states as the preeminent basis for expansion of all physical states.

1. Introduction

Non-perturbative canonical quantization of gravity attempts to overcome the perturbative non-renormalizability of Einstein's theory by treating the constraints exactly. Ashtekar's seminal simplification of the constraints through the use of gauge connection and densitized triad variables[1] bridged the distinction between geometrodynamics and gauge dynamics by identifying the densitized triad, $\tilde{E}^{ia}$ -from which the metric is a derived composite - as the momentum conjugate to an $SO(3,C)$ gauge potential A_{ia}. The introduction of spin network states[2] have also yielded discrete spectra for well defined area and volume operators[3]. To the extent that exact states and rigorous results are needed, simplifications of the classical and corresponding quantum constraints are of great importance to the program. These include Ashtekar's original simplification and also Thiemann's observation that $\epsilon_{abc}\underset{\sim}{\epsilon}_{ijk}\tilde{E}^{ia}\tilde{E}^{jb}$ in the super-Hamiltonian constraint is proportional to the Poisson bracket between the connection and the volume operator[4]. Thus it is not unreasonable to expect even more progress from further simplification of the constraints, all the more so if the simplification is naturally associated with spin networks states.

Starting with the fundamental conjugate pair and Poisson bracket,

$$\{\tilde{E}^{ia}(\vec{x}), A_{jb}(\vec{y})\}_{P.B.} = -i(\frac{8\pi G}{c^3})\delta^i_j \delta^a_b \delta^3(\vec{x}-\vec{y}), \tag{1}$$

we shall show that the super-Hamiltonian permits a further remarkable simplification: It is in fact expressible as a Poisson bracket between the volume element and other fundamental invariants, even when the cosmological constant, λ, is non-vanishing[5]. The leads naturally to an equivalent classical constraint, and to the quantum Wheeler-DeWitt Equation as a vanishing commutator relation. The preeminence of spin network states come naturally from the fact that they are eigenstates of the Hermitian volume element operator; and can therefore be used as a basis for all physical states.

2. Gauge-invariant elements of 3-geometry and their Poisson Brackets

Three physical quantities- the volume element ($\tilde{v}$), the Chern-Simons functional of the gauge potential ($C[A]$), and the integral of the mean extrinsic curvature (K)- form the basic ingredients of 4-dimensional General Relativity as a theory of the conjugate pair of densitized triad and gauge variables, $(\tilde{E}^{ia}, A_{ia})$. All are gauge-invariant, but the latter two are in addition also invariant under three-dimensional diffeomorphisms i.e. they are elements of 3-geometry.

Their definitions are as follows:

$$\tilde{v}(\vec{x}) \equiv \sqrt{\frac{1}{3!}\epsilon_{abc}\epsilon_{ijk}\tilde{E}^{ia}(\vec{x})\tilde{E}^{jb}(\vec{x})\tilde{E}^{kc}(\vec{x})} = |\det E_{ai}|. \tag{2}$$

Its integral over the Cauchy surface, M, is the volume, $V = \int_M \tilde{v}(\vec{x})d^3x$. The Chern-Simons functional of the Ashtekar connection is

$$C \equiv \frac{1}{2}\int_M (A^a \wedge dA_a + \frac{1}{3}\epsilon^{abc}A_a \wedge A_b \wedge A_c). \tag{3}$$

Its characteristic feature is that it satisfies $\frac{\delta C[A]}{\delta A_{ia}} = \tilde{B}^{ia}$ if $\partial M = 0$; wherein $\tilde{B}^{ia}$ is the non-Abelian $SO(3)$ magnetic field of A_{ia}[a]. The integral of the

[a] If M is with boundary, the imposition of appropriate boundary conditions, or the introduction supplementary boundary terms should be considered. On the other hand, one can treat $\partial M = 0$ as a predictive element of the present reformulation.

trace of the extrinsic curvature is

$$K \equiv \frac{i}{2}\int_M E^a \wedge (D_A E)_a = \int_M (\tilde{E}^{ia} k_{ia}) d^3x, \tag{4}$$

with the observation that the complex Ashtekar connection is $A_{ia} \equiv -ik_{ia} + \Gamma_{ia}$, and $(D_A E)_a = dE_a + \epsilon_a{}^{bc} A_b \wedge E_c$, and Γ_a is the torsionless connection $(dE_a + \epsilon_{ab}{}^c \Gamma_b \wedge E_c = 0)$ connection compatible with the dreibein 1-form $E_a = E_{ai} dx^i$ on M.

With the above definitions and the fundamental relation of Eq.(1), it follows that the following Poisson brackets hold:

$$\begin{aligned}
\{\tilde{v}, K\}_{P.B.} &= 3(\frac{4\pi G}{c^3})\tilde{v} \\
\{\tilde{v}, C\}_{P.B.} &= (\frac{2\pi G}{ic^3 \tilde{v}})\epsilon_{abc}\underset{\sim}{\epsilon}_{ijk}\tilde{E}^{ia}\tilde{E}^{jb}\tilde{B}^{kc} \\
\{K, C\}_{P.B.} &= (\frac{8\pi G}{c^3})\int_M (\tilde{B}^{ia} k_{ia})\, d^3x \\
\tilde{H} \equiv \{\tilde{v}, iC + \frac{\lambda}{3}K\}_{P.B.} &= (\frac{2\pi G}{c^3 \tilde{v}})[\epsilon_{abc}\underset{\sim}{\epsilon}_{ijk}\tilde{E}^{ia}\tilde{E}^{jb}(\tilde{B}^{kc} + \frac{\lambda}{3}\tilde{E}^{kc})].
\end{aligned} \tag{5}$$

3. Further simplification of the super-Hamiltonian constraint

Expressed in Ashtekar variables, the super-Hamiltonian constraint for the theory of General Relativity is *precisely*[1]

$$\tilde{\tilde{H}}_0 = \frac{c^3}{16\pi G}[\epsilon_{abc}\underset{\sim}{\epsilon}_{ijk}\tilde{E}^{ia}\tilde{E}^{jb}(\tilde{B}^{kc} + \frac{\lambda}{3}\tilde{E}^{kc})] \approx 0. \tag{6}$$

It follows that at the classical level, we may *equivalently* replace the super-Hamiltonian constraint with the vanishing of a Poisson bracket i.e.

$$\{\tilde{v}, \frac{\lambda}{3}K + iC\}_{P.B.} = 0. \tag{7}$$

The new super-Hamiltonian, $\tilde{H}$, is now a tensor density of weight 1. Using $\tilde{H} \propto \tilde{\tilde{H}}_0/\tilde{v}$ it can be demonstrated that the new super-Hamiltonian constraint together with Ashtekar's transcriptions of the Gauss' law and super-momentum constraints remain a set of first class constraints at the classical level.

Even though we may invoke Poisson bracket-quantum commutator correspondence $\{,\}_{P.B.} \mapsto (i\hbar)^{-1}[\,,\,]$, there is no unique prescription for defining a quantum theory from its classical correspondence. The previous observations naturally suggest **defining four-dimensional non-perturbative**

Quantum General Relativity as a theory of the conjugate pair $(\tilde{E}^{ia}, A_{ia})$ with super-Hamiltonian constraint imposed as the vanishing commutation relation, $[\hat{\tilde{v}}(\vec{x}), \frac{\lambda}{3}\hat{\boldsymbol{K}} + i\hat{C}] = 0$, together with the requirement of invariance under three-dimensional diffeomorphisms and internal gauge transformations.

4. Reformulation of the Wheeler-DeWitt Equation, and further comments

Physical quantum states $|\Psi\rangle$ are required to be annihilated by the constraint :

$$[\hat{\tilde{v}}(\vec{x}), \frac{\lambda}{3}\hat{K} + i\hat{C}]|\Psi\rangle = 0. \tag{8}$$

It is very noteworthy that the Wheeler-DeWitt Equation above is not merely symbolic, but is in fact *expressed explicitly in terms of gauge-invariant 3-geometry elements* C and K.

The formulation is so far not confined to a particular representation or realization of the theory. However it is most interesting that *explicit realizations* and representations of eigenstates of the volume element operator exist, and they are precisely associated with *spin network states*! On a spin network, it is known that $\tilde{v}^2$ acts in a well-defined manner[3]. Its eigenstates are linear combinations of spin network states of the same vertex valency(number of links) at $\vec{x}$, such that

$$[\hat{\tilde{v}}(\vec{x})]^2|\Psi_v\rangle = v^2|\Psi_v\rangle, \tag{9}$$

has spectrum given by $v = 0$ if valency of the vertex at $\vec{x}$ is less than 4, and v can be computed and may be non-trivial if the valency of the vertex at $\vec{x}$ is equal to or greater than four[3]. In the connection-representation of spin network states, the wave function, $\langle A|\Psi\rangle = \langle A|\mathbf{\Gamma}, \{vertices\}, \{j\}\rangle$, is such that for the spin network, $\mathbf{\Gamma}$, each link between two vertices at $\vec{x}$ and $\vec{y}$ is associated with a non-integrable phase factor $P[\exp(i\int_{\vec{x}\,p}^{\vec{y}} A)]$ with A in the spin-j representation of the Lie algebra. With appropriate assignments of combinations of Wigner symbols at the vertices, spin network states are gauge-invariant.

Simultaneous eigenstates, $|\Psi_{v,\gamma}\rangle$, with eigenvalues (v, γ) of the volume element operator and the (dimensionless) operator $\hat{\Upsilon} \equiv \frac{\lambda}{3}\hat{K} + i\hat{C}$ are solutions. But since $\hat{\Upsilon}$ does not commute off-shell with the volume element, well defined non-trivial simultaneous eigenstates may not exist.

The operator $\hat{K}$ has in fact been studied by Borrisov, De Pietri and Rovelli[6], and its action on loop or non-integrable phase factor elements of spin network states can be made well-defined. This indicates the action of the operator $\hat{\Upsilon} \equiv \frac{\lambda}{3}\hat{K} + iC$ can also be defined on spin network states. The physical meaning of $\hat{K}$ is not readily apparent in the spin network formulation, but its connection to "intrinsic time" in quantum gravity may nevertheless be deduced from a different perspective. Without resorting to particular representations, it is readily verified that, apart from a multiplicative constant, K is in fact conjugate to the intrinsic time variable i.e. $\{\ln \tilde{v}, K\}_{P.B.} = \frac{12\pi G}{c^3}$. In the quantum context $\hat{K}$ is thus proportional to the *generator of translations* of $\ln \tilde{v} = \ln |\det E_{ai}|$ which is furthermore a monotonic function of the superspace "intrinsic time variable" ($\propto \sqrt{|\det E_{ai}|}$) discovered by DeWitt in his seminal study of canonical quantum gravity[7]. The operator K is thus a Schwinger-Tomonaga "time-evolution operator" (complete with i) for an intrinsic time variable.

A few observations on the physical requirements of the inner product and the properties of the operators with respect to Hermitian conjugation are in order. At the classical level $\tilde{v}$ and K are real(according to Eqs. (2) and (4)), and it is reasonable to require that the inner product must lead to Hermitian $\hat{\tilde{v}}$ and $\hat{K}$. The Ashtekar connection on M is furthermore understood to be the pullback of the self-dual projection of the spin connection to the Cauchy surface. Thus it is also reasonable to conclude that $A^\dagger$ corresponds to the anti-self-dual projection, or the orientation-reversed transform, of A. In fact these observations suggest that $C^\dagger$ is the Chern-Simons functional for the Ashtekar connection of the manifold $\bar{M}$ with the reversed orientation with respect to M. Thus the operator $\hat{\Upsilon}$ is not in general P and CP-invariant. In this respect, spin network states are holomorphic in A and are therefore also not automatically P-invariant. Further discussions on possible P, CP and CPT violations in Ashtekar theory coupled to matter have been discussed elsewhere[8].

The observations presented here highlight many remarkable features in four-dimensional Quantum General Relativity and behoove further consideration and continued research.

Acknowledgments

This work has been supported by funds from the National Science Council of Taiwan under Grant Nos. NSC93-2112-M-006-011 and NSC94-2112-M-006-006.

References

1. A. Ashtekar, *Phys. Rev. Lett.* **57**, 2244 (1986); A. Ashtekar, *Phys. Rev. D***36**, 1587 (1987); A. Ashtekar, *Lectures on non-perturbative canonical gravity* (World Scientific, 1991).
2. C. Rovelli and L. Smolin, *Phys. Rev. D***52**, 5743 (1995).
3. C. Rovelli and L. Smolin, *Nucl. Phys. B***442**, 593 (1995), erratum-*ibid.* B**456**, 753 (1995); R. De Pietri and C. Rovelli, *Phys. Rev. D***54** 2664 (1996); A. Ashtekar and J. Lewandowski, *Class. Quantum Grav.* **14**, A55 (1997); J. Lewandowski, *Class. Quant. Grav.* **14**, 71 (1997); A. Ashtekar and J. Lewandowski, *Adv. Theor. Math. Phys.* **1**, 388 (1998); J. Brunnemann and T. Thiemann, *Simplification of the spectral analysis of the volume operator in loop quantum gravity*, gr-qc/0405060; K. A. Meissner, *Eigenvalues of the volume operator in loop quantum gravity*, gr-qc/0509049.
4. T. Thiemann, *Phys. Lett. B***380**, 257 (1996).
5. Further details can be found in C. Soo, *Further simplification of the super-Hamiltonian constraint of General Relativity, and a reformulation of the Wheeler-DeWitt Equation*, gr-qc/0512025.
6. R. Borissov, R. De Pietri and C. Rovelli, *Class. Quantum Grav.* **14**, 2793 (1997).
7. Bryce S. DeWitt, *Phys. Rev.* **160**, 1113 (1967).
8. L. N. Chang and C. Soo, *Phys. Rev. D***53**, 5682 (1996); C. Soo and L. N. Chang, *The Weyl theory of fundamental interactions: is CPT violated ?*, hep-th/9702171; L. N. Chang and C. Soo, *Class. Quantum Grav.* **20**, 1379 (2003).

Evaporation of Bardeen black holes

Sean A. Hayward
Center for Astrophysics, Shanghai Normal University,
100 Guilin Road, Shanghai 200234, China

Non-singular space-times are given which model the formation of a Bardeen black hole from an initial vacuum region and its subsequent evaporation to a vacuum region. The black hole consists of a compact space-time region of trapped surfaces, with inner and outer boundaries which join circularly as a single smooth trapping horizon.

1. Introduction

Concrete models of the formation and evaporation of non-singular black holes were recently constructed, demonstrating that anything which fell into such a black hole must eventually re-emerge[1]. There is no question of information loss, thereby resolving an apparent paradox that was widely regarded as a fundamental problem of theoretical physics. This article reports a similar construction based on the earliest non-singular black-hole model, due to Bardeen[2].

The causal structure of a Bardeen black hole is similar to that of a Reissner-Nordström black hole, with the internal singularities replaced by regular centres (Fig. 1). The idea is to remove regions to the past and future of two consecutive advanced times, then adjoin a past which describes gravitational collapse and a future which describes evaporation. This is to be done using Vaidya-like regions[3] with ingoing or outgoing radiation.

2. Bardeen black holes

Consider static, spherically symmetric metrics of the form

$$ds^2 = r^2 dS^2 + dr^2/F(r) - F(r)dt^2 \tag{1}$$

where t is the static time, r the area radius and $dS^2 = d\theta^2 + d\phi^2 \sin^2\theta$. A surface has area $4\pi r^2$, is trapped if $F(r) < 0$ and untrapped if $F(r) > 0$.

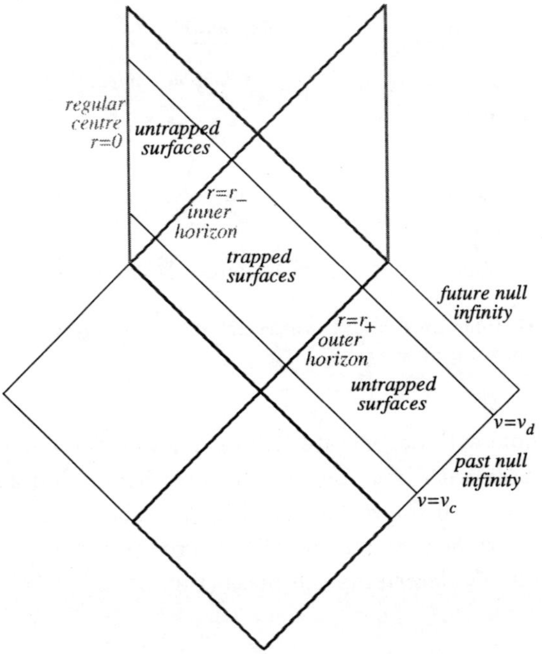

Fig. 1. Penrose diagram of a Bardeen black hole.

Trapping horizons, in this case also Killing horizons, are located at the zeros $F(r) = 0$. For an asymptotically flat space-time with total mass m,

$$F(r) \sim 1 - 2m/r \quad \text{as } r \to \infty. \tag{2}$$

Similarly, flatness at the centre requires

$$F(r) \sim 1 - \tfrac{8}{3}\pi\rho r^2 \quad \text{as } r \to 0 \tag{3}$$

where ρ is the central energy density. Bardeen's model was given by

$$F(r) = 1 - \frac{2mr^2}{(r^2 + e^2)^{3/2}} \tag{4}$$

where (m, e) are constant. The space-time reduces to the Schwarzschild solution for $e = 0$ and is flat for $m = 0$.

Elementary analysis of the zeros of $F(r)$ reveals a critical mass $m_* = 3\sqrt{3}|e|/4$ and radius $r_* = \sqrt{2}|e|$ such that, for $m > 0$, $r > 0$, $F(r)$ has no zeros if $m < m_*$, one double zero at $r = r_*$ if $m = m_*$, and two simple zeros at $r = r_\pm$ if $m > m_*$ (Fig. 2). These cases therefore describe, respectively, a space-time with the same causal structure as flat space-time, a non-singular

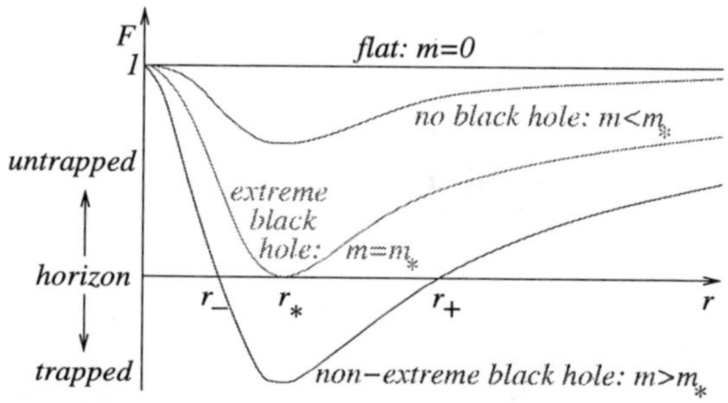

Fig. 2. The metric function $F = g^{rr}$, whose sign determines gravitational trapping, for fixed e and different total mass m.

extreme black hole with degenerate Killing horizon, and a non-singular non-extreme black hole with both outer and inner Killing horizons, located at $r_+ \approx 2m$ and $r_- \to 0$ for $m \gg m_*$ (Fig. 1). The structure of the inner horizon differs from the previous model[1], where the radius became constant. The horizon radii $r_\pm$ determine the mass (Fig. 3)

$$m(r_\pm) = \frac{(r_\pm^2 + e^2)^{3/2}}{2r_\pm^2}. \tag{5}$$

Note the existence of a mass gap: such black holes cannot form with mass $m < m_*$.

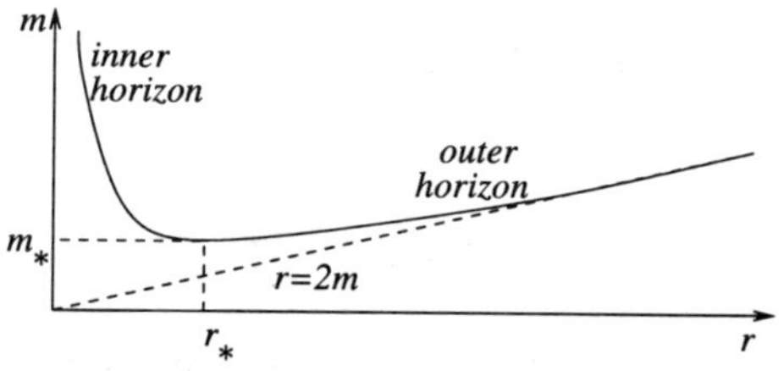

Fig. 3. Horizon mass-radius relation: a pair of horizons appears when mass m exceeds critical mass m_*.

If the Einstein equation $G = 8\pi T$ is used to interpret components of the energy tensor T, these metrics are supported by density $-T_t^t$, radial pressure T_r^r and transverse pressure $T_\theta^\theta = T_\phi^\phi$ given by

$$G_t^t = G_r^r = -\frac{6e^2 m}{(r^2+e^2)^{5/2}} \tag{6}$$

$$G_\theta^\theta = G_\phi^\phi = \frac{3(3r^2-2e^2)e^2 m}{(r^2+e^2)^{7/2}}. \tag{7}$$

They fall off rapidly at large distance, $O(r^{-5})$.

3. Adding radiation

Next rewrite the static space-times in terms of advanced time

$$v = t + \int \frac{dr}{F(r)} \tag{8}$$

so that

$$ds^2 = r^2 dS^2 + 2dvdr - Fdv^2. \tag{9}$$

Now allow the mass to depend on advanced time, $m(v)$, defining $F(r,v)$ by the same expression (4). Then the density $-T_v^v$, radial pressure T_r^r and transverse pressure T_θ^θ have the same form (6)–(7), but there is now an additional independent component, radially ingoing energy flux (or energy-momentum density) T_v^r given by

$$G_v^r = \frac{2rm'}{(r^2+e^2)^{3/2}} \tag{10}$$

where $m' = dm/dv$. This describes pure radiation, recovering the Vaidya solutions for $e = 0$ and at large radius. In the Vaidya solutions, the ingoing radiation creates a central singularity, but in these models, the centre remains regular, with the same central energy density given by (3).

The ingoing energy flux is positive if m is increasing and negative if m is decreasing. A key point is that trapping horizons still occur where the invariant $g^{rr} = F(r,v)$ vanishes[4]. Then one can apply the previous analysis to locate the trapping horizons in (v,r) coordinates parameterized by m, given by $m(r_\pm)$ in (5) and a mass profile $m(v)$; qualitatively, by inspecting Figs. 3 and 4.

Ingoing radiation. One can now model formation and evaporation of a Bardeen black hole. Introduce six consecutive advanced times $v_a < v_b < \ldots < v_f$ and consider smooth profiles of $m(v)$, meaning $m'(v)$ at least

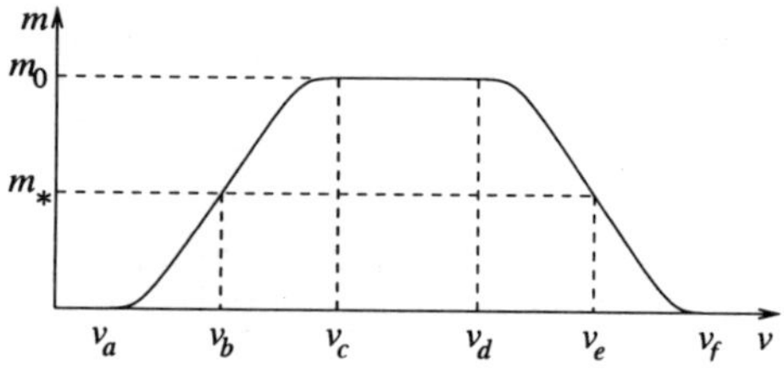

Fig. 4. A mass profile $m(v)$ in advanced time v.

continuous, such that (Fig. 4)

$$\forall v \in (-\infty, v_a) : m(v) = 0 \tag{11}$$
$$\forall v \in (v_a, v_c) : m'(v) > 0 \tag{12}$$
$$\forall v \in (v_c, v_d) : m(v) = m_0 > m_* \tag{13}$$
$$\forall v \in (v_d, v_f) : m'(v) < 0 \tag{14}$$
$$\forall v \in (v_f, \infty) : m(v) = 0. \tag{15}$$

Then

$$\exists v_b \in (v_a, v_c) : m(v_b) = m_* \tag{16}$$
$$\exists v_e \in (v_d, v_f) : m(v_e) = m_*. \tag{17}$$

These transition times mark the appearance and disappearance of a pair of trapping horizons: for $v < v_b$ and $v > v_e$, there is no trapping horizon, while for $v_b < v < v_e$, there are outer and inner trapping horizons, in the sense of a local classification[4]. These horizons join smoothly at the transitions and therefore unite as a single smooth trapping horizon enclosing a compact region of trapped surfaces (Fig. 5, for $r < r_0$).

4. Outgoing radiation

Thus far, only the ingoing Hawking radiation has been modelled, since outgoing radiation does not enter the equation of motion of the trapping horizon; in terms of retarded time u, T_{vv} and T_{uv} enter, but T_{uu} does not[4]. Outgoing Hawking radiation will now be modelled by selecting a certain radius $r_0 > 2m_0$ outside the black hole, adopting the above negative-energy radiation only inside that radius, balanced by outgoing positive-energy radiation outside that radius, with the same mass profile (Fig. 5). This is an idealized model of pair creation of ingoing particles with negative energy and outgoing particles with positive energy, locally conserving energy.

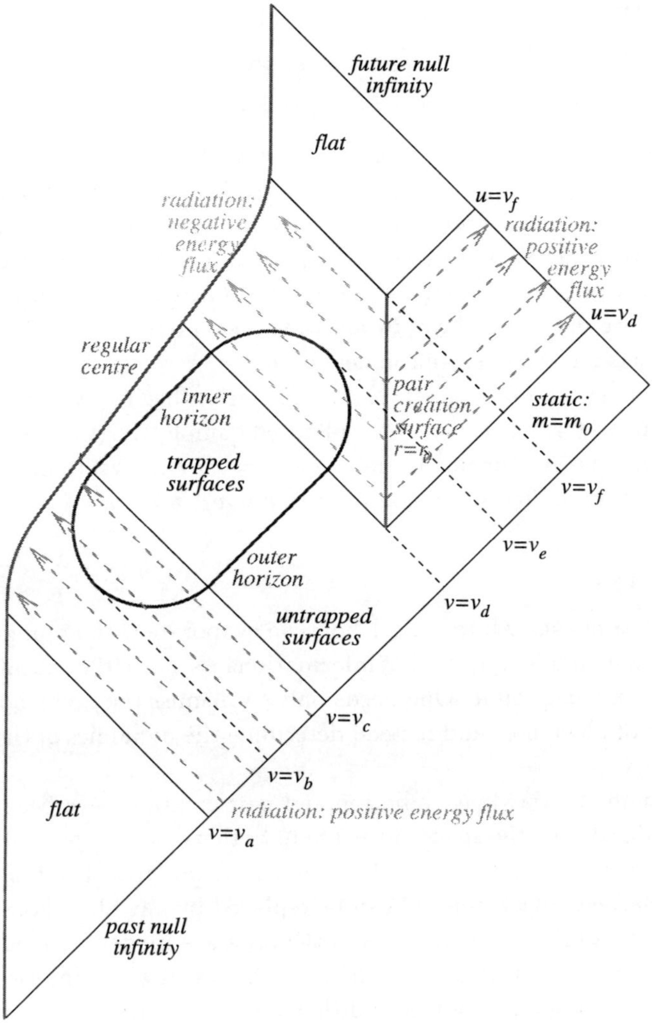

Fig. 5. Penrose diagram of formation and evaporation of a Bardeen black hole.

In more detail, consider an outgoing Vaidya-like region

$$ds^2 = r^2 dS^2 - 2dudr - F du^2 \tag{18}$$

with $F(r, u)$ as before (4), with m replaced by a mass function $n(u)$. Fix the zero point of the retarded time u so that $r = r_0$ corresponds to $u = v$. Now take the above model only for $v < v_d$ (11)–(13). For $v > v_d$, keep the profiles (14)–(15) for $r < r_0$, but for $r > r_0$, take an outgoing Vaidya-like region with

$$\forall u < v_d : n(u) = m_0 \tag{19}$$

$$\forall u > v_d : n(u) = m(u). \tag{20}$$

Then there is a static Bardeen region with total mass m_0 for $v > v_d$, $u < v_d$, and a flat region for $v > v_f$, $u > v_f$. Since the ingoing and outgoing radiation has no net energy but a net outward momentum, the pair-creation surface $r = r_0$ has a surface layer with surface pressure but no surface energy density, as can be confirmed using the Israel formalism[5]. Alternatively, one could remove the surface layer by allowing the pair-creation surface to move, in which case it would shrink under recoil.

The whole picture is given in Fig. 5. Action begins at $v = v_a$, a black hole begins to form at $v = v_b$, has collapsed completely at $v = v_c$ to a static state with mass m_0, begins to deflate at $v = v_d$ and eventually evaporates at $v = v_e$, leaving flat space finally after $v = v_f$, $u = v_f$.

5. Remarks

The space-times show how a black hole can evaporate with no singularity, no event horizon and no question of information loss, since the causal structure is that of flat space-time. One needs only to dismiss the event horizon as a definition of black hole and instead determine the dynamics of the trapping horizons.

Although the Hawking radiation and pair creation have been modelled in an idealized way, the space-times are not otherwise very special; the mass profile $m(v)$ is free up to the qualitative functional form (11)–(15), (Fig. 4), and the Bardeen black hole (4) can be replaced by any black hole with similar properties (1)–(3), (Fig. 2), as in the previous model[1]. The idealization was made only to obtain analytic models, the essential assumptions instead being that no singularity forms and that the ingoing Hawking radiation has negative energy flux[6]. Then the outer horizon shrinks and the inner horizon, which must exist to prevent a singularity, grows. Since this is generally expected to be an accelerating process, the horizons meet, the circle closes and the black hole has evaporated.

It might be interesting to reconsider semi-classical treatments of black-hole evaporation in this light, to see exactly how all the information comes out, whether steadily or mainly in the final pop.

Note: the original presentation was based on a preprint which was subsequently published separately[1].

References

1. S. A. Hayward, *Phys. Rev. Lett.* **96**, 031103 (2006).
2. J. Bardeen, Proc. GR5, Tiflis, USSR (1968). A. Borde, *Phys. Rev.* **D55**, 7615 (1997).
3. P. C. Vaidya, *Proc. Indian Acad. Sci.* **A33**, 264 (1951).
4. S. A. Hayward, *Phys. Rev.* **D49**, 6467 (1994); *Class. Quantum Grav.* **15**, 3147 (1998); *Phys. Rev.* **70**, 104027 (2004); *Phys. Rev. Lett.* **93**, 251101 (2004).
5. W. Israel, *Nuovo Cimento* **B44**, 1 (1966); **B49**, 463 (1967).
6. S. A. Hayward, gr-qc/0504037–8.

On non-uniform charged black branes

U. MIYAMOTO

Department of Physics, Waseda University,
Okubo 3-4-1, Tokyo 169-8555, Japan
E-mail: umpei @gravity.phys.waseda.ac.jp
and
Department of Physics, University of California,
Santa Barbara, CA 93106
E-mail: kudoh @physics.ucsb.edu

The final fate of Gregory-Laflamme instability is one of the most interesting problems in higher-dimensional black hole physics. In this article, non-uniform black strings with a gauge charge are constructed by static perturbations. Then, their thermodynamical stability is analyzed. It is found that in sufficiently non-extremal regions, the non-uniform black strings can be thermodynamically favored over the uniform ones in a large range of spacetime dimensions, which is realized only for higher dimensions (> 13) in vacuum spacetime.

Keywords: Black Holes; Gregory-Laflamme instability.

1. Introduction

It is known the black objects with a translational symmetry such as black branes and strings suffer from a dynamical instability to distribute their mass breaking the translational symmetry, which is called Grerogy-Laflamme (GL) instability [1]. Naive expect is that they decay into individual black holes, having larger entropy via certain topological changing. Horowitz and Maeda showed, however, that inhomogeneous black strings do not "pinch off" in finite proper time on the horizon under certain assumptions, forbidding naked singularities [2]. Although a dynamical simulation was also reported [3], the endpoint of the collapse remains the outstanding open problem.

Besides the simulation of the dynamics followed by the instability, it will be quite useful to obtain the sequence of static solutions in Kaluza-Klein (KK) compact space to know the final fate of instability and the physics of the topology change around the "pinch off" of black strings. Along this line, the whole phase diagram of black hole and black string in 5, 6-dimensionsl KK spaces were obtained in [4]. It is known that perturbative approach is also useful to see the physics changing with the spacetime dimensions. Gubser developed a method to construct perturbative non-uniform black string solutions [5]. Its generalization to arbitrary dimensions reveals the existence of the critical dimension, above which the order of phase transition from uniform to non-uniform strings is higher [6,7].

Although one can believe that charges (e.g. gauge-charge or angular momenta) play a crucial role in non-linear regime, our understanding for stability and phase structure of black branes are restricted to special cases [8,7,9], in which the systems can be reduced to a neutral case. In this article, to have an insight of the final fate of GL instability and phase structure of sensible theories beyond non-linear regime, we perform static perturbations of gauge-charged black strings up to third order. As the main result, we discover new thermodynamically favored non-uniform black strings over the uniform one for same mass (or temperature) and charge. We can understand the critical dimensions in vacuum system from a higher standpoint.

2. Magnetic black string solution

We consider the following $(d+1)$-dimensional action $(d \geq 4)$:

$$I_{d+1} = \frac{1}{16\pi G_{d+1}} \int d^{d+1}x\sqrt{-g}\left[R - \frac{1}{2(d-2)!}F_{d-2}^2\right], \tag{1}$$

where G_{d+1} is $(d+1)$-dimensional gravitational constant and F is a $(d-2)$-form field. By the dimensional reduction method, black string solution is obtained from the black hole solution in this theory [10]:

$$ds_{d+1}^2 = -e^{2a(r,z)} f_+ dt^2 + f_-\, e^{2b(r,z)}\left[\frac{dr^2}{f_+ f_-^2} + dz^2\right]$$
$$+ e^{2c(r,z)} r^2 d\Omega_{d-2}^2, \tag{2}$$

$$f_\pm(r) = 1 - \left(\frac{r_\pm}{r}\right)^{d-3}, \tag{3}$$

$$F = \sqrt{(d-1)(d-3)}\,(r_+ r_-)^{(d-3)/2}\,\varepsilon_{d-2}, \tag{4}$$

where ε_{d-2} is the volume element of a unit $(d-2)$-sphere and the uniform black string solution is obtained by setting by $a = b = c = 0$. For $0 \leq r_- \leq r_+$, this solution has an event horizon at $r = r_+$ and an inner horizon at $r = r_-$ where a curvature singularity exists.

3. Static perturbations

Gubser developed the systematic method to construct non-uniform black string solution by static perturbations [5], as well as to specify the GL critical point quite easily. The method was generalized to higher dimensions first by Sorkin [6]. Here, we introduce the perturbation scheme withought details. First, we rescale coordinates r, z and "charge" paramter r_- by the horizon radius:

$$y \equiv \frac{r}{r_+}, \quad x \equiv \frac{z}{r_+}, \quad q \equiv \frac{r_-}{r_+}. \tag{5}$$

Then, we expand the metric function $X(x, y)$ ($X = a, b, c$) around the uniform solution,

$$X(x, y) = \sum_{n=0}^{\infty} \epsilon^n X_n(y) \cos(nKx), \tag{6}$$

$$X_n(y) = \sum_{p=0}^{\infty} \epsilon^{2p} X_{n,p}(y), \quad K = \sum_{q=0}^{\infty} \epsilon^{2q} k_q, \tag{7}$$

where $X_{0,0}(y) = 0$ is imposed. Here K is the GL critical wavenumber, in other words, $L \equiv 2\pi/K$ gives the asymptotic length of the compactified space, and ϵ is an expansion parameter. Substituting these expanions into the Einstein equations, we obtain ODEs for $X_{n,p}(y)$ which should be solved order-by-order.

As we will see later, it is necessary that we integrate independent modes, $X_{1,0}$ at first order, $X_{2,0}$ and $X_{0,1}$ at second order, and $X_{1,1}$ at third order. Among these modes, $X_{0,1}$ mode is massless mode, which has power law tail at the asymptotic region and the others are massive Kaluza-Klein (KK) modes, which have rapid dacay. Note that a mode $X_{n,p}$ appears at $O(\epsilon^{n+2p})$.

4. Thermodynamics

4.1. *Thermodynamical quantities and stability*

As we mentioned in the previous section, the homogeneous perturbations (zero modes) appear at the second order, and they decay as some power law in the asymptotic region $r \gg r_+$. Suppressing all exponentially small corrections which comes from KK modes in the first and second order perturbations, leading asymptotic bahavior of the metric functions are found to be

$$
\begin{aligned}
a(r,z) &\simeq A_\infty \left(\frac{r_+}{r}\right)^{d-3}, \\
b(r,z) &\simeq B_\infty \left(\frac{r_+}{r}\right)^{d-3}, \\
c(r,z) &\simeq \begin{cases} C_\infty \dfrac{r_+}{r} \ln\left(\dfrac{r}{r_+}\right) & \text{for } d=4, \\ C_1 \dfrac{r_+}{r} + C_\infty \left(\dfrac{r_+}{r}\right)^{d-3} & \text{for } d \geq 5, \end{cases}
\end{aligned} \tag{8}
$$

where A_∞, B_∞, C_∞ and C_1 are constants. With using these asymptotics, the mass per unit length in the z-direction ($z \in [0, L]$) is computed as

$$
\frac{M}{L} = \begin{cases} \dfrac{\Omega_2 r_+}{16\pi G_5} \left[2(1 + 3B_\infty - 2C_\infty) + q\right], & \text{for } d=4 \\ \dfrac{\Omega_{d-2} r_+^{d-3}}{16\pi G_{d+1}} [2(2d-5)B_\infty & \\ \qquad +2(d-2)(d-4)C_\infty + d - 2 + q^{d-3}], & \text{for } d \geq 5. \end{cases} \tag{9}
$$

where $\Omega_{d-2} = 2\pi^{(d-1)/2}/\Gamma[(d-1)/2]$ is the surface area of a $(d-2)$-dimensional unit sphere. The entropy S, temperature T and magnetic charge Q are given by

$$
T = \frac{d-3}{4\pi r_+} \sqrt{f_-}\; e^{a-b} \Big|_{r=r_+}, \tag{10}
$$

$$
\frac{S}{L} = \frac{\Omega_{d-2}}{4G_{d+1}} r_+^{d-2} \sqrt{f_-} \int_0^L \frac{dz}{L} e^{b+(d-2)c} \Big|_{r=r_+}, \tag{11}
$$

$$
\frac{Q}{L} = \frac{(d-1)\Omega_{d-2}}{16\pi G_{d+1}} (r_+^2 q)^{(d-3)/2}, \tag{12}
$$

The z-independence of the temperature, corresponding the zeroth law of thermodynamics, is not obvious from the expression (10). The charge of the field is normalized so that $Q \to M$ in the extremal limit $q \to 1$.

The thermodynamical stability of background spacetime $(a, b, c = 0)$ is determined by the heat capacity:

$$
C_Q \equiv \left(\frac{\partial M}{\partial T}\right)_Q = \frac{L\Omega_{d-2} r_+^{d-2}}{4G_{d+1}} \frac{\sqrt{1-q^{d-3}}[(d-2) - q^{d-3}]}{(d-2)q^{d-3} - 1}. \tag{13}
$$

The Correlated Stability Conjecture (CSC) [11] asserts that the uniform black branes are locally dynamically stable if and only if both the heat capacity is positive. The heat capacity changes its sign depending on the charge parameter q. The system is thermodynamically stable when the charge parameter q is above the critical value

$$q_c = (d-2)^{1/(3-d)}. \tag{14}$$

The corresponding normalzied chrage is denoted by Q_c. As we see in Fig. 1, the GL critical static mode disappears almost exactly at this point, and will have a realization of the CSC.

4.2. *Entropy and Free-Energy*

Because the asymptotic size of the circle is not fixed, we introduce variables that are invariant under rigid rescaling of the entire solution. Such invariant quantities can be obtained by multiplying K by suitable powers. For these quantities, the differences between the critical solution and non-uniform solution are

$$\begin{aligned}
\frac{\delta\mu}{\mu} &:= \frac{\delta M}{M} + (d-3)\frac{\delta K}{K} = \mu_1\epsilon^2 + O(\epsilon^4),\\
\frac{\delta s}{s} &:= \frac{\delta S}{S} + (d-2)\frac{\delta K}{K} = s_1\epsilon^2 + O(\epsilon^4),\\
\frac{\delta\tau}{\tau} &:= \frac{\delta T}{T} - \frac{\delta K}{K} = \tau_1\epsilon^2 + O(\epsilon^4),\\
\frac{\delta\vartheta}{\vartheta} &:= \frac{\delta Q}{Q} + (d-3)\frac{\delta K}{K} = \vartheta_1\epsilon^2 + O(\epsilon^4),
\end{aligned} \tag{15}$$

where the second-order coefficients are given by

$$\begin{aligned}
\mu_1 &= \begin{cases} \dfrac{2(3b_\infty - 2c_\infty)}{2+q} + \dfrac{k_1}{k_0} & \text{for } d=4,\\[2mm] \dfrac{2(2d-5)b_\infty + 2(d-2)(d-4)c_\infty}{d-2+q^{d-3}} + (d-3)\dfrac{k_1}{k_0} & \text{for } d\geq 5, \end{cases}\\
s_1 &= b_{0,1}(1) + (d-2)c_{0,1}(1) + \frac{1}{4}[b_{1,0}^2(1) + 2(d-2)b_{1,0}(1)c_{1,0}(1)\\
&\qquad + (d-2)^2c_{1,0}^2(1)] + (d-2)\frac{k_1}{k_0},\\
\tau_1 &= a_{0,1}(1) - b_{0,1}(1) - \frac{k_1}{k_0},\\
\vartheta_1 &= (d-3)\frac{k_1}{k_0}.
\end{aligned} \tag{16}$$

Here, the asymptotic coefficients are written as $A_\infty = \epsilon^2 a_\infty$ and so on.

What we are most interested in is the difference between the entropy of a uniform black string and that of a non-uniform one of the same mass and charge. The difference is evaluated as follows:

$$\frac{S_{\rm NU} - S_{\rm U}}{S_{\rm NU}} = \sigma_2 \epsilon^4 + O(\epsilon^6),$$

$$\sigma_2 = \frac{q^{d-3}\,\vartheta_1}{d-2-q^{d-3}} \left[\frac{(d-1)(d-2)}{2(d-3)} \vartheta_1 - (d-1)s_1 \right] - \frac{s_1}{2} \left[\frac{1-(d-2)q^{d-3}}{d-2-q^{d-3}} s_1 + \tau_1 \right], \qquad (17)$$

where we have used the first law of thermodynamics. We can discuss the thermodynamical stability also in the canonical ensumble, i.e., the comparison of Helmholtz free-energy ($F = M - TS$) with fixed mass and charge. The difference of free energy between the critical uniform solution and non-uniform solutions is evaluated as,

$$\frac{F_{\rm NU} - F_{\rm U}}{F_{\rm NU}} = \rho_2 \epsilon^4 + O(\epsilon^6),$$

$$\rho_2 = -\frac{(d-3)(1-q^{d-3})}{2[1+(d-2)q^{d-3}]} \left\{ \frac{(d-1)q^{d-3}\vartheta_1}{[1-(d-2)q^{d-3}]} \left[\frac{\vartheta_1}{d-3} + 2\tau_1 \right] + \tau_1 \left[s_1 + \frac{(d-2)-q^{d-3}}{1-(d-2)q^{d-3}} \tau_1 \right] \right\}, \qquad (18)$$

where we have used the first law of thermodynamics again. These formulae can be derived by starting from the frame with $\delta K = 0$ and re-expressing the results in terms of the invariant quantities.

For neutral case $Q = 0$, it is known that σ_2 is negative for $5 \leq D < 14$ and positive for $D \geq 14$ as first shown by [6]. In addition, it is known that ρ_2 is positive for $5 \leq D < 13$ and negative $D \geq 13$ [7]. This means that the phase transition in microcanonical (canonical) ensemble is first order (discontinuous) in lower dimensions $4 \leq D < 14$ ($4 \leq D < 13$) and higher order (continuous) in higher dimensions $D \geq 14$ ($D \geq 13$) in neutral case.

The quantities σ_2 and ρ_2 are numerically plotted as functions of Q in Fig. 2. We see that there exists the range of charge in which σ_2 (ρ_2) is positive (negative). This means that the order of phase transition change even in a certain fixed spacetime dimension. According to the discussion in Refs. [6,9], it suggests that the final fate of GL instability would be non-uniform black string rather than black hole in the region of $\sigma_2 > 0$ ($\rho_2 < 0$). In the canonical ensemble, this type of inversion phenomenon occurs, i.e., ρ_2 becomes positive and the non-uniform black string is disfavored. These change of sign in σ_2 and ρ_2 can be seen every case we examined. In other words, the order of phase transition in higher dimensions than 13 can be low order (discontinuous) for charged cases.

5. Conclusion

We have constructed gauge-charged non-uniform black strings by static perturbations up to third order and investigated their thermodynamical properties in the large range of spacetime dimensions. At the first order of perturbation, we see a realization of the correlated stability conjecture, i.e., we see that the dynamical instability disappears almost exactly at the background solution becomes thermodynamically stable. The behavior of critical wave length around the boundary of thermodynamical stability was discussed in [12]. It is interesting to investigate in this line. With using the results of higher-order perturbations, we have found that there exist non-uniform black strings being thermodynamically favored over the uniform black strings. It was known that *neutral* black strings have larger entropy (lower free-energy) than uniform ones only for large spacetime dimensions, $D \geq 14$ ($D \geq 13$). We have seen that non-uniform theremodynamically favored black strings/branes are possible in lower spacetime dimensions $6 \leq D \leq 14$ for charged cases.

The discovery of thermodynamically stable phase of non-uniform solutions obtained in this article is followed by a natural question whether they are dynamically stable or not. Unfortunately, there is no simple criterion of dynamical stability for black strings without translational symmetry. Dynamical stability have to be examined independently, which is quite interesting.

References

1. R. Gregory and R. Laflamme, *Phys. Rev. Lett.* **70**, 2837 (1993).
2. G. T. Horowitz and K. Maeda, *Phys. Rev. Lett.* **87**, 131301 (2001).
3. M. W. Choptuik, L. Lehner, I. Olabarrieta, R. Petryk, F. Pretorius and H. Villegas, *Phys. Rev. D* **68**, 044001 (2003).
4. H. Kudoh and T. Wiseman, *Phys. Rev. Lett.* **94**, 161102 (2005).
5. S. S. Gubser, *Class. Quant. Grav.* **19**, 4825 (2002).
6. E. Sorkin, *Phys. Rev. Lett.* **93**, 031601 (2004).
7. H. Kudoh and U. Miyamoto, *Class. Quant. Grav.* **22**, 3853 (2005).
8. T. Harmark and N. A. Obers, JHEP **0409**, 022 (2004).
9. J. L. Hovdebo and R. C. Myers, Phys. Rev. D **73**, 084013 (2006).
10. G. T. Horowitz and A. Strominger, Nucl. Phys. B **360**, 197 (1991).
11. S. S. Gubser and I. Mitra, arXiv:hep-th/0009126.
12. S. S. Gubser and A. Ozakin, JHEP **0305**, 010 (2003).

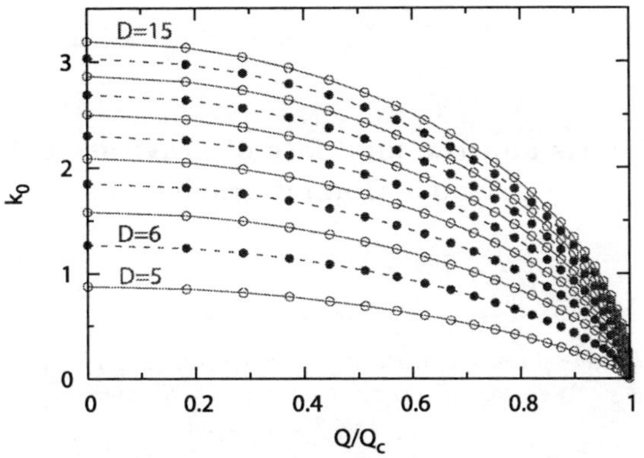

Fig. 1. Charge dependence of GL critical wave number k_0, normalized by the inverse of the horizon radius. We see that the critical wavenumbers decrease monotonically as the background charge increases. As expected the correlated stability conjecture, this decreasing wavenumber almost exactly vanishes at the critical charge Q_c, above which the background uniform black strings are thermodynamically stable.

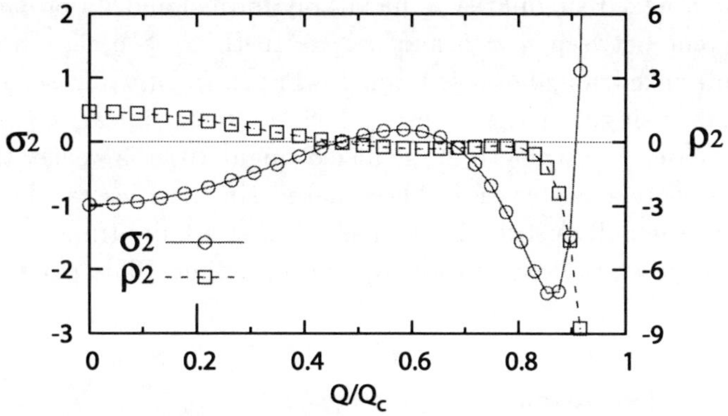

Fig. 2. The difference of entropy and free energy between the uniform and non-uniform black string solutions, (17) and (18), in $D = 6$. One can see that there are regions in which that the non-uniform black string is favored ($\sigma_2 > 0$ or $\rho_2 < 0$) over the uniform solution.

Quasi-local mass and the final fate of gravitational collapse in Gauss-Bonnet gravity

H. MAEDA

Advanced Research Institute for Science and Engineering,
Waseda University, Okubo 3-4-1, Shinjuku,
Tokyo 169-8555, Japan

E-mail: hideki@gravity.phys.waseda.ac.jp

We obtain a general spherically symmetric solution of a null dust fluid in $n(\geq 4)$-dimensions in Gauss-Bonnet gravity. This solution is a generalization of the n-dimensional Vaidya-(anti)de Sitter solution in general relativity. For $n = 4$, the Gauss-Bonnet term in the action does not contribute to the field equations, so that the solution coincides with the Vaidya-(anti)de Sitter solution. Using the solution for $n \geq 5$ with a specific form of the mass function, we present a model for a gravitational collapse in which a null dust fluid radially injects into an initially flat and empty region. It is found that a naked singularity is inevitably formed and its properties are quite different between $n = 5$ and $n \geq 6$. In the $n \geq 6$ case, a massless ingoing null naked singularity is formed, while in the $n = 5$ case, a massive timelike naked singularity is formed, which does not appear in the general relativistic case. The strength of the naked singularities is weaker than that in the general relativistic case. These naked singularities can be globally naked when the null dust fluid is turned off after a finite time and the field settles into the empty asymptotically flat spacetime. This paper is based on [1].

1. Model and solution

We begin with the following n-dimensional $(n \geq 5)$ action:

$$S = \int d^n x \sqrt{-g}\left[\frac{1}{2\kappa_n^2}(R - 2\Lambda + \alpha L_{GB})\right] + S_{\rm matter}, \tag{1}$$

where R and Λ are the n-dimensional Ricci scalar and the cosmological constant, respectively. κ_n is defined by $\kappa_n \equiv \sqrt{8\pi G_n}$, where G_n is the n-dimensional gravitational constant. The Gauss-Bonnet term L_{GB} is the combination of the Ricci scalar, Ricci tensor $R_{\mu\nu}$, and Riemann tensor $R^{\mu}{}_{\nu\rho\sigma}$ as

$$L_{GB} = R^2 - 4R_{\mu\nu}R^{\mu\nu} + R_{\mu\nu\rho\sigma}R^{\mu\nu\rho\sigma}. \tag{2}$$

In 4-dimensional spacetime, the Gauss-Bonnet term does not contribute to the field equations. α is the coupling constant of the Gauss-Bonnet term. This type of action is derived in the low-energy limit of heterotic superstring theory [2]. In that case, α is regarded as the inverse string tension and positive definite. Therefore, only the case where $\alpha \geq 0$ is considered in this paper. We consider a null dust fluid as a matter field, whose action is represented by $S_{\rm matter}$ in Eq. (1).

The gravitational equation of the action (1) is

$$G^{\mu}{}_{\nu} + \alpha H^{\mu}{}_{\nu} + \Lambda \delta^{\mu}{}_{\nu} = \kappa_n^2 T^{\mu}{}_{\nu}, \tag{3}$$

where

$$G_{\mu\nu} = R_{\mu\nu} - \frac{1}{2}g_{\mu\nu}R, \tag{4}$$

$$H_{\mu\nu} = 2\Big[RR_{\mu\nu} - 2R_{\mu\alpha}R^{\alpha}_{\ \nu} - 2R^{\alpha\beta}R_{\mu\alpha\nu\beta} + R_{\mu}^{\ \alpha\beta\gamma}R_{\nu\alpha\beta\gamma}\Big] - \frac{1}{2}g_{\mu\nu}L_{GB} \tag{5}$$

$H_{\mu\nu} \equiv 0$ holds for $n = 4$. The energy-momentum tensor of a null dust fluid is

$$T_{\mu\nu} = \rho l_{\mu} l_{\nu}, \tag{6}$$

where ρ is the non-zero energy density and l_{μ} is a null vector.

We obain the general spherically symmetric solution:

$$ds^2 = -f(v,r)dv^2 + 2dvdr + r^2 d\Omega_{n-2}^2, \tag{7}$$

$$f \equiv 1 + \frac{r^2}{2\tilde{\alpha}}\left\{1 \mp \sqrt{1 + 4\tilde{\alpha}\left[\frac{m(v)}{r^{n-1}} + \tilde{\Lambda}\right]}\right\}, \tag{8}$$

$$\rho = \frac{n-2}{2\kappa_n^2 r^{n-2}}\dot{m}, \tag{9}$$

$$l_{\mu} = -\partial_{\mu} v, \tag{10}$$

where $\tilde{\alpha} \equiv (n-3)(n-4)\alpha$ and $\tilde{\Lambda} \equiv 2/[(n-1)(n-2)]\Lambda$ and the dot denotes the derivative with respect to v. v is the advanced time coordinate and hence a curve $v =$ const denotes a radial ingoing null geodesic. $d\Omega_{n-2}^2$ is the line element of the $(n-2)$-dimensional unit sphere and $m(v)$ is an arbitrary function of v. In order for the energy density to be non-negative, $\dot{m} \geq 0$ must be satisfied.

In the general relativistic limit $\tilde{\alpha} \to 0$, the minus-branch solution is reduced to

$$f = 1 - \frac{m(v)}{r^{n-3}} - \tilde{\Lambda} r^2, \tag{11}$$

which is the n-dimensional Vaidya-(anti)de Sitter solution, while there is no such limit for the plus-branch solution. In the static case $\dot{m} = 0$, the solution (8) is reduced to the solution independently discovered by Boulware and Deser [3] and Wheeler [4].

2. Naked singularity formation

In this section, we study the gravitational collapse of a null dust fluid in Gauss-Bonnet gravity and compare it with that in general relativity by use of the solution obtained in the previous section. We consider the minus-branch solution for $n \geq 5$ in order to compare with the general relativistic case. For simplicity, we do not consider a cosmological constant, i.e., $\tilde{\Lambda} = 0$. We consider the situation in which a null dust fluid radially injects at $v = 0$ into an initially Minkowski region ($m(v) = 0$ for $v < 0$). The form of $m(v)$ is assumed to be

$$m = m_0 v^{n-3}, \tag{12}$$

where m_0 is a positive constant. In this case, the solution is reduced to the n-dimensional self-similar Vaidya solution in the general relativistic limit $\tilde{\alpha} \to 0$.

From Eq. (9), it is seen that there is a central singularity at $r = 0$ for $v > 0$. The point of $v = r = 0$ is more subtle but will be shown to be singular as well below. We will study the nature of the singularity.

We have

$$\frac{dr}{dv} = \frac{f}{2} \tag{13}$$

along a future-directed outgoing radial null geodesic, so that the region with $f < 0$ is the trapped region. A curve $f = 0$ represents the trapping horizon, i.e., the orbit of the apparent horizon, which is

$$m(v) = m_0 v^{n-3} = \tilde{\alpha} r^{n-5} + r^{n-3} \tag{14}$$

and shown to be spacelike for $v > 0$ and $r > 0$, i.e., it is a future outer trapping horizon, which is a local definition of black hole. From Eq. (14), only the point $v = r = 0$ may be a naked singularity in the case of $n \geq 6$, while the central singularity may be naked for $0 \leq v \leq v_{\rm AH}$ in the case of $n = 5$, where $v_{\rm AH}$ is defined by

$$m_0 v_{\rm AH}^2 = \tilde{\alpha}. \tag{15}$$

In order to determine whether or not the singularity is naked, we investigate the future-directed outgoing geodesics emanating from the singularity. It is shown that if a futre-directed radial null geodesic does not emanate from the singularity, then a future-directed causal (excluding radial null) geodesic does not also. So we consider here only the future-directed outgoing radial null geodesics. We found the asymptotic solution of Eq. (13)

$$v \simeq 2r \tag{16}$$

for $n \geq 6$, while

$$v \simeq v_0 + \frac{2}{1 - \sqrt{m_0 v_0^2/\tilde{\alpha}}} r \tag{17}$$

for $n = 5$, where $0 \leq v_0 < v_{\rm AH}$ is satisfied. Along these geodesics, the Kretschmann invariant $K \equiv R_{\mu\nu\rho\sigma} R^{\mu\nu\rho\sigma}$ diverges for $r \to 0$, so that they are actually singular null geodesics. We have now shown that at least a locally naked singularity is formed, and consequently the strong version of cosmic censorship hypothesis (CCH) is violated [5].

For $n \geq 6$, the naked singularity at $v = r = 0$ is massless and has the ingoing-null structure. In the case of $n = 5$, on the other hand, the singularity at $r = 0$ and $v = v_0$ with $0 \leq v_0 < v_{\rm AH}$ is massive and has the timelike structure. These naked singularities can be globally naked if we consider the situation in which the null dust fluid is turned off at a finite time $v = v_{\rm f} > 0$, whereupon the field settles into the Boulware-Deser-Wheeler spacetime with $m = m_0 v_{\rm f}^{n-3}$, which is asymptotically flat.

Next we consider the general relativistic case, i.e., the n-dimensional Vaidya solution. In this case, the trapping horizon is represented by $v = v_{\rm h} r$, where

$$v_{\rm h} \equiv \frac{1}{m_0^{1/(n-3)}}, \tag{18}$$

and hence only the point $v = r = 0$ may be naked. We find the exact power-law solution of the null geodesic equation (13):

$$v = v_1 r \tag{19}$$

for

$$m_0 \leq \frac{(n-3)^{n-3}}{2^{n-3}(n-2)^{n-2}}, \tag{20}$$

where v_1 satisfies

$$m_0 v_1^{n-2} - v_1 + 2 = 0. \tag{21}$$

Along these geodesics, the Kretschmann invariant diverges for $r \to 0$, so that they are actually singular null geodesics. The naked singularity at $r = v = 0$ is massless and has the ingoing-null structure. We can show that when Eq. (20) is satisfied, the singularity is always globally naked, i.e., the weak CCH is violated.

Next, we investigate the curvature strength of the naked singularity. We define

$$\psi \equiv R_{\mu\nu} k^\mu k^\nu, \tag{22}$$

where $k^\mu \equiv dx^\mu/d\lambda$ is the tangent vector of the future-directed outgoing radial null geodesic which emanates from the singularity and is parameterized by an affine parameter λ. We evaluate the strength of the naked singularity by the dependence of ψ on λ near the singularity.

After straightforward calculations, we show that

$$\lim_{\lambda \to 0} \psi \propto \frac{1}{\ln(\lambda + 1)} \tag{23}$$

for $n \geq 5$ in Gauss-Bonnet gravity, while

$$\psi \propto \frac{1}{\lambda^2}. \tag{24}$$

for $n \geq 5$ in general relativity, where $\lambda = 0$ corresponds to the singularity. Consequently, it is concluded by comparing Eq. (23) with Eq. (24) that the strength of the naked singularity in Gauss-Bonnet gravity is weaker than that in general relativity.

3. Mass of naked singularity

In this section, we show that the higher-dimensional Misner-Sharp mass is not an appropriate quasi-local mass in Gauss-Bonnet gravity.

Firstly, we give a definition of the higher-dimensional Misner-Sharp mass. We consider the n-dimensional spherically symmetric spacetime $M^n \approx M^2 \times S^{n-2}$ with the general metric

$$g_{\mu\nu} = \mathrm{diag}(g_{AB}, r^2 \gamma_{ij}), \tag{25}$$

where g_{AB} is an arbitrary Lorentz metric on M^2, r is a scalar function on M^2 with $r = 0$ defining the boundary of M^2, and γ_{ij} is the unit curvature metric on S^{n-2}. The higher-dimensional Misner-Sharp mass $m_{\rm MS}$ is a scalar on M^2 defined by

$$m_{\rm MS} \equiv \frac{(n-2)V_{n-2}}{2\kappa_n^2} r^{n-3}(1 - r_{,A}r^{,A}). \tag{26}$$

There are several properties which a well-defined quasi-local mass should satisfy [6,7]. One of them states that there is an invariant mass function in spherically symmetric spacetime, that any definition of the quasi-local mass should be reduced to in the special case with spherical symmetry. In particular, in the Schwarzschild spacetime of mass $M_{\rm Sch}$, the invariant mass function should coincide with $M_{\rm Sch}$ because it is the unique vacuum solution in general relativity. In Gauss-Bonnet gravity, on the other hand, the unique spherically symmetric vacuum solution is the GB-Schwarzschild solution. Therefore, a well-defined quasi-local mass in Gauss-Bonnet gravity should be reduced to the mass for this solution in the spherically symmetric vacuum case.

The metric of the GB-Schwarzschild solution is given by

$$ds^2 = -f dv^2 + 2dvdr + r^2 d\Omega_{n-2}^2 \tag{27}$$

with

$$f = 1 + \frac{r^2}{2\tilde{\alpha}}\left(1 - \sqrt{1 + 4\tilde{\alpha}\frac{m}{r^{n-1}}}\right), \tag{28}$$

where m is a constant. The n-dimensional Schwarzschild solution is obtained in the limit of $\tilde{\alpha} \to 0$ as

$$f = 1 - \frac{m}{r^{n-3}}, \tag{29}$$

of which mass is given by $M_{\rm Sch} \equiv (n-2)V_{n-2}m/(2\kappa_n^2)$, where V_{n-2} is the volume of an $(n-2)$-dimensional unit sphere. On the other hand, the mass for the GB-Schwarzschild solution also coincides with $M_{\rm Sch}$ [8]. Thus, a well-defined quasi-local mass should be reduced to $M_{\rm Sch}$ for the spherically symmetric vacuum case in Gauss-Bonnet gravity as well as in general relativity.

From Eq. (26), we can show $m_{\rm MS} = M_{\rm Sch}$ for the n-dimensional Schwarzschild solution, while we obtain

$$m_{\rm MS} = -\frac{(n-2)V_{n-2}}{4\tilde{\alpha}\kappa_n^2} r^{n-1}\left(1 - \sqrt{1 + 4\tilde{\alpha}\frac{m}{r^{n-1}}}\right) \tag{30}$$

for the GB-Schwarzschild solution. Indeed, $m_{\rm MS}$ does not coincide with $M_{\rm Sch}$ for $n \geq 5$ for the latter case. Consequently, $m_{\rm MS}$ is not an appropriate quasi-local mass in Gauss-Bonnet gravity.

In this paper, $m(v)$ in Eq. (8) is adopted in order to evaluate the mass of the singularity. $m(v)$ is preferable rather than $m_{\rm MS}$ in Gauss-Bonnet gravity because it is reduced to $2\kappa_n^2 M_{\rm Sch}/[(n-2)V_{n-2}]$ for the static limit, i.e., in the GB-Schwarzschild solution.

4. Discussion and conclusions

We have obtained an exact solution in Gauss-Bonnet gravity, which represents the spherically symmetric gravitational collapse of a null dust fluid in $n(\geq 5)$-dimensions. For $n \geq 5$, the solution is reduced to the n-dimensional Vaidya-(anti)de Sitter solution in the general relativistic limit. For $n = 4$, the Gauss-Bonnet term does not contribute to the field equations, so that the solution coincides with the Vaidya-(anti)de Sitter solution. Applying the solution to the situation in which a null dust fluid radially injects into an initially Minkowski region, we have investigated the effects of the Gauss-Bonnet term on the final fate of the gravitational collapse.

We have assumed that the mass function has the form of $m(v) = m_0 v^{n-3}$. Then, it has been found that there always exists a future-directed outgoing radial null geodesic emanating from the singularity in Gauss-Bonnet gravity, i.e., a naked singularity is inevitably formed. On the other hand, in the general relativistic case, there exists such a null geodesic only when m_0 takes a sufficiently small value. This result implies that the effects of the Gauss-Bonnet term on gravity worsen the situation from the viewpoint of CCH rather than prevent naked singularity formation.

Furthermore, the Gauss-Bonnet term drastically changes the nature of the singularity and the whole picture of gravitational collapse. The picture of the gravitational collapse for $n = 5$ is quite different from that for $n \geq 6$. For $n \geq 6$, as well as the general relativistic case for $n \geq 4$, a massless ingoing null naked singularity is formed. On the other hand, for the special case $n = 5$, a massive timelike naked singularity is formed. It is found from Eq. (14) that the formation of a massive timelike singularity in 5-dimensions is generic for the general mass function $m(v)$ which satisfies $m(0) = 0$ and $\dot{m} \geq 0$.

Although naked singularities are inevitably formed in Gauss-Bonnet gravity, the Gauss-Bonnet term makes the strength of the naked singularity weaker than that in the general relativistic case. In association with this, there does exist a possible formulation of CCH, which asserts that the formation of weak naked singularities need not be ruled out [9]. In this sense, the Gauss-Bonnet term works well in the spirit of CCH.

Acknowledgments

I would like to thank T. Torii, U. Miyamoto, T. Harada, and M. Nozawa for discussion and comments. I would also like to thank N. Ohta for comments on references.

References

1. H. Maeda, *Class. Quantum Grav.* **23**, 2155 (2006).
2. D.J. Gross and J.H. Sloan, *Nucl. Phys.* **B291**, 41 (1987); M.C. Bento and O. Bertolami, *Phys. Lett.* **B368**, 198 (1996).
3. D.G. Boulware and S. Deser, *Phys. Rev. Lett.* **55**, 2656 (1985).
4. J.T. Wheeler, *Nucl. Phys.* **B268**, 737 (1986).
5. R. Penrose, *Riv. Nuovo Cim.* **1**, 252 (1969); R. Penrose, in *General Relativity, an Einstein Centenary Survey*, edited by S.W. Hawking and W. Israel (Cambridge University Press, Cambridge, England, 1979), p. 581.
6. S.A. Hayward, *Phys. Rev.* **D53**, 1938 (1996).
7. L.B. Szabados, *Living Rev. Relativity* **7**, 4 (2004).
8. Y.M. Cho and I.P. Neupane, *Phys. Rev.* **D66**, 024044 (2002); A. Padilla, *Class. Quantum Grav.* **20**, 3129 (2003); N. Deruelle, J. Katz, and S. Ogushi, *Class. Quantum Grav.* **21**, 1971 (2004).
9. F.J. Tipler, C.J.S. Clarke and G.F.R. Ellis, in *General Relativity and Gravitation*, edited by A. Held (Plenum, New York, 1980), vol. 2, p. 97.

Gyraton Solutions in Einstein-Maxwell Theory and Supergravity

Valeri P. Frolov

Theoretical Physics Institute, Department of Physics, University of Alberta, Edmonton, AB, Canada, T6G 2J1

E-mail: frolov@phys.ualberta.ca

A gyraton is an object moving with the speed of light and having finite energy and internal angular momentum (spin). We study gyraton solutions of the Einstein-Maxwell gravity and in supergravity. We demonstrate that in the both cases the solutions in 4 and higher dimensions reduce to linear problems in a Euclidean space.

1. Introduction

A gyraton is an object moving with the speed of light and having finite energy and internal angular momentum (spin). A physically interesting example of a gyraton-like object is a spinning (circular polarized) beam-pulse of the high-frequency electromagnetic or gravitational radiation. Studies of the gravitational fields of beams and pulses of light have a long history. Tolman [1] found a solution in the linear approximation. Peres [2,3] and Bonnor [4] obtained exact solutions of the Einstein equations for a pencil of light. Polarization effects were studied in [5]. The generalization of these solutions to the case where the beam of radiation carries angular momentum and the number of spacetime dimensions is arbitrary was obtained in [6,7]. Such solutions are important for study mini black hole formation in a high-energy collision of two particles with spin. In the present paper we obtain generalisation of these solutions to the case when a gyraton has an electric or Kalb-Ramon charge.

2. Electrically charged gyratons

We start by considering the gyraton solutions of the Einstein-Maxwell equations [8]. Our starting point is the following ansatz for the metric in the $D = n + 2$ dimensional spacetime

$$ds^2 = d\bar{s}^2 + 2\left(a_u du + a_a dx^a\right) du\,. \tag{1}$$

Here

$$d\bar{s}^2 = -2\,du\,dv + d\mathbf{x}^2 \tag{2}$$

is a flat D–dimensional metric, and $a_u = a_u(u, x^a)$, $a_a = a_a(u, x^a)$. The spatial part of the metric $d\mathbf{x}^2 = \delta_{ab}dx^a dx^b$ in the n–dimensional space R^n is flat. Here and later the Greek letters μ, $\nu\ldots$ take values $1, \ldots, D$, while the Roman low-case letters a, $b, \ldots$ take values $3, \ldots, D$. We denote

$$a_\mu = a_u \delta^u_\mu + a_a \delta^a_\mu .$$

The form of the metric (1) is invariant under the following coordinate transformation

$$v \to v + \lambda(u, \mathbf{x})\,, \quad a_\mu \to a_\mu + \lambda_{,\mu}\,. \tag{3}$$

This transformation for a_μ reminds the gauge transformation for the electromagnetic potential. The quantity

$$f_{\mu\nu} = \partial_\mu a_\nu - \partial_\nu a_\mu ,$$

which is the gravitational analogue of the electromagnetic strength tensor, is gauge invariant.

The metric (1) admits a parallelly propagating null Killing vector $l = l^\mu \partial_\mu = \partial_v$. It is the most general D-dimensional null Brinkmann metrics [9] with flat transverse space. The metric for the gravitational field of a relativistic neutral gyratons is of the form (1) [6,7].

Let us denote $l_\mu = -\delta^u_\mu$. Then it is easy to check that

$$l^\epsilon a_\epsilon = l^\epsilon f_{\epsilon\mu} = 0\,. \tag{4}$$

Now let us consider the electromagnetic field in the spacetime (1). We choose its vector potential A_μ in the form $A_\mu = A_u \delta^u_\mu + A_a \delta^a_\mu$, where the functions A_u and A_a are independent of the null coordinate v. The electromagnetic gauge transformations $A_\mu \to A_\mu + \Lambda_{,\mu}$ with $\Lambda = \Lambda(u, \mathbf{x})$ preserve the form of the potential. The Maxwell strength tensor for this vector potential is

$$F_{\mu\nu} = \partial_\mu A_\nu - \partial_\nu A_\mu\,.$$

The quantities A_μ and $F_{\mu\nu}$ obey the relations

$$l^\epsilon A_\epsilon = l^\epsilon F_{\epsilon\mu} = 0\,. \tag{5}$$

3. Reduction of the Einstein-Maxwell equations

The Einstein-Maxwell action in higher dimensions reads

$$S = \frac{1}{16\pi G}\int d^D\mathbf{x}\sqrt{|g|}\left[R - \frac{1}{4}F^{\mu\nu}F_{\mu\nu} + A_\mu J^\mu\right].$$

Here G is the gravitational coupling constant in D−dimensional spacetime. The stress-energy tensor for the electromagnetic field is

$$T_{\mu\nu} = \frac{1}{16\pi G}\left[F_\mu{}^\epsilon F_{\nu\epsilon} - \frac{1}{4}g_{\mu\nu}\mathbf{F}^2\right], \tag{6}$$

where $\mathbf{F}^2 = F_{\epsilon\sigma}F^{\epsilon\sigma}$.

Our aim now is to find solutions of the system of Einstein-Maxwell equations

$$R_{\mu\nu} - \frac{1}{2}g_{\mu\nu}R = 8\pi G T_{\mu\nu}\,, \tag{7}$$

$$F_\mu{}^\nu{}_{;\nu} = J_\mu\,, \tag{8}$$

for the adopted field ansatz (1) and (5). Direct calculations [7] show that for the metric (1) the scalar curvature vanishes $R = 0$ and the only nonzero components of the Ricci tensor are

$$R_{ua} = \frac{1}{2}f_{ab}{}^{,b}, \tag{9}$$

$$R_{uu} = -(a_u)_{,a}^{;a} + \frac{1}{4}f_{ab}f^{ab} + \partial_u(a_a{}^{,a})\,. \tag{10}$$

Since $R = \delta^{ab}R_{ab} = 0$ one has $\mathbf{F}^2 = 0$, and hence $F_{ab} = 0$. Thus in a proper gauge the transverse components of the electromagnetic vector potential vanish, $A_a = 0$. Let us denote $\mathcal{A} = A_u$, then the only non-vanishing components of $F_{\mu\nu}$ and $T_{\mu\nu}$ are

$$F_{ua} = -F_{au} = -\mathcal{A}_{,a}\,, \qquad T_{uu} = \frac{1}{16\pi G}(\nabla\mathcal{A})^2\,. \tag{11}$$

where $(\nabla\mathcal{A})^2 = \delta^{ab}\mathcal{A}_{,a}\mathcal{A}_{,b}$.

Thus the requirement that the electromagnetic field is consistent with the Einstein equations for the gyraton metric ansatz (1) implies that the vector potential a_μ can be chosen $a_\mu \sim l_\mu$, i.e., to be aligned with the null Killing vector. In this case $F_{\mu\nu}$ and all its covariant derivatives are also aligned with l_μ. Together with the orthogonality conditions (5) these properties can be used to prove that all local scalar invariants constructed from the Riemann tensor, the Maxwell tensor (11) and their covariant derivatives vanish. This property generalizes the analogous property for non-charged gyratons [7] and gravitational shock waves. This property can also be used to prove that the charged gyraton solutions of the Einstein-Maxwell equations are also exact solutions of any other nonlinear electrodynamics and the Einstein equations [8].

Eventually the Einstein equations reduce to the following two sets of equations in n-dimensional flat space R^n

$$(a_u)_{,a}^{;a} - \partial_u(a_a{}^{,a}) = \frac{1}{4}f_{ab}f^{ab} - \frac{1}{2}(\nabla\mathcal{A})^2\,, \tag{12}$$

$$f_{ab}{}^{,b} = 0\,, \qquad f_{ab} = a_{b,a} - a_{a,b}\,. \tag{13}$$

We are looking for the field outside the region occupied by the gyraton, where $J_\mu = 0$. The Maxwell equations then reduce to the relation

$$\triangle\mathcal{A} = 0\,. \tag{14}$$

Here $\triangle$ is a flat n-dimensional Laplace operator. The electric charge of the gyraton is determined by the total flux of the electric field across the surface $\partial\Sigma$ surrounding it

$$Q = \frac{1}{16\pi G}\int_{\partial\Sigma} F^{\mu\nu}d\sigma_{\mu\nu}. \tag{15}$$

For a charged gyraton one has

$$Q \equiv \int_{u_1}^{u_2} du\,\rho(u). \tag{16}$$

where $\rho(u)$ is a linear charge density. By comparing (15) and (16) one can conclude that in the asymptotic region $r \to \infty$ the following relations are valid

$$F_{ur} \approx \begin{cases} \dfrac{16\pi G(n-2)g_n\rho(u)}{r^{n-1}}\,, & \text{for } n > 2, \\ \\ \dfrac{8G\rho(u)}{r}\,, & \text{for } n = 2. \end{cases}$$

Here $g_n = \Gamma(\frac{n-2}{2})/(4\pi^{n/2})$.

Now we return to the Einstein equations. The combination which enters the left hand side of (12) is invariant under the transformation (3). One can use this transformation to put $a_a{}^{,a} = 0$. We shall use this "gauge" choice and denote $a_u = \frac{1}{2}\Phi$ in this "gauge". The equations (12)-(13) take the form

$(\Phi = \varphi + \psi)$

$$\triangle\varphi = 0\,, \tag{17}$$

$$\triangle\psi = \frac{1}{2}f_{ab}f^{ab} - (\nabla\mathcal{A})^2\,, \tag{18}$$

$$\triangle a_a = 0\,. \tag{19}$$

The set of equations (14) and (17)–(19) determines the metric

$$ds^2 = d\bar{s}^2 + \Phi du^2 + 2a_a dx^a\, du$$

and the electromagnetic field $\mathcal{A}$ of a gyraton. (Let us emphasize again that these equations are valid only outside the region occupied by the gyraton.) These equations are linear equations in an Euclidean n-dimensional space R^n. The equations (14) and (17) coincide with the equations for electric potential created by a point like source, while the equation (19) formally coincides with the equation for the magnetic field. The last equation (18) is a linear equation in the Euclidean space which can be solved after one finds solutions for f_{ab} and $\mathcal{A}$. Thus for a chosen ansatz for the metric and the electromagnetic fields, the solution of the Einstein-Maxwell equations in $D-$dimensional spacetime reduce to linear problems in an Euclidean $n-$dimensional space ($n = D-2$). Special solutions of these equations can be found in [8].

4. Gyratons in supergravity

We discuss now the gyratonic solutions in the supergravity [10]. We consider the massless bosonic sector of supergravity. We restrict ourselves by discussing what is called the common sector. The fields in the common sector are the metric $g_{\mu\nu}$, the Kalb-Ramond antisymmetric field $B_{\mu\nu}$ and the dilaton field ϕ. The corresponding action, which is also the low-energy superstring effective action, is

$$S = \frac{1}{16\pi G}\int d^Dx\,\sqrt{|g|}e^{-2\phi}[R - 4(\nabla\phi)^2$$

$$-\frac{1}{12}H_{\mu\nu\lambda}H^{\mu\nu\lambda}] + \frac{1}{2}\int d^Dx\,\sqrt{|g|}B_{\mu\nu}J^{\mu\nu} + \mathcal{S}_m\,. \tag{20}$$

Here G is the $D-$dimensional gravitational (Newtonian) coupling constant, and $\mathcal{S}_m$ is the action for the string matter source. The string coupling constant g_s is determined by the vacuum expectation value of the dilaton field ϕ_0, $g_s = \exp(\phi_0)$. The 3-form flux

$$H_{\mu\nu\lambda} = \partial_\mu B_{\nu\lambda} + \partial_\nu B_{\lambda\nu} + \partial_\lambda B_{\mu\nu} \tag{21}$$

is the Kalb-Ramond (KR) field strength and $B_{\mu\nu}$ is its anti-symmetric 2-form potential. The field $H_{\mu\nu\lambda}$ is invariant under the gauge transformation

$$B_{\mu\nu} \to B_{\mu\nu} + \partial_\mu \Lambda_\nu - \partial_\nu \Lambda_\mu . \tag{22}$$

$J^{\mu\nu}$ is the antisymmetric tensor of the current which plays the role of a source for the KR field.

For example, for the interaction of the KR field with a fundamental string described by the action

$$S_{int} = -\frac{q}{2} \int d^2\zeta \epsilon^{ab} B_{\mu\nu} \frac{\partial X^\mu}{\partial \zeta^a} \frac{\partial X^\nu}{\partial \zeta^b} \tag{23}$$

this current is

$$J^{\mu\nu}(x) = \frac{q}{2} \int d^2\zeta \frac{\delta^D(x - X(\zeta))}{\sqrt{|g|}} \epsilon^{ab} \frac{\partial X^\mu}{\partial \zeta^a} \frac{\partial X^\nu}{\partial \zeta^b} . \tag{24}$$

Here ϵ^{ab} is the antisymmetric symbol, $\zeta^a = (\tau, \sigma)$ are parameters on the string surface and the functions $X^\mu = X^\mu(\zeta)$ determine the embedding of the string worldsheet in the bulk (target) spacetime. The parameter q is the "string charge". The current $J^{\mu\nu}$ is tangent to the worldsheet of the string, $J^{[\mu\nu} X^{\lambda]}_{,c} = 0$.

We shall study a special class of gyraton solutions for which the dilaton field is constant, i.e., $e^\phi = g_s$. In this case the field equations are

$$R_{\mu\nu} - \frac{1}{2} g_{\mu\nu} R = T_{\mu\nu} + g_s^2 \kappa \mathcal{T}_{\mu\nu} , \tag{25}$$

$$H_{\mu\nu}{}^{l}{}_{;l} = 8\kappa J_{\mu\nu} . \tag{26}$$

Here the stress-energy tensor for the 3-form flux is

$$T_{\mu\nu} = \frac{1}{12} (3 H_{\mu\lambda\rho} H_\nu{}^{\lambda\rho} - \frac{1}{2} g_{\mu\nu} H_{\rho\sigma\lambda} H^{\rho\sigma\lambda}) , \tag{27}$$

and $\kappa = 8\pi g_s^2 G$. $\mathcal{T}_{\mu\nu}$ which enters the equation (25) is the stress-energy of the matter (string) which we shall specify later.

Let Σ be a $(D-2)$–dimensional spacelike surface, and $\partial\Sigma$ be its boundary. We define the charge of the fundamental string intersecting Σ by Gauss's law as

$$Q := \int_{\partial\Sigma} d\sigma_{D-3} *_D H_3 = \int d\sigma_{\mu\nu\lambda} H_{\mu\nu\lambda} . \tag{28}$$

By using the Stoke's theorem and (26) one has

$$Q = \int d\sigma_{\mu\nu} J^{\mu\nu} = 8\kappa q . \tag{29}$$

Here $d\sigma_{\mu_1\cdots\mu_n} := i_{\mu_1}\cdots i_{\mu_n}(*1)$ in which $*1$ is the volume form of D-dimensional spacetime and $i_{\mu u} : \Lambda^p T^* \to \Lambda^{p-1}T^*$, $i_\mu dx^{\nu_1}\wedge\cdots\wedge dx^{\nu_p} = p\delta_\mu^{[\nu_1} dx^{\nu_2}\wedge\cdots\wedge dx^{\nu_p]}$.

In what follows, we consider the gravitational and KR fields outside the sources, that is in the region where $\mathcal{T}_{\mu\nu} = 0$ and $J_{\mu\nu} = 0$. The relation (29) will be used to relate the parameters which enter a solution to the charge of the string.

5. Ansatz for Supergravity Gyraton

We use the same ansatz (1) for the metric as in the case of an electrically charged gyraton. For the KR field potential we use the ansatz similar to the one adopted for the electromagnetic gyratons (5). Namely we postulate that $B_{\mu\nu} = B_{\mu\nu}(u, \boldsymbol{x})$ and

$$l^\mu B_{\mu\nu} = 0\,. \tag{30}$$

It is easy to check that

$$l^\mu H_{\mu\nu\lambda} = 0. \tag{31}$$

The imposed constraints imply that the only non-vanishing components of $B_{\mu\nu}$ are $B_{ua}(u,\boldsymbol{x})$ and $B_{ab}(u,\boldsymbol{x})$, and of $H_{\mu\nu l}$ are $H_{uab}(u,\boldsymbol{x})$ and $H_{abc}(u,\boldsymbol{x})$. Moreover, to preserve the constraint (30) under gauge transformation, we should impose

$$\partial_v \Lambda_\mu - \partial_\mu \Lambda_v = 0. \tag{32}$$

The equation (31) implies that

$$H^2 = \boldsymbol{H}^2 \equiv H_{abc}H^{abc}. \tag{33}$$

We discuss now the ansatz for $\mathcal{T}_{\mu\nu}$ which enters the equation (26). We require that this tensor obeys the conservation low

$$\mathcal{T}^{\mu\nu}{}_{;\nu} = 0\,, \tag{34}$$

and is aligned to the null Killing vector l_μ

$$\mathcal{T}_{\mu\nu} = l_{(\mu}p_{\nu)}\,, \qquad l^\mu p_\mu = 0\,. \tag{35}$$

The last condition guarantees that the trace of $\mathcal{T}^{\mu\nu}$ vanishes, $\mathcal{T}^\mu_\mu = 0$. For the metric (1) these conditions are satisfied when

$$p_\mu = p_\mu(u, x^a)\,, \qquad p^a{}_{,a} = 0\,. \tag{36}$$

This can be checked by using the condition $l_{\mu;\nu} = 0$. Bonnor [5] called such matter in 4-dimensional spacetime spinning null fluid.

6. Reduction of supergravity equations

The only non-vanishing components of the Ricci tensor for the metric (1) are given by equations (9) and (10). These relations imply $R = 0$. Since $\mathcal{T}^{\mu}_{\mu} = 0$ the field equation (25) yields $T^{\mu}_{\mu} = 0$. This equation implies $\boldsymbol{H}^2 = 0$, and hence $H_{abc} = 0$. Therefore, the only non-vanishing components of $H_{\mu\nu\lambda}$ are $H_{uab}(u, \boldsymbol{x})$. This means that $H_{\mu\nu\lambda} = l_{[\mu} P_{\nu\lambda]}$, so that the field strength $H_{\mu\nu\lambda}$ is aligned to the null Killing vector l_{μ}.

The field equations (25)–(26) reduce to

$$(a_u)_{;a}^{;a} - \partial_u (a_a^{;a}) = \frac{1}{4}\left(f_{ab} f^{ab} - H_{uab} H_u{}^{ab}\right) + \kappa p_u \,, \tag{37}$$

$$f_{ab}{}^{,b} = -\kappa p_a \,, \tag{38}$$

$$H_{ua}{}^{b}{}_{,b} = 8\kappa J_{ua} \,. \tag{39}$$

The last two relations are linear differential equation in the n–dimensional Euclidean space ($n = D-2$). They can be solved for f_{ab} and H_{uab} once the source J_{ua} and the distribution for the source for gravito-magnetic field f_{ab} is given. After this we can solve the first equation for a_u, which for a given right-hand-side is also linear.

On the other hand, we can also solve the constraint $H_{abc} = 0$ by the following ansatz for the 2-form potential

$$B_{\mu\nu} = A_{\mu} l_{\nu} - A_{\nu} l_{\mu} \,. \tag{40}$$

From $l^{\mu} B_{\mu\nu} = 0$, we have $l^{\mu} A_{\mu} = 0$. This is equivalent to choose a gauge so that the only non-vanishing component of $B_{\mu\nu}$ is $B_{ua} = A_a(u, \boldsymbol{x})$, and of $H_{\mu\nu\lambda}$ is

$$H_{uab} = \partial_b A_a - \partial_a A_b \equiv F_{ba}. \tag{41}$$

Let us denote

$$\Phi = 2a_u \,, \quad \boldsymbol{a} = a_a \,, \quad \boldsymbol{f} = f_{ab} \tag{42}$$

$$\boldsymbol{A} = A_a \,, \quad \boldsymbol{F} = F_{ab} \,, \quad \boldsymbol{J} = J_{ua} \,, \quad \boldsymbol{p} = p_a \,. \tag{43}$$

In these notations the gyraton metric is

$$ds^2 = d\bar{s}^2 + \Phi du^2 + 2(\boldsymbol{a}, d\boldsymbol{x}) du \,, \tag{44}$$

and the field equations (37) and (39) reduce to

$$\triangle\Phi - 2\partial_u (\boldsymbol{\nabla} \cdot \boldsymbol{a}) = \frac{1}{2}\left(\boldsymbol{f}^2 - \boldsymbol{F}^2\right) + 2\kappa p_u \,, \tag{45}$$

$$\triangle \boldsymbol{A} + \boldsymbol{\nabla}(\boldsymbol{\nabla} \cdot \boldsymbol{A}) = 8\kappa \boldsymbol{J}. \tag{46}$$

Here $\nabla = \partial_a$, and $\triangle$ is the Laplacian operator in the n-dimensional Euclidean space.

Using the coordinate and electromagnetic gauge transformations one can put

$$\nabla \cdot \boldsymbol{A} = 0\,, \qquad \nabla \cdot \boldsymbol{a} = 0\,. \tag{47}$$

For these gauge fixing conditions the equations (45), (38) and (46) take the form ($\Phi = \varphi + \psi$)

$$\triangle \varphi = 2\kappa p_u\,, \tag{48}$$

$$\triangle \psi = \frac{1}{2}\left(\boldsymbol{f}^2 - \boldsymbol{F}^2\right)\,, \tag{49}$$

$$\triangle \boldsymbol{a} = \kappa \boldsymbol{p}\,, \tag{50}$$

$$\triangle \boldsymbol{A} = 8\kappa \boldsymbol{J}\,. \tag{51}$$

It is interesting to note that the magnetic and gravitomagnetic terms enter the right hand side of (49) with the opposite signs. A special type of solutions is the case when these terms cancel one another, so that the equation for ψ outside the matter source becomes homogeneous. We call such solutions *saturated.* The condition of saturation is $\boldsymbol{f}^2 = \boldsymbol{F}^2$. This condition can be achieved by letting $\boldsymbol{p} = 8\boldsymbol{J}$ as suggested by (50) and (51).Solutions of the obtained field equations for a ring-like string configuration of gyratons are obtained in [10].

7. Summary and discussions

We demonstrated that the vacuum Einstein equations for the gyraton-type metric (1) in an arbitrary number of spacetime dimensions D can be reduced to linear problems in the Euclidean $(D-2)$-dimensional space. These problems are: (1) To find a static electric field created by a point-like source; (2) To find a magnetic field created by a point-like source. The retarded time u plays the role of an external parameter. One can include u-dependence by making the coefficients in the harmonic decomposition for φ and $\mathbf{A}$ to be arbitrary functions of u. After choosing the solutions of these two problems one can define ψ by means of equations (18) and (49).

It should be emphasized that the point- or line-like sources are certainly an idealization. In [6] it was shown that gyraton solutions can describe the gravitational field of beam-pulse spinning radiation. In such a description one uses the geometric optics approximation. For its validity the size of the cross-section of the beam must be much larger than the wave-length of the radiation. In the presence of spin J one can expect additional restrictions on the minimal size of both, the cross-section size and the duration of the pulse. As usual in physics, one must have in mind that in the possible physical applications the obtained solution is valid only outside some region surrounding the immediate neighborhood of the singularity.

The obtained gyraton solutions might be used, for example, for study the gravitational interaction of ultrarelativistic particles with spin and charge. The gyraton metrics might be also interesting as possible exact solutions in the string theory. The generalization of the gyraton-type solutions to the case when a spacetime is asymptotically AdS was obtained recently in [11].

Acknowledgments

This work was supported by NSERC and by the Killam Trust.

References

1. R. C. Tolman, *Relativity, Thermodynamics, and Cosmology*, p.272, Oxford, Clarendon Press (1934).
2. A. Peres, Phys. Rev. Lett. **3**, 571 (1959).
3. A. Peres, Phys. Rev. **118**, 1105 (1960).
4. W. B. Bonnor, Commun. Math. Phys. 13, 163 (1969).
5. W. B. Bonnor, Int. J. Theor. Phys. **3**, 257 (1970).
6. V. P. Frolov and D. V. Fursaev, Phys.Rev. **D71**,104034 (2005).
7. V. P. Frolov, W. Israel and A. Zelnikov, Phys.Rev. **D72** 084031 (2005).
8. V. P. Frolov and A. Zelnikov , *Class.Quant. Gravity* **23** 2119 (2006).
9. H.W. Brinkmann, Math. Ann. **94**, 119 (1925).
10. V. P. Frolov and F.-L. Lin, "String gyratons in supergravity", hep-th/0603018; Phys. Rev. **D73** (accepted for publication) (2006).
11. V.P. Frolov and A. Zelnikov, hep-th/0509044; Phys.Rev. **D72** 104005 (2005).

Stationary Spacetime with Intersecting Branes in M/Superstring Theory*

Kei-ichi MAEDA†
Dept. of Physics, Waseda University,
Shinjuku, Tokyo 169-8555, JAPAN
E-mail: maeda@waseda.jp

We study a stationary "black" brane in M/superstring theory. Assuming BPS-type relations between the first-order derivatives of metric functions, we present general stationary black brane solutions with a traveling wave for the Einstein equations in D-dimensions. The solutions are given by a few independent harmonic equations (and plus the Poisson equation). Using the hyperspherical coordinate system for a flat base space, we explicitly give the solutions in 11-dimensional M theory for the case with M2⊥M5 intersecting branes and a traveling wave. Compactifying these solutions into five dimensions, we present general solutions, which include the BMPV black hole and the Brinkmann wave solution. We prove that the solutions preserve the 1/8 supersymmetry if the gravi-electromagnetic field $\mathcal{F}_{ij}$, which is a rotational part of gravity, is self-dual. We also construct a supersymmetric rotating black hole in a compactified 4-dimensional spacetime by superposing an infinite number of BMPV black holes.

1. Introduction

A black hole is now one of the most important subjects in string theory. The Beckenstein-Hawking black hole entropy of an extreme black hole is obtained in string theory by statistical counting of the corresponding microscopic states. The fundamental unified theory is constructed in either

*This manuscript is based on the papers by K. Maeda and M. Tanabe [1] (paper I), and by K. Maeda, N. Ohta, and M. Tanabe [2] .

†This work is partially partially supported by the Grant-in-Aid for Scientific Research Fund of the JSPS (No. 17540268) and for the Japan-U.K. Research Cooperative Program, and by the Waseda University Grants for Special Research Projects and for The 21st Century COE Program (Holistic Research and Education Center for Physics Self-organization Systems) at Waseda University.

ten or eleven dimensions. M-theory is the best candidate for such a unified theory.

So far, we know several interesting black hole solutions in supergravity theories, which are obtained as an effective theory of a superstring model in a low energy limit. We also have black hole solutions in a higher-dimensional spacetime. In higher dimensions, there is no uniqueness theorem of black holes. In fact, we know a variety of black objects such as a black ring, which horizon has a topology of $S^1 \times S^2$.

When we discuss the entropy of black holes, we have to show the relation between black holes in four dimensions and more fundamental "black" branes either in ten or eleven dimensions, from which we obtain "black" holes (or rings) via compactification. Since the low energy limit of M/superstring theory coincides with a higher-dimensional supergravity model, it provides a natural framework to study "black" brane or BPS brane solutions. Therefore, here we study a class of intersecting brane solutions in higher dimensions.

In §2, we consider the dilaton coupling gauged supergravity actions in D dimensions, and derive the basic equations for a stationary "black" brane. The extra $(D-d)$-dimensions are filled by several branes with a traveling wave. In §3, we construct the solution in eleven (or ten) dimensions. In §4, we present the explicit solutions for $d=5$ by use of a hyperspherical coordinate system. We recover the BMPV solution as a special case. We also construct a supersymmetric rotating black hole in a compactified 4-dimensional spacetime by superposing an infinite number of BMPV black holes in §5.

2. Basic equations for a stationary spacetime with branes

We first present the basic equations for a stationary spacetime with intersecting branes and describe how to construct generic solutions. We consider the following bosonic sector of a low energy effective action of superstring theory or M-theory in D dimensions ($D \leq 11$):

$$S = \frac{1}{16\pi G_D} \int d^D X \sqrt{-g} \left[\mathcal{R} - \frac{1}{2}(\nabla\varphi)^2 - \sum_A \frac{1}{2 \cdot n_A!} e^{a_A \varphi} F_{\mathbf{n_A}}^2 \right], \qquad (1)$$

where $\mathcal{R}$ is the Ricci scalar of a spacetime metric $g_{\mu\nu}$, $F_{\mathbf{n_A}}$ is the field strength of an arbitrary form with a degree $n_A (\leq D/2)$, and a_A is its coupling constant with a dilaton field φ. Each index A describes a different type of brane.

The equations of motion are written in the following forms:

$$\mathcal{R}_{\mu\nu} = \frac{1}{2}\partial_\mu\varphi\partial_\nu\varphi + \sum_A \Theta_{\mathbf{n_A}\mu\nu},$$
$$\nabla^2\varphi = \sum_A \frac{a_A}{2\cdot n_A!} e^{a_A\varphi} F^2_{\mathbf{n_A}},$$
$$\partial_{\mu_1}(\sqrt{-g}\; e^{a_A\varphi} F_{\mathbf{n_A}}^{\ \mu_1\cdots\mu_{n_A}}) = 0\,, \tag{2}$$

where $\Theta_{\mathbf{n_A}\mu\nu}$ is the stress-energy tensor of the n_A-form, which is given by

$$\Theta_{\mathbf{n_A}\mu\nu} = \frac{1}{2\cdot n_A!} e^{a_A\varphi}\left[n_A F_{\mathbf{n_A}\mu}{}^{\rho\cdots\sigma} F_{\mathbf{n_A}\nu\rho\cdots\sigma} - \frac{n_A-1}{D-2} F^2_{\mathbf{n_A}} g_{\mu\nu}\right]. \tag{3}$$

We also have an additional equation, which is the Bianchi identity for the n_A-form, i.e.,

$$\partial_{[\mu} F_{\mathbf{n_A}\mu_1\cdots\mu_{n_A}]} = 0\,. \tag{4}$$

This is automatically satisfied if we introduce the potentials of n_A-form.

As for a metric of a spacetime with intersecting branes, we assume the following form:

$$ds^2 = 2e^{2\xi}du\left(dv + f du + \frac{\mathcal{A}_i}{\sqrt{2}}dx^i\right) + e^{2\eta}\sum_{i=1}^{d-1}(dx^i)^2 + \sum_{\alpha=2}^{p} e^{2\zeta_\alpha}(dy^\alpha)^2\,, \tag{5}$$

where $D = d + p$. Here we have used null coordinates; $u = -(t - y_1)/\sqrt{2}$ and $v = (t + y_1)/\sqrt{2}$. This metric form includes rotation of spacetime and a traveling wave. Since we are interested in a stationary solution, we assume that the metric components f, $\mathcal{A}_i$, ξ, η and ζ_α depend only on the spatial coordinates x^i in d-dimensions, which coordinates are given by $\{t, x^i(i = 1, 2, \cdots, d-1)\}$. In this setting, we set each brane A in a submanifold of p-spatial dimensions, which coordinates are given by $\{y_\alpha(\alpha = 1, 2, \cdots, p)\}$.

As for the n_A-form field with a q_A-brane, we assume that the source brane exists in the coordinates $\{y_1, y_{\alpha_2}, \cdots, y_{\alpha_{q_A}}\}$. The form field generated by an "electric" charge is given by the following form:

$$\begin{aligned} F_{\mathbf{n_A}} &= \partial_j E_A dx^j \wedge du \wedge dv \wedge dy_2 \wedge \cdots \wedge dy_{q_A} \\ &+ \frac{1}{\sqrt{2}}\partial_i B^A_j dx^i \wedge dx^j \wedge du \wedge dy_2 \wedge \cdots \wedge dy_{q_A}\,, \end{aligned} \tag{6}$$

where $n_A = q_A + 2$ and E_A and B^A_j are scalar and vector potentials. This setting automatically guarantees the Bianchi identity (4).

We can also discuss the form field generated by a "magnetic" charge by use of a dual ${}^*n_A$-field with ${}^*q_A$-brane, which is obtained by a dual transformation of the n_A-field with a q_A-brane ($^*n_A \equiv D - n_A, {}^*q_A \equiv {}^*n_A - 2$). In other words, the field components of $F_{\mathbf{n_A}}$ generated by a "magnetic" charge are described by the same form of (6) of the dual field $^*F_{\mathbf{n_A}} = F_{^*\mathbf{n_A}}$. When we sum up by the types of branes A, we treat $F_{^*\mathbf{n_A}}$, which is generated by a "magnetic" charge, as another independent form field with a different brane from $F_{\mathbf{n_A}}$, which is generated by an "electric" charge.

Setting

$$H_A \equiv \exp\left[-\left(2\xi + \sum_{\alpha=\alpha_2}^{\alpha_{q_A}} \zeta_\alpha - \frac{1}{2}\epsilon_A a_A \varphi\right)\right] \tag{7}$$

$$V \equiv \exp\left[2\xi + (d-3)\eta + \sum_{\alpha=2}^{p} \zeta_\alpha\right] = 1 \text{ (gauge condition)}, \tag{8}$$

where $\epsilon_A = +1$ or for n_A-form field ($F_{\mathbf{n_A}}$) or -1 for the dual field ($^*F_{\mathbf{n_A}}$), we find the basic equations as follows:

$$\partial^2 f = \frac{1}{8} e^{2(\xi-\eta)}\left[\mathcal{F}_{ij}^2 - \frac{1}{2}\sum_A \left(\mathcal{F}_{ij}^{(A)}\right)^2\right], \tag{9}$$

$$\partial^2 \xi = \frac{1}{2(D-2)}\sum_A (D - q_A - 3) H_A^2 (\partial E_A)^2, \tag{10}$$

$$\partial^j \mathcal{F}_{ij} + 2\mathcal{F}_{ij}\partial^j(\xi - \eta) = \sum_A H_A \mathcal{F}_{ij}^{(A)} \partial^j E_A, \tag{11}$$

$$\begin{aligned} &\partial^2\eta\delta_i^j + 2\partial_i\xi\partial^j\xi + (d-3)\partial_i\eta\partial^j\eta + \sum_{\alpha=2}^{p}\partial_i\zeta_\alpha\partial^j\zeta_\alpha \\ &= -\frac{1}{2}\partial_i\varphi\partial^j\varphi + \frac{1}{2}\sum_A H_A^2\left[\partial_i E_A \partial^j E_A - \frac{q_A+1}{(D-2)}(\partial E_A)^2\delta_i^j\right], \end{aligned} \tag{12}$$

$$\partial^2\zeta_\alpha = \frac{1}{2(D-2)}\sum_A \delta_{\alpha A} H_A^2 (\partial E_A)^2, \tag{13}$$

$$\partial^2\varphi = -\frac{1}{2}\sum_A \epsilon_A a_A H_A^2 (\partial E_A)^2, \tag{14}$$

$$\partial_j(H_A^2 \partial^j E_A) = 0, \tag{15}$$

$$\partial^j \mathcal{F}_{ij}^{(A)} = 0, \tag{16}$$

where ∂_i is a partial derivative $\partial/\partial x^i$ in a flat $(d-1)$-space, $\partial^2 \equiv \partial_i\partial^i$, and

$\mathcal{F}_{ij}$, $\mathcal{F}_{ij}^{(A)}$, and $\delta_{\alpha A}$ for each coordinate α are defined by

$$\begin{aligned} \mathcal{F}_{ij} &\equiv \partial_i \mathcal{A}_j - \partial_j \mathcal{A}_i \\ \mathcal{F}_{ij}^{(A)} &\equiv 2H_A \left(\mathcal{A}_{[i}\partial_{j]} E_A - \partial_{[i} B_{j]}^A \right) \\ \delta_{\alpha A} &\equiv \begin{cases} D - q_A - 3 & \alpha = \alpha_2, \ldots, \alpha_{q_A} \\ -(q_A + 1) & \text{otherwise} \end{cases} . \end{aligned} \tag{17}$$

Setting $\tilde{E}_A = E_A - E_A^{(0)}$, where $E_A^{(0)}$ is a constant, which is fixed by a boundary condition, and using Eqs. (15) and (16), we obtain from Eqs. (10), (13) and (14)

$$\partial^j \left[\partial_j \xi - \frac{1}{2(D-2)} \sum_A (D - q_A - 3) H_A^2 \tilde{E}_A \partial_j \tilde{E}_A \right] = 0, \tag{18}$$

$$\partial^j \left[\partial_j \zeta_\alpha - \frac{1}{2(D-2)} \sum_A \delta_{\alpha A} H_A^2 \tilde{E}_A \partial_j \tilde{E}_A \right] = 0, \tag{19}$$

$$\partial^j \left[\partial_j \varphi + \frac{1}{2} \sum_A \epsilon_A a_A H_A^2 \tilde{E}_A \partial_j \tilde{E}_A \right] = 0 . \tag{20}$$

This set of equations is a coupled system of elliptic-type differential equations, for which it is very difficult to find general solutions. Hence, we assume the following special relations:

$$\partial_j \xi = \frac{1}{2(D-2)} \sum_A (D - q_A - 3) H_A^2 \tilde{E}_A \partial_j \tilde{E}_A, \tag{21}$$

$$\partial_j \zeta_\alpha = \frac{1}{2(D-2)} \sum_A \delta_{\alpha A} H_A^2 \tilde{E}_A \partial_j \tilde{E}_A, \tag{22}$$

$$\partial_j \varphi = -\frac{1}{2} \sum_A \epsilon_A a_A H_A^2 \tilde{E}_A \partial_j \tilde{E}_A , \tag{23}$$

which guarantee that Eqs. (18), (19) and (20) are exact.

These equations are relations between the first-order derivatives of variables, which may be related to a BPS condition. In fact, if $\mathcal{F}_{ij}^{(A)}$ is proportional to $\mathcal{F}_{ij}$ and $\mathcal{F}_{ij}$ is self-dual, we prove that 1/8 supersymmetry remains in the solutions with M2$\perp$M5 branes for $D = 11$ supergravity theory (see Appendix A in the paper I [1]).

η is obtained from the gauge condition (8) and Eqs. (21)-(23) as

$$\partial^j \eta = -\frac{1}{2(D-2)} \sum_A (q_A + 1) H_A^2 \tilde{E}_A \partial^j \tilde{E}_A . \tag{24}$$

We have, however, another equation for η, i.e., Eq. (12), which should be satisfied as well. This consistency leads to two results: (1) a crossing rule of intersecting branes, that is, the crossing dimensions $\bar{q}_{AB}$ of q_A and q_B branes must satisfies

$$\bar{q}_{AB} = \frac{(q_A+1)(q_B+1)}{D-2} - 1 - \frac{1}{2}\epsilon_A a_A \epsilon_B a_B \,, \tag{25}$$

and (2) the relation between $\tilde{E}_A$ and H_A, that is,

$$E_A = -\sqrt{\frac{2(D-2)}{\Delta_A}}\left(1 - \frac{1}{H_A}\right), \quad (\text{or} \quad E_A = 0)\,. \tag{26}$$

Inserting this relation into Eq. (15), we obtain the equation for H_A as

$$\partial^2 H_A = 0\,. \tag{27}$$

From the relation (26) with Eqs. (10), (13) and (14), we obtain the solutions for metric functions in terms of the harmonic function H_A:

$$\begin{aligned} \xi &= -\sum_A \frac{D-q_A-3}{\Delta_A}\ln H_A, & \eta &= \sum_A \frac{q_A+1}{\Delta_A}\ln H_A, \\ \zeta_\alpha &= -\sum_A \frac{\delta_{\alpha A}}{\Delta_A}\ln H_A, & \varphi &= (D-2)\sum_A \frac{\epsilon_A a_A}{\Delta_A}\ln H_A\,. \end{aligned} \tag{28}$$

Next we analyze two equations (11) and (16) for $\mathcal{A}_i$ ($\mathcal{F}_{ij}$) and one Poisson equation (9) for f. We expect that each brane A has a charge $\mathcal{Q}_H^{(A)}$ (either electric or magnetic type), and then E_A becomes non-trivial, i.e., $H_A \neq 1$[a]. In this case, if we set

$$B_i^A = -\tilde{E}_A\mathcal{A}_i = -\sqrt{\frac{2(D-2)}{\Delta_A}}\frac{\mathcal{A}_i}{H_A}\,, \tag{29}$$

then we have

$$\mathcal{F}_{ij}^{(A)} = \sqrt{\frac{2(D-2)}{\Delta_A}}\mathcal{F}_{ij}\,. \tag{30}$$

Inserting Eqs. (28), we can show that two equations (11) and (16) are reduced to the following one equation:

$$\partial^j\mathcal{F}_{ij} = 0\,. \tag{31}$$

[a]When H_A=1, we find Brinkmann type wave solution[1]

Table 1. Examples of intersecting branes for $d = 5$. M2, M5 and W denote the location where the M2 brane, the M5 brane and a wave exist, respectively.

$d = 5$

y_1	y_2	y_3	y_4	y_5	y_6
M2	M2				
		M2	M2		
				M2	M2

$d = 5$

y_1	y_2	y_3	y_4	y_5	y_6
M2					M2
M5	M5	M5	M5	M5	
W					

B_i^A describes a magnetic-type field produced by a current appearing through rotation of a charged brane. It turns out that the condition (30) plays a key role for the system to keep supersymmetry (see the paper I[1]).

Finally we discuss the last equation (9) for f, i.e.

$$\partial^2 f = \frac{\beta}{8} \prod_A H_A^{-\frac{2(D-2)}{\Delta_A}} \mathcal{F}_{ij}^2 , \qquad (32)$$

where $\beta \equiv 1 - (D-2)\sum_A \Delta_A^{-1}$ is just a constant. If $(D-2)\sum_A \Delta_A^{-1} = 1$, then β vanishes, and f is given by an arbitrary harmonic function on $\mathbb{E}^{d-1}$.

3. "Black" brane solutions in M-theory

In this section, we show how to construct the exact stationary solutions in M-theory (or in type IIB superstring theory), and present concrete examples.

(1) M2 and M5-brane solutions in M-theory

In 11-dimensional supergravity, we have a 4-form field ($n_A = 4$) and no dilaton φ ($a_A = 0$). Setting $D = 11$ and $a_A = 0$, we have $\Delta_A = (q_A + 1)(8 - q_A)$. The form field produced by an "electric" charge is related to the M2-brane, i.e., $q_A = n_A - 2 = 2$. This gives $\Delta_A = 18$. Similarly, the field with a "magnetic" charge is related to the M5-brane because ${}^*q_A = {}^*n_A - 2 = D - n_A - 2 = 5$. This also gives $\Delta_A = 18$.

These two branes (M2 and M5) can intersect if and only if

$$M2 \cap M2 \to \bar{q}_{22} = 0, \quad M2 \cap M5 \to \bar{q}_{25} = 1, \quad M5 \cap M5 \to \bar{q}_{55} = 3 . \qquad (33)$$

The crossing rule leads that there exists a four-dimensional (4D) "black" object with four independent branes or three M5 branes and one wave, or a five-dimensional (5D) "black" object with three independent M2 branes or two branes and one wave (see some examples in Table 1).

(2) Compactification of a black brane

Compactifying eleven (or ten) dimensional spacetime, we obtain an effective d-(four or five) dimensional world. Performing a conformal transformation of our metric as

$$ds_D^2 = \Omega^{-\frac{2}{d-2}} d\bar{s}_d^2 + \Omega_1^2 \left[dy_1 - \frac{1}{1+f}\left(fdt - \frac{A_i}{2}dx^i\right)\right]^2 + \sum_{\alpha=2}^{p} \Omega_\alpha^2 dy_\alpha^2 \,, \quad (34)$$

where the conformal factors Ω_1, Ω_α and Ω are defined by

$$\Omega_1^2 = (1+f)\prod_A H_A^{-2\frac{D-q_A-3}{\Delta_A}} \,, \qquad \Omega_\alpha^2 = \prod_A H_A^{-2\frac{\delta_{\alpha A}}{\Delta_A}} \qquad (\alpha = 2, \cdots, p)$$

$$\Omega^2 = \prod_{\alpha=1}^{p} \Omega_\alpha^2 = (1+f)\prod_A H_A^{\frac{2}{\Delta_A}[D-d-q_A(d-2)]} \,, \quad (35)$$

we obtain the Einstein gravity in d-dimensions, which metric is given by

$$d\tilde{s}_d^2 = -\Xi^{d-3}\left(dt + \frac{A_i}{2}dx^i\right)^2 + \Xi^{-1}\sum_{i=1}^{d-1} dx_i^2 ,$$

$$\Xi \equiv (1+f)^{-1/(d-2)} \prod_A H_A^{-\frac{2(D-2)}{(d-2)\Delta_A}} \,. \quad (36)$$

If the compactified space is sufficiently small, we find the effective d-dimensional world with the metric (36).

If this spacetime is asymptotically flat, i.e.,

$$H_A \to 1 + \frac{Q_H^{(A)}}{\rho^{d-3}}, \quad f \to \frac{Q_0}{\rho^{d-3}} \,, \quad (37)$$

where $Q_H^{(A)}$ and Q_0 are conserved charges, we obtain the ADM mass as

$$M_{\rm ADM} = \frac{(d-3)\pi^{\frac{d-3}{2}}}{8G_d\Gamma\left(\frac{d-1}{2}\right)}\left[Q_0 + \sum_A \frac{2(D-2)}{\Delta_A} Q_H^{(A)}\right] . \quad (38)$$

In what follows, we discuss the details of the 5D "black" object with M2⊥M5⊥W branes for $D = 11$ and $d = 5$ and present the exact solutions. For $D = 10$, the construction of solutions is almost the same.

4. "Black" brane solutions with M2-M5 branes : the case of $d = 5$

We consider the case of $d = 5$. There are two branes (M2 and M5). The metric in the Einstein frame in five-dimensions is written by

$$d\tilde{s}_5^2 = -\Xi^2\left(dt + \frac{A_i}{2}dx^i\right)^2 + \Xi^{-1} ds_{\mathbb{E}^4}^2 \,, \quad (39)$$

where $\Xi = [H_2 H_5 (1+f)]^{-1/3}$. The unknown functions $H_A (A = 2, 5)$, $\mathcal{A}_i$ and f satisfy the following equations:

$$\partial^2 f = 0 \quad , \quad \partial^2 H_A = 0 \ \ (A = 2, 5) \tag{40}$$

$$\partial_j \mathcal{F}^{ij} = 0 \,. \tag{41}$$

In order to find the exact solutions, we assume that the 4-dimensional x-space has two rotation symmetries which Killing vectors ($\xi^i_{(\phi)}$ and $\xi^i_{(\psi)}$) commute each other. In this case, Eq. (41) is reduced to two uncoupled equations for two scalar fields, $\mathcal{A}_\phi = \mathcal{A}_i \xi^i_{(\phi)}$ and $\mathcal{A}_\psi = \mathcal{A}_i \xi^i_{(\psi)}$, as

$$\partial^2 \mathcal{A}_\phi - \partial_i \ln \left(\xi_{(\phi)} \cdot \xi_{(\phi)} \right) \, \partial^i \mathcal{A}_\phi = 0 \,, \tag{42}$$

$$\partial^2 \mathcal{A}_\psi - \partial_i \ln \left(\xi_{(\psi)} \cdot \xi_{(\psi)} \right) \, \partial^i \mathcal{A}_\psi = 0 \,. \tag{43}$$

Here we have assumed that the other components of $\mathcal{A}_i$ vanish.

We now have a set of Laplace equations and the similar equations (Eqs. (42) and (43)) for several scalar functions (H_A, $\mathcal{A}_\phi$, $\mathcal{A}_\psi$, and f). Each equation is linear and uncoupled. Hence it is very easy to find general solutions because the Laplace-Beltrami operator is defined on the flat Euclidian space. Once we obtain a complete set of solutions in an appropriate curvilinear coordinate system, we can construct any solutions by superposing them.

Here we adopt the hyperspherical coordinates:

$$x_1 + i x_2 = \rho \cos\theta e^{i\phi}, \quad x_3 + i x_4 = -\rho \sin\theta e^{i\psi} \,, \tag{44}$$

where $0 \le \phi, \psi < 2\pi$ and $0 \le \theta \le \pi/2$. The line element of 4D flat space is

$$ds^2_{\mathbb{E}^4} = d\rho^2 + \rho^2 \left(d\theta^2 + \cos^2\theta d\phi^2 + \sin^2\theta d\psi^2 \right) \,. \tag{45}$$

Eqs. (40) in this coordinate system are

$$\frac{1}{\rho} \partial_\rho \left(\rho^3 \partial_\rho \mathcal{H}_a \right) + \frac{1}{\sin\theta \cos\theta} \partial_\theta \left(\sin\theta \cos\theta \partial_\theta \mathcal{H}_a \right) = 0 \,, \tag{46}$$

where $\mathcal{H}_a = (f, H_A)$. With an asymptotically flatness and regularity conditions on the axis, the solutions are given by

$$f = \sum_{\ell=0}^{\infty} \mathcal{Q}_\ell \, \rho^{-2(\ell+1)} P_\ell(\cos 2\theta) \tag{47}$$

$$H_A = 1 + \sum_{\ell=0}^{\infty} h^{(A)}_\ell \rho^{-2(\ell+1)} P_\ell(\cos 2\theta) \,, \tag{48}$$

where $\mathcal{Q}_\ell$ and $h^{(A)}_\ell$ are coefficients of multipole moments.

Next, we discuss Eqs. (42) and (43), which are written as

$$\rho\partial_\rho\left(\rho\partial_\rho \mathcal{A}_\phi\right) + \cot\theta\partial_\theta\left(\tan\theta\partial_\theta \mathcal{A}_\phi\right) = 0\,, \tag{49}$$

$$\rho\partial_\rho\left(\rho\partial_\rho \mathcal{A}_\psi\right) + \tan\theta\partial_\theta\left(\cot\theta\partial_\theta \mathcal{A}_\psi\right) = 0\,. \tag{50}$$

Assuming asymptotically flatness and regularity on the axis, we find general solution for $\mathcal{A}_i$ as

$$\mathcal{A}_\phi = \sum_{m=1}^{\infty} \frac{b_m^{(\phi)}}{\rho^{2m}} F(-m, m, 1, \sin^2\theta)\,, \quad \mathcal{A}_\psi = \sum_{n=1}^{\infty} \frac{b_n^{(\psi)}}{\rho^{2n}} F(-n, n, 1, \cos^2\theta)\,. \tag{51}$$

The solution (Eqs. (47)-(51)) with the lowest multipole moment is given by

$$f = \frac{\mathcal{Q}_0}{\rho^2}\,, \quad H_A = 1 + \frac{\mathcal{Q}_H^{(A)}}{\rho^2} \quad (A = 2, 5)\,,$$

$$\mathcal{A}_\phi = \frac{J_\phi \cos^2\theta}{\rho^2}\,, \quad \mathcal{A}_\psi = \frac{J_\psi \sin^2\theta}{\rho^2}\,, \tag{52}$$

where J_ϕ and J_ψ are angular momenta.

Supersymmetry implies $J_\phi = -J_\psi = J$, which corresponds to the BMPV solution. In what follows, we assume this relation. The mass and the entropy of this spacetime are given as

$$M_{\mathrm{ADM}} = \frac{\pi}{4G_5}(\mathcal{Q}_0 + \mathcal{Q}_H^{(2)} + \mathcal{Q}_H^{(5)}), \tag{53}$$

$$S = \frac{A_h}{4G_5} = \frac{\pi^2}{2G_5}\sqrt{\mathcal{Q}_0\mathcal{Q}_H^{(2)}\mathcal{Q}_H^{(5)} - \frac{J^2}{8}}\,. \tag{54}$$

5. Supersymmetric rotating black hole in compactified spacetime

Since the origin of a flat space can be shifted by any distance, we can move such a black hole to any position. Since the basic equations (40) and (41) are linear, we can also superpose those black hole solutions at different positions. In particular, if we superpose an infinite number of black holes, each of which is separated by the same distance in one direction, we obtain a periodic BMPV black hole solution. It can be regarded as a deformed BMPV black hole in a compactified four-dimensional spacetime. We call it a compactified BMPV black hole (a CBMPV black hole).

Here we present its explicit solution. We assume that an infinite number of black holes exist along the w-axis with the same coordinate distance ℓ.

By superposing those black hole solutions, we find

$$f = Q_0 \sum_{n=-\infty}^{\infty} \frac{1}{r^2+(w+n\ell)^2}, \quad H_A = 1+Q_H^{(A)} \sum_{n=-\infty}^{\infty} \frac{1}{r^2+(w+n\ell)^2}, \tag{55}$$

$$\mathcal{A}_\phi = J \sum_{n=-\infty}^{\infty} \frac{x^2+y^2}{[r^2+(w+n\ell)^2]^2}, \quad \mathcal{A}_\psi = -J \sum_{n=-\infty}^{\infty} \frac{z^2+(w+n\ell)^2}{[r^2+(w+n\ell)^2]^2}, \tag{56}$$

where $r^2 = x^2 + y^2 + z^2$.

Using the formula

$$\sum_{n=-\infty}^{\infty} \frac{1}{r^2+(w+n\ell)^2} = \frac{2\pi^2}{\ell^2} \frac{\sinh \bar{r}}{\bar{r}\,(\cosh \bar{r} - \cos \bar{w})}, \tag{57}$$

where $\bar{r} = 2\pi\rho/\ell$ and $\bar{w} = 2\pi w/\ell$, we find

$$f = \frac{2\pi^2 Q_0}{\ell^2} \frac{\sinh \bar{r}}{\bar{r}\,(\cosh \bar{r} - \cos \bar{w})}, \quad H_A = 1 + \frac{2\pi^2 Q_H^{(A)}}{\ell^2} \frac{\sinh \bar{r}}{\bar{r}\,(\cosh \bar{r} - \cos \bar{w})}, \tag{58}$$

$$\mathcal{A}_\phi = \frac{\pi^2 J}{\ell^2} \frac{(\bar{x}^2+\bar{y}^2)}{\bar{r}^2} \left[\frac{(\cosh \bar{r} \cos \bar{w} - 1)}{(\cosh \bar{r} - \cos \bar{w})^2} + \frac{\sinh \bar{r}}{\bar{r}\,(\cosh \bar{r} - \cos \bar{w})} \right], \tag{59}$$

$$\mathcal{A}_\psi = -\frac{\pi^2 J}{\ell^2} \left[-\frac{(\bar{x}^2+\bar{y}^2)}{\bar{r}^2} \frac{(\cosh \bar{r} \cos \bar{w} - 1)}{(\cosh \bar{r} - \cos \bar{w})^2} + \frac{(\bar{r}^2+\bar{z}^2)}{\bar{r}^2} \frac{\sinh \bar{r}}{\bar{r}\,(\cosh \bar{r} - \cos \bar{w})} \right] \tag{60}$$

We analyze spacetime structures of this solution in order.

(1) Horizon

The horizon exists at $(r, w) = (0, 0)$. Setting $\bar{x} = \epsilon \cos\theta \cos\phi, \bar{y} = \epsilon \cos\theta \sin\phi, \bar{z} = \epsilon \sin\theta \cos\psi, \bar{w} = \epsilon \sin\theta \cos\psi$ ($\epsilon \ll 1$), we find near-horizon structure as

$$f = \frac{4\pi^2 Q_0}{\ell^2 \epsilon^2}, \quad H_A = 1 + \frac{4\pi^2 Q_H^{(A)}}{\ell^2 \epsilon^2}, \mathcal{A}_\phi = \frac{4\pi^2 J}{\ell^2 \epsilon^2} \cos^2\theta, \quad \mathcal{A}_\psi = -\frac{4\pi^2 J}{\ell^2 \epsilon^2} \sin^2\theta \tag{61}$$

This shows exactly the same near-horizon structure as that of the BMPV black hole. The horizon structure is not modified by superposition of an infinite number of black holes. This is because of supersymmetry.

(2) Asymptotic structure

The metric describes the 5-dimensional spacetime, but it behaves effectively as the 4-dimensional one as the observer is far from the black hole. In fact, w-direction is compactified as $0 \le w < \ell$, while the other spatial directions are not $(0 \le r < \infty)$.

We show how the asymptotic 4-dimensional spacetime looks like. Then we introduce the 3-dimensional spherical coordinates $(\bar{r}, \Theta, \Phi)$, which are defined by the transformations $\bar{x} = \bar{r}\sin\Theta\cos\Phi\,, \bar{y} = \bar{r}\sin\Theta\sin\Phi\,, \bar{z} = \bar{r}\cos\Theta\,.$ Using this coordinate system, dropping the w-components of the metric, and taking an average of the other metric components over the period ℓ for the w coordinate, we find the effective 4-dimensional metric in the Einstein frame as

$$d\bar{s}_4^2 = -\Xi^{3/2}\left(d\bar{t} + \frac{\bar{A}_\phi}{2}d\Phi\right)^2 + \Xi^{-3/2}\left[d\bar{r}^2 + \bar{r}^2\left(d\Theta^2 + \sin^2\Theta d\Phi^2\right)\right]\,, \quad (62)$$

where

$$f = 2\left(\frac{\pi}{\ell}\right)^2\frac{Q_0}{\bar{r}}\,, \quad H_A = 1 + 2\left(\frac{\pi}{\ell}\right)^2\frac{Q_H^{(A)}}{\bar{r}}\,, \quad \bar{A}_\phi = 2\left(\frac{\pi}{\ell}\right)^3\frac{J}{\bar{r}}\sin^2\Theta \quad (63)$$

Comparing this with the asymptotic form of the 4-dimensional Kerr-Newman metric, we find a gravitational mass of the black hole M and a rotation parameter a, which is the angular momentum per unit mass, as follows.

$$G_4 M = \frac{\pi(Q_0 + Q_H^{(2)} + Q_H^{(5)})}{4\ell}\left(= \frac{G_5 M_{\rm ADM}}{\ell}\right)\,, \quad G_4 M a = \frac{\pi J}{8\ell} \quad (64)$$

Since $G_4 = G_5/\ell$, $M = M_{\rm ADM}$. The Kerr rotation parameter is given as $a = J/[2(Q_0 + Q_H^{(2)} + Q_H^{(5)})]$.

From the asymptotic form of the electric fields, we also find charges in four-dimensions as

$$Q_A \sim \left(\frac{\pi}{\ell}\right)Q_H^{(A)}\,. \quad (65)$$

(3) Entropy

The entropy of the present black hole solution is given by that of the BMPV black hole (Eq. (54)). Setting $Q^{(2)} = Q^{(5)} = \alpha Q_0$, we rewrite it as

$$S_{\rm CBMPV} = \frac{\pi r_g^2}{2G_4}\sqrt{\frac{1}{2}\left(\frac{1}{\lambda^2} - q^2\right)} \quad (66)$$

where $r_g \equiv 2G_4 M$ is the Schwarzschild radius of a black hole mass M, and

$$\lambda^2 \equiv \frac{\pi(1+2\alpha)^3}{16\alpha^2}\times\frac{r_g}{\ell} \quad \text{(ratio of the BH size to the compactification scale)}$$

$$q^2 \equiv \frac{a^2}{G_4^2 M^2} \quad \text{(normalized Kerr rotation parameter)}$$

$$e_A \equiv \frac{Q_A}{G_4 M} = \frac{4\alpha}{1+2\alpha} \quad (A = 2, 5\,,\ \text{normalized charges})\,. \quad (67)$$

For a CBMPV black hole, we have a constraint of $\lambda^2 \leq 1/q^2$, that is, the size of a black hole must be smaller than some critical value, i.e. $r_g \leq r_g^{(\mathrm{cr})}$, where

$$r_g^{(\mathrm{cr})} = 2G_4 M^{(\mathrm{cr})} \equiv \frac{16\alpha^2}{\pi(1+2\alpha)^3 q^2}\,\ell\,. \qquad (68)$$

This means that if a black hole is rapidly rotating, i.e. $q \sim O(1)$ and all charges are almost the same, i.e. $\alpha \sim O(1)$, then we find $r_g^{(\mathrm{cr})} \sim 5\ell$. The scale of extra-dimension is constraint by the experiment of Newtonian gravity as $\ell < 1$ mm . Hence, if a black hole mass is smaller than 10^{28} g, which is about the Earth's mass scale, a rapidly rotating CBMPV black hole could be realized.

References

1. K. Maeda and M. Tanabe, Nucl.Phys. B738 (2006) 184 [hep-th/0510082] (paper I).
2. K. Maeda, N. Ohta, and M. Tanabe, in preparation.

Wave maps on black holes in any dimensions

Makoto Narita
Department of Physics, National Central University, Jhongli 320, Taiwan
E-mail: narita@phy.ncu.edu.tw

We study global properties for wave maps on black holes.

Keywords: Cosmic censorship; Wave maps.

1. Introduction

Let us consider $D+1$-dimensional spherically symmetric spcetimes ($D \geq 3$). It is known already that one have black hole solutions to the Einstein equations under the stationary and some asymptotic conditions. From the viewpoint of Penrose's cosmic censorship, which states that physically reasonable spacetimes are globally hyperbolic, it is important to study whether such black hole spacetimes are stationary limit after dynamical evolution (i.e. naked singularities never appear by generic gravitational collapse) and/or the spacetimes are stable or not. In the $D = 3$ case, Christodoulou has shown that generic spherically symmetric gravitational collapse leads to the Schwarzschild spacetime as the final state [8]. It has been shown that $D+1$-dimensional static and spherical symmetric black hole spacetimes are stable against linear perturbation [1,12]. These results support the validity of the cosmic censorship conjecture, that is, the stability of the black holes. Next, nonlinear perturbation should be considered. However, we have no mathematical tool to analyze full nonlinear perturbation for curved spacetimes at the present time[a] and the only one results of stability against nonlinear perturbation is for the Minkowski spacetime [9,3]. Then, we will consider nonlinear scalar fields (as test fields) on curved spacetimes. Our choice for the test fields are wave maps, which play an important role in general relativity. For example, dynamical evolution equations of the (four dimensional) vacuum Einstein equations for spacelike $U(1)$-symmetric spacetimes

[a] Recently, nonlinear orbital stability of five-dimensional static and spherically symmetric black hole (Schwarzschild-Tangherlini) spacetimes has been shown [10].

can be written as a wave map. Also, it is known that nonlinearity of wave maps is similar with one of the Einstein equations, that is "null form" [7,13]. This nonlinearity is a key to prove global existence theorems for the Einstein equations and wave maps in small initial data in four dimensions. Thus, wave maps are better choices as nonlinear test fields.

In this paper, to investigate nonlinear stability of curved spacetimes, as a simple model, we will survey global properties of solutions to wave maps on static and spherically symmetric black hole spacetimes in higher-dimensions.

2. Wave maps on curved manifolds

2.1. *Definitions*

According to Choquet-Bruhat [2,4,5,6], we will use the following definitions for our problems. Let (V, g) and (M, h) be $D+1$-dimensional Lorentzian and d-dimensional Riemannian manifolds. A mapping $u : (V, g) \to (M, h)$ is called a *wave map* if it satisfies the following PDE in local coordinates on V and M:

$$g^{\alpha\beta}\nabla_\alpha\partial_\beta u^a = g^{\alpha\beta}\left(\partial^2_{\alpha\beta}u^a - \Gamma^\lambda_{\alpha\beta}\partial_\lambda u^a + \Gamma^a_{bc}\partial_\alpha u^b\partial_\beta u^c\right) = 0, \tag{1}$$

where $\Gamma^\lambda_{\alpha\beta}$ and Γ^a_{bc} are the Christoffel symbols of the base V and target M, respectively. Here, Greek indices runs from 0 to D and Roman indices run from 1 to d. This wave map system is a critical point of the following action:

$$S_{\rm WM} = \int d^{D+1}x\sqrt{-g}g^{\alpha\beta}h_{ab}\partial_\alpha u^a\partial_\beta u^b. \tag{2}$$

Then, the wave map equations read as a semilinear quasidiagonal system of second order partial differential equations for d scalar functions u^a. The principal part is the wave operator for the Lorentzian metric g and the nonlinear terms are quadratic forms in the derivative of u with cofficients u, which satisfy the null condition [7,13].

Now, let us consider the Cauchy problem, that is, given values on a given spacelike hypersurface S_0 in V, $u(x)|_{S_0} = \phi(x) \in M$ and $\partial_t u(x)|_{S_0} = \psi(x) \in T_{\phi(x)}M$, the construction of a wave map. Here, $x \in S_0$ and ϕ is a map from the initial hypersurface S_0 into the target M. To do this, we need some assumptions for target and base manifolds

2.1.1. *Assumptions for target manifolds*

Now we put an assumption for target manifolds.

Definition 2.1. The manifold (M, h) is said to be regularly embedded in the Euclidean manifold (R^N, Q) if it is defined by P smooth scalar equations $F^{(P)} = 0$ on R^N, of rank P on M, and h is the pullback of the metric Q under this embedding. We denote by $\nu^{(P)}$ the normal to M in R^N defined by the gradient of $F^{(P)}$. We set $\nu_{(I)} = m_{IJ}\nu^{(J)}$, where m_{IJ} is the inverse matrix with elements $m^{IJ} = Q^{AB}\nu_A^{(I)}\nu_B^{(J)}$.

Under the assumption, the equations for U read as a semilinear quasidiagonal system of second order partial differential equations for N scalar functions on V.

2.1.2. *Assumptions for base manifolds*

To consider the (well-posed) Cauchy problem for the wave map, we need to suppose conditions on the base manifold.

Definition 2.2. The manifold (V, g) is said to be regularly hyperbolic if:

(1) It is globally hyperbolic. Then V is of the type $S \times \mathrm{R}$, with S a D-dimensional oriented smooth spacelike manifold, the past of any compact subset of V intersects any $S_t \equiv S \times \{t\}$ along a compact set.
(2) The metric g, assumed here for simplicity to be smooth, can be written

$$g = -N^2 dt^2 + g_{ij}\theta^i\theta^j, \qquad \theta^i \equiv dx^i + \beta^i dt, \tag{3}$$

with $0 < B_1 \le N \le B_2$ on V, where B_1 and B_2 are positive and continuous functions of t.
(3) The metrics $g_t = g_{ij}dx^i dx^j$, induced by g on S_t, are uniformly equivalent to a given smooth Riemannian metric e on S, that is, there exist positive and continuous functions of t, A_1 and A_2 which, for any vector field ξ on S and t, satisfy

$$A_1 e(\xi, \xi) \le g_t(\xi, \xi) \le A_2 e(\xi, \xi) \tag{4}$$

on S .

To show global existence theorem to our wave map, we should specify the base manifold further. Let us consider the base manifold $V \sim \mathrm{R}_t \times (0, \infty)_r \times S^n$ (thus, $n = D - 1$) endowed with the Lorentzian metric

$$g = -f_n(r)dt^2 + \frac{dr^2}{f_n(r)} + r^2 d\sigma_n^2, \tag{5}$$

where $f_n \in C^\infty((0,\infty))$. Hereafter, we shall consider higher-dimensional black hole spacetimes which are solutions of the vacuum Einstein equations, $G_{\alpha\beta} = 0$, where $G_{\alpha\beta}$ is the Einstein tensor:

$$f_n(r) = 1 - \frac{C}{r^{n-1}}, \tag{6}$$

where C is a integration constant. Unfortunately, since this coordinate is singular at finite radius r_g which is a root of $f_n = 0$, we will use the Regge-Wheeler coordinate:

$$\frac{dr}{d\rho} = f_n(r). \tag{7}$$

Then we have the following metric

$$g = f_n(r)\left(-dt^2 + d\rho^2\right) + r^2 d\sigma_n^2, \tag{8}$$

with $-\infty < \rho < +\infty$, $r = r(\rho)$ is a C^∞ function of ρ, increasing from r_g to $+\infty$.

3. Global existence and asymptotic behavior of wave maps on black holes in higher dimension

The followings are main results and the proofs are given in author's paper [15]:

Theorem 3.1. *Let (M,h) be a Riemannian manifold C^∞ regularly embedded into a Euclidean space. If $\Phi \in W^{2,2}$ and $\Psi \in W^{1,2}$, then there exists a global spherically symmetric wave map $u \in C^k([0,\infty), W^{2-k,2}), 0 \leq k \leq 2$, from the exterior of a higher dimensional spherically symmetric black hole into (M,h) taking these Cauchy data. The solution is unique. The solution in a compact set depends only on the data in the intersection of the initial line with the past of this set. In particular, if the Cauchy data Φ and Ψ belong to $C_0^\infty(\mathrm{R}_\rho)$, then there exists a unique global solution which belongs to $C^\infty(\mathrm{R}_t \times \mathrm{R}_\rho)$.* □

Theorem 3.2. *Given Cauchy data $\Phi, \Psi \in C_0^\infty(\mathrm{R}_\rho)$ which are bounded by $\mathcal{E} > 0$, for $t - |\rho| > 0$, the solution has the following decay property*

$$|u(t,\rho)| \leq C\mathcal{E}(t+\rho)^{-n/2}(t-|\rho|)^{-n/2}, \tag{9}$$

where $\mathcal{E}$ depends only on initial data. □

Arguments for Theorem 3.1 are similar with Choquet-Bruhat's [2,4,5,6]. In the proof for Theorem 3.2, Christodoulou's idea [7,11,14], which is that a global problem is transformed into a local problem by conformal transformations [16], is used.

References

1. Chandrasekhar, S., *The mathematical theory of black holes*, International Series of Monographs on Physics, **69**, Oxford University Press, New York, (1983).
2. Choquet-Bruhat, Y., *Global existence theorems for hyperbolic harmonic maps*, Ann. Inst. Henri Poincaré **46**, (1987) 97-111.
3. Choquet-Bruhat, Y., *Global existence theorems for Einstein equations in high dimensions*, in Gravity, Particles and Space-Time, edited by Pronin, P. and Sardanashvily, G., World Sci. Publishing, River Edge, NJ, (1996) 19-28.
4. Choquet-Bruhat, Y., *Wave maps in general relativity*, in On Einstein path, edited by Harvey, A., Springer-Verlag, New York, (1999) 147-169.
5. Choquet-Bruhat, Y., *Global wave maps on curved space times*, in Mathematical and quantum aspects of relativity and cosmology, Lecture Notes in Physics **537**, edited by Cotsakis, S. and Gibbons, G. W., Springer-Verlag, Berlin, (2000) 1-29.
6. Choquet-Bruhat, Y., *Global wave maps on black holes*, in Proceedings of Jean Leray '99 Conference, edited by de Gosson, M., Kluwer Academic Publishers, Netherlands, (2003) 469-482.
7. Christodoulou, D., *Global solutions of nonlinear hyperbolic equations for small initial data*, Commun. Pure Appl. Math. **39**, (1986) 267-282.
8. Christodoulou, D., *On the global initial value problem and the issue of singularities*, Class. Quantum Grav. **16**, (1999) A23-A35.
9. Christodoulou, D. and Klainerman, S., *The global nonlinear stability of the Minkowski space*, Princeton Mathematical Series, **41**, Princeton University Press, Princeton, (1993).
10. Dafermos, M. and Holzegel G., *On the nonlinear stability of higher-dimensional triaxial Bianchi IX black holes*, gr-qc/0510051.
11. Godin, P., *Global existence of solutions to some exterior radial quasilinear Cauchy-Dirichlet problems*, Amer. J. Math. **117**, (1995) 1475-1505.
12. Ishibashi, A. and Kodama, H., *Stability of higher-dimensional Schwarzschild black holes*, Progr. Theoret. Phys. **110** (2003) 901-919.
13. Klainerman, S., *The null condition and global existence to nonlinear wave equations*, Lectures in Appl. Math. **23**, (1986) 293-326.
14. Lindblad, H., *A remark on global existence for small initial data of the minimal surface equation in Minkowskian space time*, Proc. Amer. Math. Soc. **132**, (2004) 1095-1102.
15. Narita, M., *Wave maps in gravitational theory*, to appear in the Proceedings of The 14th International Research Institute, "Asymptotic Analysis and Singularity" (Advanced Studies in Pure Mathematics, Mathematical Society of Japan).
16. Penrose, R., *Confomal treatment of infinity*, in Relativity, Groups and Topology, Les Houches 1963, edited by DeWitt, C. and DeWitt, B., Gordon and Breach, New York, (1964) 563-584.

Geometric characterization of purely in- or out-going gravitational radiation at a finite distance

Jong Hyuk Yoon

Department of Physics, Konkuk University,
Seoul 143-701, Korea
E-mail: yoonjh@konkuk.ac.kr

By studying the canonical expression of quasilocal energy-flux that follows from the Einstein's equations, I find geometric conditions for purely in- or out-going gravitational radiation of the most general type at a finite distance. These conditions are the vanishing of the transverse traceless parts of the second fundamental forms of a 2-surface with respect to the in- or out-going null vector fields normal to the surface. I also discuss the quasilocal momentum conservation equation, which has a remarkably similar structure with the Navier-Stokes equation for a viscous fluid. The deviation from the affinity of the parameter of the in-going null geodesics turns out to play the role of the coefficient of viscosity, whereas the scalar curvature of 2-surface is like a local pressure. Thus, Einstein's field equations, which consist of the Hamilton's equations of motion and a set of quasilocal conservation equations, describe a dissipative system of the Einstein's gravitation.

Keywords: Conservation equation; Gravitational radiation; Energy-flux; Dissipative Hamiltonian.

1. Review of (2,2) fibre bundle formalism

Let us begin with the following metric of (3+1) dimensional spacetimes,

$$ds^2 = -2dudv - 2hdu^2 + \phi_{ab}\left(dy^a + A_+^{\ a}du + A_-^{\ a}dv\right)\left(dy^b + A_+^{\ b}du + A_-^{\ b}dv\right), \tag{1}$$

where $+, -$ stands for u, v, respectively. In order to understand the geometry of this metric[1,2,3,4,5,6,7,8], it is convenient to introduce the vector fields $\{\hat{\partial}_\pm\}$ defined as

$$\hat{\partial}_+ := \partial_+ - A_+^{\ a}\partial_a, \tag{2}$$

$$\hat{\partial}_- := \partial_- - A_-^{\ a}\partial_a, \tag{3}$$

where

$$\partial_+ = \frac{\partial}{\partial u}, \qquad \partial_- = \frac{\partial}{\partial v}, \qquad \partial_a = \frac{\partial}{\partial y^a} \qquad (a = 2, 3). \tag{4}$$

The inner products of the vector fields $\{\hat{\partial}_\pm, \partial_a\}$ are given by

$$\begin{aligned} &< \hat{\partial}_+,\, \hat{\partial}_+ >= -2h, \qquad < \hat{\partial}_+,\, \hat{\partial}_- >= -1, \qquad < \hat{\partial}_-,\, \hat{\partial}_- >= 0, \\ &< \hat{\partial}_\pm,\, \partial_a >= 0, \qquad < \partial_a,\, \partial_b >= \phi_{ab}. \end{aligned} \tag{5}$$

The hypersurface u = constant is an out-going null hypersurface generated by $\hat{\partial}_-$ whose norm is zero. The hypersurface v = constant is generated by $\hat{\partial}_+$ whose norm is $-2h$, which can be either negative, zero, or positive, depending on whether $\hat{\partial}_+$ is timelike, null, or spacelike, respectively. The horizontal vector fields $\{\hat{\partial}_\pm\}$ orthogonal to $\{\partial_a\}$ span a two dimensional section of Lorentzian signature. The intersection of two hypersurfaces $u, v =$ constant defines a spacelike 2-surface N_2 coordinatized by $\{y^a\}$, which we assume compact with a positive-definite metric ϕ_{ab}(see FIG. 1). The metric ϕ_{ab} can be written as a product of the area element e^σ and the conformal metric ρ_{ab} normalized to have a unit determinant

$$\phi_{ab} = e^\sigma \rho_{ab} \qquad (\det \rho_{ab} = 1). \tag{6}$$

It is useful to write the in-going null vector field n and out-going null vector field l in term of $\{\hat{\partial}_\pm\}$. They are given by

$$n = \hat{\partial}_+ - h\hat{\partial}_-, \tag{7}$$

$$l = \hat{\partial}_-, \tag{8}$$

and satisfy the normalization condition

$$< n,\, l >= -1. \tag{9}$$

The coordinate v increases uniformly as l evolves, since we have

$$\pounds_l v = 1. \tag{10}$$

In the gauge where $A_-{}^a = 0$, l is given by

$$l = \frac{\partial}{\partial v}, \tag{11}$$

which tells us that v becomes the affine parameter of the out-going null geodesic l.

The spacetime integral of the scalar curvature of the metric (1) is given by

$$I = \int du\, dv\, d^2y\, L + \text{surface integrals}, \tag{12}$$

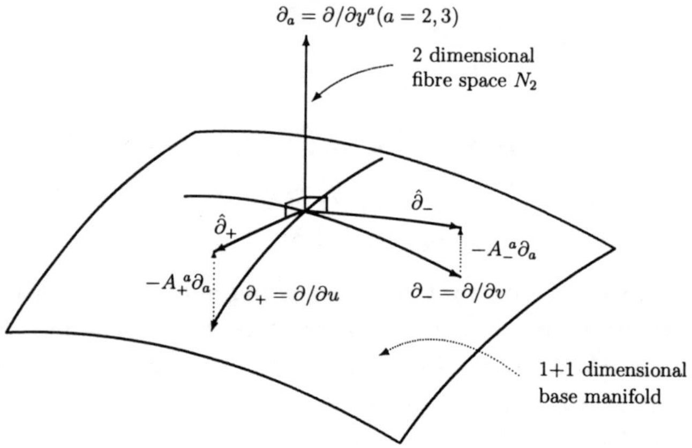

Fig. 1. This figure shows the geometry of the (2,2) fibre bundle splitting of 3+1 dimensional spacetime. The 1+1 dimensional base manifold is spanned by $\{\partial_\pm\}$ and the 2 dimensional fibre space N_2 by $\{\partial_a\}$. The horizontal vector fields $\{\hat{\partial}_\pm\}$ are orthogonal to N_2, and $A_\pm{}^a$ are the connections valued in the Lie algebra of the diffeomorphisms of N_2.

where L is given by[5,6,7,8]

$$\begin{aligned} L = &-\frac{1}{2}\mathrm{e}^{2\sigma}\rho_{ab}F_{+-}{}^{a}F_{+-}{}^{b} + \mathrm{e}^{\sigma}(D_+\sigma)(D_-\sigma) - \frac{1}{2}\mathrm{e}^{\sigma}\rho^{ab}\rho^{cd}(D_+\rho_{ac})(D_-\rho_{bd}) \\ &-2\mathrm{e}^{\sigma}(D_-h)(D_-\sigma) - h\mathrm{e}^{\sigma}(D_-\sigma)^2 + \frac{1}{2}h\mathrm{e}^{\sigma}\rho^{ab}\rho^{cd}(D_-\rho_{ac})(D_-\rho_{bd}) \\ &-\mathrm{e}^{\sigma}R_2. \end{aligned} \tag{13}$$

Here R_2 is the scalar curvature of N_2, and the covariant derivatives are defined by

$$F_{+-}{}^{a} = \partial_+ A_-{}^{a} - \partial_- A_+{}^{a} - [A_+,\ A_-]_{\mathrm{L}}^{a}, \tag{14}$$

$$D_\pm\sigma = \partial_\pm\sigma - [A_\pm,\ \sigma]_{\mathrm{L}}, \tag{15}$$

$$D_\pm h = \partial_\pm h - [A_\pm,\ h]_{\mathrm{L}}, \tag{16}$$

$$D_\pm\rho_{ab} = \partial_\pm\rho_{ab} - [A_\pm,\ \rho]_{\mathrm{L}ab}, \tag{17}$$

where the Lie brackets $[A_\pm, *]$ are the Lie derivatives of $*$ along the vector field $A_\pm := A_\pm{}^{a}\partial_a$.

2. Hamilton's equations of motion and a set of quasilocal conservation equations

The Einstein's equations in this formalism split into 12 first-order Hamilton's equations of motion and 4 equations of divergence-type from which quasilocal conservation equations follow[6,7,8]. The Hamilton's equations of motion are given by

$$D_- q^I = \frac{\delta K}{\delta \pi_I}, \tag{18}$$

$$D_- \pi_I = -\frac{\delta K}{\delta q^I}, \tag{19}$$

where K is the Hamiltonian given by

$$K = \int du \oint d^2y \Big\{ H + \lambda\,(\det\,\rho_{ab} - 1) \Big\}, \tag{20}$$

and the Hamiltonian density H is

$$\begin{aligned} H = &-\frac{1}{2}\mathrm{e}^{-\sigma}\pi_\sigma\pi_h + \frac{1}{4}h\mathrm{e}^{-\sigma}\pi_h^2 - \frac{1}{2}\mathrm{e}^{-2\sigma}\rho^{ab}\pi_a\pi_b + \frac{1}{2}\pi_h(D_+\sigma) \\ &+\frac{1}{2h}\mathrm{e}^{-\sigma}\rho_{ac}\rho_{bd}\pi^{ab}\pi^{cd} + \frac{1}{2h}\pi^{ab}(D_+\rho_{ab}) + \frac{1}{8h}\mathrm{e}^{\sigma}\rho^{ab}\rho^{cd}(D_+\rho_{ac})(D_+\rho_{bd}) \\ &+\mathrm{e}^{\sigma}R_2, \end{aligned} \tag{21}$$

and λ is the Lagrange multiplier. In the above variation we assume the boundary conditions

$$\delta\sigma = \delta\rho_{ab} = 0 \tag{22}$$

at the endpoints of u-integration in K. The momentum variables $\pi_I = \{\pi_h, \pi_\sigma, \pi_a, \pi^{ab}\}$ conjugate to the configuration variables $q^I = \{h, \sigma, A_+^{\ a}, \rho_{ab}\}$ are defined as

$$\pi_I := \frac{\partial L}{\partial(D_- q^I)}, \tag{23}$$

where L is given by (13). Explicitly, they are found to be

$$\pi_h = -2\mathrm{e}^{\sigma}(D_-\sigma), \tag{24}$$

$$\pi_\sigma = -2\mathrm{e}^{\sigma}(D_- h) - 2h\mathrm{e}^{\sigma}(D_-\sigma) + \mathrm{e}^{\sigma}(D_+\sigma), \tag{25}$$

$$\pi_a = \mathrm{e}^{2\sigma}\rho_{ab}F_{+-}^{\quad b}, \tag{26}$$

$$\pi^{ab} = h\mathrm{e}^{\sigma}\rho^{ac}\rho^{bd}(D_-\rho_{cd}) - \frac{1}{2}\mathrm{e}^{\sigma}\rho^{ac}\rho^{bd}(D_+\rho_{cd}). \tag{27}$$

By integrating the 4 divergence-type equations over N_2 which we assume compact, then we obtain the following quasilocal conservation equations

$$\frac{\partial}{\partial u}U = \frac{1}{16\pi}\oint d^2y\,(\pi^{ab}D_+\rho_{ab} + \pi_\sigma D_+\sigma - hD_+\pi_h), \tag{28}$$

$$\frac{\partial}{\partial u}P = \frac{1}{16\pi}\oint d^2y\,H, \tag{29}$$

$$\frac{\partial}{\partial u}L(\xi) = \frac{1}{16\pi}\oint d^2y\,(\pi^{ab}\pounds_\xi\rho_{ab} + \pi_\sigma\pounds_\xi\sigma - h\pounds_\xi\pi_h - A_+^{\ a}\pounds_\xi\pi_a), \tag{30}$$

where U, P, and $L(\xi)$ are 2-surface integrals defined as

$$U := \frac{1}{16\pi}\oint d^2y\,(h\,\pi_h + 2e^\sigma D_+\sigma) + \bar{U}, \tag{31}$$

$$P := \frac{1}{16\pi}\oint d^2y\,(\pi_h) + \bar{P}, \tag{32}$$

$$L(\xi) := \frac{1}{16\pi}\oint d^2y\,(\xi^a\pi_a) + \bar{L}. \tag{33}$$

Here $\bar{U}$, $\bar{P}$, and $\bar{L}$ are undetermined subtraction terms that satisfy the conditions

$$\frac{\partial\bar{U}}{\partial u} = \frac{\partial\bar{P}}{\partial u} = \frac{\partial\bar{L}}{\partial u} = 0, \tag{34}$$

and ξ^a are u-independent,

$$\frac{\partial\xi^a}{\partial u} = 0, \tag{35}$$

but otherwise arbitrary functions.

3. Geometric characterization of purely in- or out-going wave conditions

We wish to express the quasilocal energy-flux of gravitational radiation such that the in- and out-going energy-fluxes are manifest. Recall that there are two kinds of the second fundamental forms of a 2-surface N_2, since there exist two null vector fields normal to N_2. Let us introduce tensor densities $\pi_{\pm\,ab}$ proportional to the traceless parts of the second fundamental forms of N_2 associated with the in- and out-going null vector field n and l, respectively. Explicitly, we define $\pi_{\pm\,ab}$

$$\pi_{+\,ab} := -\frac{1}{2}\,e^{3\sigma}\,\pounds_n\rho_{ab}, \tag{36}$$

$$\pi_{-\,ab} := \frac{h}{2}\,e^{3\sigma}\,\pounds_l\rho_{ab}. \tag{37}$$

In components, they are given by

$$\pi_{+\,ab} = -\frac{1}{2}\,\mathrm{e}^{3\sigma} D_{+}\rho_{ab} + \frac{h}{2}\,\mathrm{e}^{3\sigma} D_{-}\rho_{ab}, \tag{38}$$

$$\pi_{-\,ab} = \frac{h}{2}\,\mathrm{e}^{3\sigma} D_{-}\rho_{ab}, \tag{39}$$

and $\pi_{\pm}^{ab}$ are naturally defined as

$$\pi_{\pm}{}^{ab} = \mathrm{e}^{-2\sigma} \rho^{ac}\rho^{bd}\pi_{\pm\,cd}. \tag{40}$$

Thus, from (27), (38), and (39), we find that

$$\pi^{ab} = \pi_{+}{}^{ab} + \pi_{-}{}^{ab}. \tag{41}$$

Conversely, one can write $D_{\pm}\rho_{ab}$ in terms of $\pi_{\pm\,ab}$ as follows,

$$D_{+}\rho_{ab} = -2\,\mathrm{e}^{-3\sigma}(\pi_{+\,ab} - \pi_{-\,ab}), \tag{42}$$

$$D_{-}\rho_{ab} = \frac{2}{h}\,\mathrm{e}^{-3\sigma}\pi_{-\,ab}. \tag{43}$$

3.1. *Splitting quasilocal energy-flux into in- and out-flux*

Now we will express the quasilocal energy-flux given by the right hand side of (28) using the splitting described above. The term $\pi^{ab}D_{+}\rho_{ab}$ can be written as

$$\pi^{ab}D_{+}\rho_{ab} = 2\,\mathrm{e}^{-3\sigma}(\pi_{-}{}^{ab}\pi_{-\,ab} - \pi_{+}{}^{ab}\pi_{+\,ab}). \tag{44}$$

If we introduce the supermetric G_{abcd} as

$$G_{abcd} := \rho_{ac}\rho_{bd} + \rho_{ad}\rho_{bc}, \tag{45}$$

such that

$$G_{abcd} = G_{(ab)(cd)}, \qquad G_{abcd} = G_{(cd)(ab)}, \tag{46}$$

and define the positive-definite norm of $\pi_{\pm}{}^{ab}$ as

$$|\pi_{\pm}{}^{ab}|^2 := G_{abcd}\,\pi_{\pm}^{ab}\pi_{\pm}^{cd} \geq 0, \tag{47}$$

then (44) becomes

$$\pi^{ab}D_{+}\rho_{ab} = \mathrm{e}^{-\sigma}(|\pi_{-}{}^{ab}|^2 - |\pi_{+}{}^{ab}|^2). \tag{48}$$

Therefore the quasilocal energy conservation equation (28) becomes

$$\frac{\partial}{\partial u}U = \frac{1}{16\pi}\oint d^2y\left\{\mathrm{e}^{-\sigma}(|\pi_{-}{}^{ab}|^2 - |\pi_{+}{}^{ab}|^2) + \pi_{\sigma}D_{+}\sigma - hD_{+}\pi_{h}\right\}. \tag{49}$$

The last two-terms in the right hand side of (49) are *work terms* of the type[8,9]

$$p_i \dot{q}^i, \tag{50}$$

each term being associated with the deforming the kinematic elements of the geometry. The first two terms in the right hand side of (49) are quadratic in $\pi_{\pm}{}^{ab}$, traceless parts of the second fundamental forms of N_2, and represent energy-flux carried by the transverse traceless degrees of freedom of gravitational radiation[1,2,3,4] along the in- or out-going null direction. When

$$\pi_{-}{}^{ab} = 0 \tag{51}$$

on N_2, then the net gravitational energy-flux becomes negative-definite, representing the situation that the radiation is purely outward. Likewise, when

$$\pi_{+}{}^{ab} = 0, \tag{52}$$

the radiation is purely inward, as the net energy-flux of the gravitational radiation is positive-definite. Thus,

$$\pi_{\pm}{}^{ab} = 0 \tag{53}$$

is the condition for gravitational radiation to be purely inward or outward on N_2, respectively.

3.2. *Splitting quasilocal momentum-flux into physical modes*

One can also split the quasilocal momentum-flux H given by (21) into distinct physical modes characteristic of gravitational fields. Let us collect terms in H which contain π^{ab} and $D_+\rho_{ab}$, and define ζ as

$$\begin{aligned}
\zeta &:= \frac{1}{2h} \mathrm{e}^{-\sigma} \rho_{ac}\rho_{bd}\pi^{ab}\pi^{cd} + \frac{1}{2h}\pi^{ab} D_+\rho_{ab} + \frac{1}{8h}\mathrm{e}^{\sigma} \rho^{ab}\rho^{cd}(D_+\rho_{ac})(D_+\rho_{bd}) \\
&= \frac{h}{2}\, \mathrm{e}^{\sigma} \rho^{ac}\rho^{bd}(D_-\rho_{ab})(D_-\rho_{cd}) \\
&= \frac{2}{h}\, \mathrm{e}^{-\sigma} \rho_{ac}\rho_{bd}\pi_{-}{}^{ab}\pi_{-}{}^{cd} \\
&= \frac{1}{h}\, \mathrm{e}^{-\sigma} |\pi_{-}{}^{ab}|^2 .
\end{aligned} \tag{54}$$

Thus, ζ is simply quadratic in $\pi_{-}{}^{ab}$. Likewise, let us collect terms in H which contain π_σ, π_h and define ξ as

$$\xi := -\frac{1}{2}\mathrm{e}^{-\sigma}\pi_\sigma \pi_h + \frac{h}{4}\, \mathrm{e}^{-\sigma}\pi_h^2 + \frac{1}{2}\pi_h D_+\sigma. \tag{55}$$

This quantity ξ can be written in a more suggestive way as follows. Recall that, due to the nullity of n, the covariant derivatives of n with respect to itself is orthogonal to n, so that one may write without loss of generality as[10]

$$n^B \nabla_B n^A = \kappa\, n^A + \eta^a m_a{}^A, \tag{56}$$

where $m_a{}^A := (\partial/\partial y^a)^A$. When $\eta^a = 0$, Eq. (56) is a null geodesic equation, and κ measures the deviation from the affinity of the parameter ω defined by

$$n^A \nabla_A \omega = 1. \tag{57}$$

The function η^a measures the rate of turning of n along $m_a{}^A$. Thus, the magnitude $|\eta|$ of η^a is a measure of rotation of n^A, which is given by

$$|\eta|^2 := \eta^a \eta_a = (n^B \nabla_B n^A)^2. \tag{58}$$

If we contract both sides of (56) by l_A, then we have

$$\begin{aligned} -\kappa = {} & l_A n^B \nabla_B n^A = n^B \Big\{ \nabla_B (l_A n^A) - n^A \nabla_B l_A \Big\} \\ = {} & -n^A n^B \nabla_B l_A \\ = {} & -n^A \Big\{ [n,\ l]_A + l^B \nabla_B n_A \Big\} \\ = {} & -n^A [n,\ l]_A \\ = {} & - < n, [n,\ l] > . \end{aligned} \tag{59}$$

where we used the identities

$$l_A n^A = -1, \qquad n^2 = 0. \tag{60}$$

Thus, κ is given by

$$\kappa = < n, [n,\ l] > . \tag{61}$$

In components, $[n,\ l]$ is given by

$$[n,\ l] = -F_{+-}{}^a \partial_a + (D_- h) l, \tag{62}$$

so that κ becomes

$$\kappa = -D_- h. \tag{63}$$

Therefore, if we use Eqs. (24), (25), and (63) in (55), we find that ξ becomes haha

$$\begin{aligned}\xi = & -\frac{1}{2}\,e^{-\sigma}\pi_h(\pi_\sigma - \frac{h}{2}\pi_h - e^{\sigma}D_+\sigma)\\ = & -\frac{1}{2}\,e^{-\sigma}\pi_h(2\kappa\, e^{\sigma} + \frac{h}{2}\,\pi_h)\\ = & -\frac{h}{4}\,e^{-\sigma}\pi_h^2 - \kappa\,\pi_h.\end{aligned} \tag{64}$$

Thus, from Eqs. (54) and (64), H becomes

$$H = \frac{2}{h}\,e^{-\sigma}\rho_{ac}\rho_{bd}\pi_-^{\ ab}\pi_-^{\ cd} - \frac{1}{2}\,e^{-2\sigma}\rho^{ab}\pi_a\pi_b - \frac{h}{4}\,e^{-\sigma}\pi_h^2 - \kappa\,\pi_h + e^{\sigma}R_2, \tag{65}$$

where H represents the net momentum-flux in the quasilocal momentum conservation equation (29).

4. Discussions

It seems appropriate to mention that, with the momentum-flux H given by (65), the quasilocal momentum conservation equation has a similar structure to the Navier-Stokes equation for a viscous fluid[7,11], which is given by

$$\frac{\partial P_i}{\partial u} = -\oint dS^k\,(p\delta_{ik} + \rho v_i v_k - \sigma'_{ik}). \tag{66}$$

Here P_i and σ'_{ik} are the total momentum and the viscous term,

$$P_i = \int dV\,(\rho v_i), \tag{67}$$

$$\sigma'_{ik} = \mu\Big(\frac{\partial v_i}{\partial x^k} + \frac{\partial v_k}{\partial x^i} - \frac{2}{3}\delta_{ik}\frac{\partial v_l}{\partial x^l}\Big) + \nu\delta_{ik}\frac{\partial v_l}{\partial x^l}, \tag{68}$$

and μ and ν are the coefficients of shear and bulk viscosity, respectively. This equation tells us that the rate of the net momentum change of a fluid within a given volume is determined by the net momentum-flux across 2-surface enclosing the volume. Notice that the momentum-flux H in (65) is at most quadratic in the conjugate momenta π_I. From this point of view, terms quadratic in π_I are responsible for direct momentum transfer, terms linear in π_I may be regarded as viscosity terms, and terms independent of π_I as pressure terms. Thus, the deviation κ from the affinity of the parameter of the in-going null geodesics plays the role of the coefficient of viscosity, whereas the scalar curvature of 2-surface is like a local pressure, so that the following correspondence holds,

$$\kappa \longleftrightarrow \mu \text{ (and/or) } \nu,$$
$$R_2 \longleftrightarrow p \text{ (pressure)}. \qquad (69)$$

Therefore, the Hamilton's equations of motion (18) and (19), together with the quasilocal conservation equations (28), (29), and (30), describe a dissipative system of the Einstein's gravitation.

Acknowledgments and Appendices

It is a great pleasure to thank Profs. J.M. Nester and Chiang-Mei Chen of National Central University, Taiwan, the organizers of the 7th Asia-Pacific International Conference on Gravitation and Astrophysics, where I presented a lecture based on this article. This work is supported by Konkuk University Research Grant 2004.

References

1. R. A. d'Inverno and J. Stachel, *J. Math. Phys.* **19**, 2447 (1978).
2. R. A. d'Inverno and J. Smallwood, *Phys. Rev. D* **22**, 1233 (1980).
3. E. T. Newman and T. Unti, *J. Math. Phys.* **3**, 891 (1962).
4. E. T. Newman and K. P. Tod, in *Asymptotically Flat Space-Times in General Relativity and Gravitation*, vol.2, ed. A. Held (Plenum Press, New York, 1980).
5. J. H. Yoon, *Phys. Lett. B* **308**, 240 (1993).
6. J. H. Yoon, *J. Korean Phys. Soc.* **34**, 108 (1999).
7. J. H. Yoon, *Phys. Lett. A* **292**, 166 (2001).
8. J. H. Yoon, *Phys. Rev. D* **70**, 084037 (2004).
9. J. H. Yoon, *in preparation.*
10. R. Penrose and W. Rindler, *Spinors and Space-time*, vol. 2, p.173 (Cambridge University Press, 1986).
11. L. D. Landau and E. M. Lifshitz, *The Classical Theory of Fields* (Pergamon Press, 1989).

Radiation reaction in curved space-time: local method.*

Dmitri Gal'tsov, Pavel Spirin and Simona Staub
Department of Theoretical Physics, Moscow State University, 119899, Moscow, Russia
E-mail: galtsov@phys.msu.ru

Although consensus seems to exist about the validity of equations accounting for radiation reaction in curved space-time, their previous derivations were criticized recently as not fully satisfactory: some ambiguities were noticed in the procedure of integration of the field momentum over the tube surrounding the world-line. To avoid these problems we suggest a purely local derivation dealing with the field quantities defined only *on* the world-line. We consider point particle interacting with scalar, vector (electromagnetic) and linearized gravitational fields in the (generally non-vacuum) curved space-time. To properly renormalize the self-action in the gravitational case, we use a manifestly reparameterization-invariant formulation of the theory. Scalar and vector divergences are shown to cancel for a certain ratio of the corresponding charges. We also report on a modest progress in extending the results for the gravitational radiation reaction to the case of non-vacuum background.

1. Introduction

Study of the radiation reaction problem in classical electrodynamics initiated by Lorentz and Abraham by the end of the 19-th century [1], remained an area of active research during the whole 20-th century. Although with the development of quantum electrodynamics this problem became somewhat academic, it still attracts attention in connection with new applications and new ideas in fundamental theory. Current understanding of the radiation reaction has emerged in the classical works of Dirac [2], Ivanenko and Sokolov [3], Rohrlich [4], Teitelboim [5] and some other (for a recent discussion see [6]).

*This work was supported in part by the RFBR grant 02-04-16949.

The Lorentz-Dirac equation was covariantly generalized to curved space-time with arbitrary metric by DeWitt and Brehme in 1960. In their paper [7] an elegant technique of covariant expansion of two-point tensor quantities was introduced which later became a basis of the perturbative quantum field theory calculations in curved space-time. It was used for calculation of the field momentum within a small tube surrounding the world-line of a point charge which resulted in the charge equation of motion with the radiation damping term. The main difference with the flat space case was the presence of the tail term depending on the entire history of a charge. Its presence signals violation of the Huygens' principle in curved space: a sharp pulse of light does not in general remain sharp, but gradually develops a "tail". The equation (as extended by Hobbs to non-vacuum metrics) [8]) reads

$$\begin{aligned} m\ddot{z}^{\alpha} = \; & eF^{\text{in}\,\alpha}_{\;\;\;\;\beta}\dot{z}^{\beta} + \frac{2}{3}e^2(\dddot{z}^{\alpha} - \ddot{z}^2\dot{z}^{\alpha}) \\ + \; & \frac{e^2}{3}\left(R^{\alpha}{}_{\beta}\dot{z}^{\beta} + R_{\gamma\beta}\dot{z}^{\gamma}\dot{z}^{\beta}\dot{z}^{\alpha}\right) + e^2\,\dot{z}^{\beta}\int_{-\infty}^{\tau} f^{\alpha}{}_{\beta\gamma}\dot{z}^{\gamma}(\tau')d\tau'. \end{aligned} \quad (1)$$

where $F^{\text{in}\,\alpha}_{\;\;\;\;\beta}$ is the incoming electromagnetic field $R^{\alpha}{}_{\beta}$ is the Ricci tensor and $f^{\alpha}{}_{\beta\gamma}$ is some two-point function taken on the world-line. This results was later generalized to other massless fields — scalar and linearized gravitational. The gravitational radiation reaction was discussed long ago by C. DeWitt-Morette and Jing [9], and more recently reconsidered in detail by Mino, Sasaki and Tanaka [10] for vacuum background metrics. In [10] an extension of DeWitt-Brehme method was used consisting in integration of the gravitational field momentum flux across the small world-tube surrounding the particle world-line. The equation of motion obtained in [10] can be presented as

$$\ddot{z}^{\mu} = -\frac{1}{2}(g^{\mu\nu} + \dot{z}^{\mu}\dot{z}^{\nu})(2h^{\text{tail}}_{\nu\lambda\rho} - h^{\text{tail}}_{\lambda\rho\nu})\dot{z}^{\lambda}\dot{z}^{\rho}. \quad (2)$$

where (contrary to the case of 1) the motion is supposed to be geodesic, so the local higher-derivative terms vanishes within the linear approximation, and the reaction force is given entirely by the tail term

$$h^{\text{tail}}_{\mu\nu\lambda} = 4mG\int_{-\infty}^{\tau^-}\left(G^{\;\text{ret}}_{\mu\nu\tau\sigma;\lambda} - \frac{1}{2}g_{\mu\nu}G^{\rho\;\text{ret}}_{\;\;\rho\nu\tau\sigma;\lambda}\right)z(\tau), z(\tau')\dot{z}^{\tau}(\tau')\dot{z}^{\sigma}(\tau'). \quad (3)$$

Although there is consensus about the *validity* of these equations, their previous *derivation* is somewhat problematic. In fact, as was discussed recently [11], all calculations involving integration of the field momentum located the small tube surrounding the particle world-line contain some yet unsolved problems. One such problem consists in computing the contributions of "caps" at the ends of the chosen tube segment: contributions of "caps" were rather *conjectured* than rigorously calculated. Another problem constitutes the singular integral over the internal boundary of the tube which was simply discarded in these calculations with no clear justification. In addition, the usual mass-renormalization is not directly applicable in the gravitational case since, due to the equivalence principle, the mass does not enter at all into the geodesic equations (in [10] the mass parameter was actually reintroduced by hand).

Several new derivations were suggested during past few years. One is based on the redefinition of the Green's functions of massless fields in curved space-time proposed by Detweiler and Whiting [12]. But this scheme involves an axiomatic assumption about the nature of the singular term and so it does not help to solve the above problems. Another scheme was suggested by Quinn and Wald [13] under the name of an "axiomatic approach to radiation reaction". This scheme makes use of some intuitive "comparison axioms", which do not follow from the first principles either. A recent attempt by Sanchez and Poisson [11] to overcome the above difficulties in fact is tight to a particular model of an extended particle (a "dumbbell" model). Thus, though the results obtained within several different approaches agree in the final form of non-divergent terms, their consistent derivation *from the first principles* is still lacking. Also, the derivation of the gravitation radiation reaction force for non-vacuum background metrics remains an open problem.

Here we briefly report on the new derivation (more detailed version will be published elsewhere) of the radiation reaction in curved space-time which has an advantage to deal with the fields *only on the world-line* and not to involve the volume integrals over the world-tube at all. This provides a more economic calculation and at the same time removes objections raised in [11]. The problem of divergences is relocated to the definition of the delta function with the support lying on the boundary of the integration domain, but this is exactly the same problem which is encountered in the flat space case where the approbated prescription amounts to the point-splitting procedure. Thus the local method is free from ambiguities which were met in the previous curved space calculations and can be considered as an adequate solution of the radiation reaction problem in curved space-time from the first principles. In general, our final results are conformal with those derived previously; in addition we partly remove the restriction by vacuum metrics in the gravitational case. Our signature is $(-+++)$.

2. Non-geodesic motion

Here we consider the non-geodesic motion of point particle along the affinely parametrized world-line $x^\mu = z^\mu(\tau)$ interacting with the scalar ϕ and vector A^μ fields. The total action is the sum $S = S_p + S_f$, where the world-line part reads

$$S_{\rm p} = -m_0 \int (1+q\phi)\sqrt{-\dot{z}^2}d\tau + e\int A^\mu \dot{z}_\mu d\tau, \tag{4}$$

(m_0 being the bare mass, q, e – the scalar and electric charges) while the volume part is

$$S_{\rm f} = -\frac{1}{4\pi}\int \left(\frac{1}{2}\partial_\mu\phi\partial^\mu\phi + \frac{1}{4}F^2\right)\sqrt{-g}d^4x. \tag{5}$$

Our local approach to radiation reaction simply consists in the substitution of the proper fields into the particle equation of motion (for brevity we omit the external force)

$$m_0(1+q\phi)\ddot{z}^\mu = eF^\mu{}_\nu\dot{z}^\nu - m_0 q\Pi^{\mu\nu}\phi_{;\nu} \tag{6}$$

where the velocity-transverse projector in the gauge $\dot{z}^2 = -1$ reads

$$\Pi^{\mu\nu} = g^{\mu\nu} + \dot{z}^\mu\dot{z}^\nu.$$

The equations for the scalar field and the 4-potential ($F = dA$) read

$$\Box\phi = 4\pi\rho \tag{7}$$

$$\Box A_\mu - R^\nu_\mu A_\nu = -4\pi j^\mu \tag{8}$$

where the covariant D'Alembert operator is understood $\Box = D_\mu D^\mu$, and the source terms are standard. The Green's functions are defined in the usual way starting with the Hadamard solution (in the scalar case):

$$G^{(1)}(x,z) = \frac{1}{(2\pi)^2}\left[\frac{\Delta^{1/2}}{\sigma} + v\ln\sigma + w\right], \tag{9}$$

where $\sigma(x,z)$ is the Synge's world function, Δ is van Vleck determinant, and v, w satisfy the system

$$\Box v = 0,$$

$$2v + (2v_{;\mu} - v\Delta^{-1}\Delta_{;\mu})\sigma^{;\mu} + \Box\Delta^{1/2} + \sigma\Box w = 0.$$

In the vector case

$$G^{(1)}_{\mu\alpha} = \frac{1}{(2\pi)^2}\left(\frac{u_{\mu\alpha}}{\sigma} + v_{\mu\alpha}\ln\sigma + w_{\mu\alpha}\right), \tag{10}$$

where $u_{\mu\alpha}(x,z)$, $v_{\mu\alpha}(x,z)$ and $w_{\mu\alpha}(x,z)$ are bi-vectors, satisfying a similar system. In particular, $u_{\mu\alpha} = \bar{g}_{\mu\alpha}\Delta^{1/2}$, where $\bar{g}_{\mu\alpha}$ is the bi-vector of parallel transport. Notation is that of DeWitt and Brehme: indices $\alpha, \beta, \ldots$ are associated with the "emission" point z, while $\mu, \nu, \ldots$ — with the "observation" point x. When both points are taken on the world-line, we use the first set to denote $z^\alpha(\tau')$ associated with the integration variable τ', and the second set to denote $z^\mu(\tau)$, where τ is the proper time in the resulting equation. In terms of these quantities the retarded solutions of the field equation read

$$\phi^{\rm ret}(x) = m_0 q \int\limits_{-\infty}^{\tau_{\rm ret}(x)} \left[-u\delta(\sigma) + v\theta(-\sigma)\right] d\tau', \tag{11}$$

$$F_{\mu\nu}^{\rm ret}(x) = -2e \int\limits_{-\infty}^{\tau_{\rm ret}(x)} \left[u_{[\mu\alpha}\sigma_{\nu]}\delta'(\sigma) + (u_{[\mu\alpha;\nu]} + v_{[\mu\alpha}\sigma_{\nu]})\delta(\sigma) + v_{[\nu\alpha;\mu]}\right] \dot{z}^\alpha d\tau', \tag{12}$$

where an anti-symmetrization over the indices μ and ν is used.

When substituted into the equations of motion, two-point functions become localized on the world-line, so we are led to use the expansions of the Synge's function $\sigma(z(\tau), z(\tau'))$(in terms containing delta-function and its derivative) as follows

$$\sigma(z(\tau), z(\tau')) = \sum_{k=0}^{\infty} \frac{1}{k!}\frac{D^k}{d\tau^k}\sigma(\tau,\tau)(\tau-\tau')^k \tag{13}$$

In what follows we will denote by dots the quantities

$$\dot{\sigma} = \sigma_\alpha \dot{z}^\alpha \quad , \quad \ddot{\sigma} = \sigma_{\alpha\beta}\dot{z}^\alpha\dot{z}^\beta + \sigma_\alpha \ddot{z}^\alpha, \quad \text{etc.} \tag{14}$$

Denoting the difference as $s = \tau - \tau'$, we get for σ an expansion similar to that in the flat space

$$\sigma(s) = -\frac{s^2}{2} - \ddot{z}^2(\tau)\frac{s^4}{24} + \mathcal{O}(s^5), \tag{15}$$

but with dots corresponding to covariant derivatives along the world-line. Similarly, for the derivative of σ with respect to $z^\mu(\tau)$ one finds

$$\sigma^\mu(s) = s\left(\dot{z}^\mu - \ddot{z}^\mu\frac{s}{2} + \dddot{z}^\mu\frac{s^2}{6}\right) + \mathcal{O}(s^4). \tag{16}$$

This quantity is a vector at the point $z(\tau)$ and a scalar at the point $z(\tau')$ where the index μ now corresponds to the point $z(\tau)$: $\sigma^\mu = \partial\sigma(z,z')/\partial z_\mu$.

An expansion for the delta-function reads

$$\delta(-\sigma) = \delta(s^2/2) + s^4 \frac{\ddot{z}^2(\tau)}{24} \delta'(s^2/2) + \ldots \tag{17}$$

and an analogous expansion is easily obtained for its derivative. Since the most singular term is $u\delta'(\sigma)\sigma_\nu$ the maximal expansion order to be retained is s^3. This also means that we have to retain for the delta-function only the leading term: $\delta(\sigma) = \delta(s^2/2) + \mathcal{O}(s^4)$. The delta-function of the squared argument can be regularized in exactly the same way as in the flat space as

$$\theta(s)\delta(s^2/2) \to \theta(s)\delta([s^2 - \varepsilon^2]/2) = \delta(s - \varepsilon)/\varepsilon,$$

where the positive regularization parameter $\varepsilon \to 0$ of the dimension of length is introduced. One finds in particular

$$\theta(s)\delta'(\sigma) \to \theta(s) \frac{1}{s} \frac{d}{ds} \delta\left(\frac{s^2 - \varepsilon^2}{2}\right) = \frac{1}{\varepsilon s} \delta'(s - \varepsilon)$$

We have to perform the expansions up to the third order in s in all terms containing delta-function and its derivative (local terms). For the function u we will have Ricci-terms involved into expansions, up to third order one finds:

$$u = 1 + \frac{1}{12} R_{\sigma\tau} \dot{z}^\tau \dot{z}^\sigma s^2, \quad u_{;\nu} = \frac{1}{6} R_{\nu\tau} \dot{z}^\tau s, \tag{18}$$

and similarly in the vector case

$$u_{\nu\alpha}\sigma^{;\mu}\dot{z}^\nu\dot{z}^\alpha = -s\dot{z}^\mu + \frac{s^2}{2}\ddot{z}^\mu - s^3\left(\frac{1}{6}\dddot{z}^\mu + \frac{1}{12}R_{\lambda\nu}\dot{z}^\lambda\dot{z}^\nu\dot{z}^\mu + \frac{1}{2}\ddot{z}^2\dot{z}^\mu\right). \tag{19}$$

The functions of v-type in the tail term can not be found in a closed form.

Collecting all the contributions, one obtains the following expressions for the scalar and electromagnetic reaction forces:

$$f^\mu_{\text{sc}} = m^2 q^2 \left[\frac{\ddot{z}^\mu}{2\epsilon} + \Pi^{\mu\nu}\left(\frac{1}{3}\dddot{z}_\nu + \frac{1}{6}R_{\nu\tau}\dot{z}^\tau - \int\limits_{-\infty}^{\tau} v_{;\nu}\, d\tau'\right) - \ddot{z}^\mu \int\limits_{-\infty}^{\tau} v d\tau'\right], \tag{20}$$

$$f^\mu_{\text{em}} = e^2 \left[-\frac{\ddot{z}^\mu}{2\epsilon} + \Pi^{\mu\nu}\left(\frac{2}{3}\dddot{z}_\nu + \frac{1}{3}R_{\nu\alpha}\dot{z}^\alpha\right) + 2\dot{z}^\nu(\tau) \int\limits_{-\infty}^{\tau} v^{[\mu}{}_{\alpha;\nu]}\dot{z}^\alpha(\tau')d\tau'\right]. \tag{21}$$

Divergent terms can be absorbed by the renormalization of mass.

$$m = m_0 + \frac{1}{2\epsilon}(m_0^2 q^2 - e^2) \tag{22}$$

Note different signs of the divergent terms for scalar and vector self-forces, so for a "BPS" particle with the ratio of charges $m|q| = |e|$ the model is free of divergencies. Finite contributions coincide with the previous results obtained via (somewhat questionable) world-tube derivations. Thus, in view of the criticisms in [11], our derivation may be considered as confirmation of these results. Technically, the local calculation is substantially simpler than the world-tube one.

3. Neutral particle at geodesic motion

In the case of gravitational radiation reaction we have the only one parameter — particle mass, which actually does not enter into the geodesic equation. Thus it is not possible to use the above renormalization scheme. This problem can be remedied using a manifestly reparametrization invariant treatment. This is done by introducing the einbein $e(\tau)$ on the world line acting as a Lagrange multiplier:

$$S[z^\mu, e] = -\frac{1}{2}\int \left[e(\tau)\, g_{\mu\nu}\dot{z}^\mu \dot{z}^\nu + \frac{m^2}{e(\tau)}\right] d\tau, \tag{23}$$

Varying (23) with respect to $z^\mu(\lambda)$ and $e(\tau)$ gives

$$\frac{D}{d\tau}(e\dot{z}^\mu) = 0, \quad e = \frac{m}{\sqrt{-g_{\mu\nu}\dot{z}^\mu \dot{z}^\nu}}, \tag{24}$$

and we obtain the geodesic equation in a manifestly reparametrization invariant form

$$\frac{D}{d\tau}\left(\frac{\dot{z}^\lambda}{\sqrt{-g_{\mu\nu}\dot{z}^\mu \dot{z}^\nu}}\right) = 0. \tag{25}$$

We split the total metric into background and perturbation due to point particle $g_{\mu\nu} \to \hat{g}_{\mu\nu} = g_{\mu\nu} + \kappa h_{\mu\nu}$. Assuming now that the particle motion with no account for radiation reaction is geodesic on the background metric, the perturbed equation in the leading order in κ will read

$$\ddot{z}^\mu = \frac{\kappa}{2}\left(g^{\mu\nu} - \frac{\dot{z}^\mu \dot{z}^\nu}{\dot{z}^2}\right)(h_{\lambda\rho;\nu} - 2h_{\nu\lambda;\rho})\dot{z}^\lambda \dot{z}^\rho, \tag{26}$$

where contractions are with the background metric.

The derivation of the field equations for the metric perturbation in the general case of non-vacuum background is non-trivial [14]. It is expected that the particle stress-tensor $\overset{(P)}{T^{\mu\nu}}$ has to be divergence-free with respect to the background metric, this allows one to derive the equation for the metric perturbation in a harmonic gauge, which leads to a manageable Green's function. But expanding the Bianchi identities for the full metric one finds that this is only possible if the Einstein tensor $G_{\mu\nu}$ and metric perturbation $h_{\mu\nu}$ satisfies an additional equation

$$G^{\lambda}_{\mu} h_{;\lambda} - G_{\lambda\rho} h^{\lambda\rho;\mu} = \mathcal{O}(\ddot{z}^{\mu}). \tag{27}$$

Otherwise, the "naive" particle stress tensor on a given background is not enough, and construction of a reliable source term for metric perturbation becomes a complicated problem. Clearly, there is no general solution to the constraint (27), but some particular cases can be found. One can notice that this equation is identically satisfied for Einstein metrics $R_{\mu\nu} = -\Lambda g_{\mu\nu}$. Another case is conformally-flat metrics with some special conformal factor (details will be given elsewhere). Provided the Eq. (27) holds, one can derive (using the results of Sciama et al. [14]) the following equation for the trace-reversed perturbation $\psi_{\mu\nu} = h_{\mu\nu} - g_{\mu\nu} h^{\lambda}_{\lambda}/2$:

$$\psi^{\mu\nu;\xi}{}_{;\xi} + 2R^{\mu}{}_{\xi}{}^{\nu}{}_{\rho}\psi^{\xi\rho} - 2\psi^{(\mu}{}_{\sigma}R^{\nu)\sigma} + \psi^{\mu\nu}R - g^{\mu\nu}R_{\alpha\beta}\psi^{\alpha\beta} = -2\kappa^2 \overset{(P)}{T}{}^{\mu\nu}. \tag{28}$$

The corresponding Hadamard's function is similar to (10) with four-index bi-tensors $u_{\lambda\rho\alpha\beta}$, $v_{\lambda\rho\alpha\beta}$. The retarded solutions is constructed in an analogous way and substituted into the equation of motion (26) leading to an integral involving bi-tensor quantities depending on two points on the world-line τ, τ'. As before, one can distinguish four different contributions: terms proportional to the delta function, the derivative of the delta function, the derivative of the Heaviside function, and the tail term. Performing series expansions in the first three cases up to the third order in $\tau - \tau' = s$ one finds a local contribution to the self-force. To facilitate expansions of bi-tensors one has first to reduce them to scalars at the point of expansion, e.g. quantities like $u_{\nu\lambda\alpha\beta} z^{\alpha}(\tau') z^{\beta}(\tau')$ behave as a scalar with respect to the point $z(\tau')$ and tensors with respect to $z(\tau)$. Then expansion in s around τ is straightforward.

Contrary to the non-geodesic case, now we have to drop all (covariant) derivatives with respect to the proper time of the second and higher order, with the only exception of divergent term. Actually, the divergent term is proportional to the acceleration, so formally it vanishes for the geodesic motion. But still it may be argued that its infinite value demands some renormalization to be made, and we have to find an appropriate quantity to be renormalized . Since the mass does not enter the equation, we invoke the einbein $e(\tau)$, and insert as the corresponding background quantity

e_0 =const. Collecting the expansions we obtain

$$\ddot{z}^\mu = \kappa^2 e_0 \left[\frac{7}{2\epsilon} \ddot{z}^\mu - \frac{11}{6} \Pi^{\mu\nu} R_{\nu\lambda} \dot{z}^\lambda + F^\mu_{\text{tail}} \right]. \tag{29}$$

Recall, that Ricci term is not arbitrary here, but has to be proportional to the metric for consistency. So actually the only local term here is proportional to the four-velocity $\dot{z}_\nu$ and vanishes by virtue of the projector. The tail term is

$$F^\mu_{\text{tail}} = \frac{\kappa^2 e_0}{4} \Pi^{\mu\nu} \int_{-\infty}^{\tau} [4 v_{\nu\lambda\alpha\beta;\rho} - 2(g_{\nu\lambda} v^\sigma_{\sigma\alpha\beta;\rho} + v_{\lambda\rho\alpha\beta;\nu}) -$$
$$- g_{\lambda\rho} v^\sigma_{\sigma\alpha\beta;\nu})] \dot{z}'^\alpha \dot{z}'^\beta \dot{z}^\lambda \dot{z}^\rho d\tau'. \tag{30}$$

Renormalization of the einbein is performed as

$$\left(\frac{1}{e_0} - \frac{7\kappa^2}{2\epsilon} \right) \ddot{z}^\mu = \frac{1}{e} \ddot{z}^\mu. \tag{31}$$

Finally the choice of the affine parameter $\dot{z}^2 = -1$, equivalent to setting $e = m$, leads to the final form

$$\ddot{z} = m\kappa^2 \Pi^{\mu\nu} \left[- \int_{-\infty}^{\tau} [(2 v_{\nu\lambda\alpha\beta;\rho} - g_{\nu\lambda} g^{\sigma\tau} v_{\sigma\tau\alpha\beta;\rho}) - \right.$$
$$\left. - (v_{\lambda\rho\alpha\beta;\nu} - 1/2\, g_{\lambda\rho} g^{\sigma\tau} v_{\sigma\tau\alpha\beta;\nu})] \dot{z}'^\alpha \dot{z}'^\beta d\tau' \dot{z}^\lambda \dot{z}^\rho \right]. \tag{32}$$

This coincides with the result obtained in [10,13]. We have thus shown that this equation remains valid for a class on non-vacuum metrics, in particular, for Einstein spaces.

4. Discussion

We have presented a local calculation of radiation reaction in both geodesic and non-geodesic cases which do not involve any ambiguous integrals outside the world-line and thus is free from the corresponding problems. It is based on the expansion of bi-tensor quantities only on the world line and technically is much simpler than the DeWitt-Brehme type approach. We have also generalized the gravitational radiation reaction to some non-vacuum metrics, however, satisfying an additional condition. The general non-vacuum case can not be treated within the test body/external field approximation because the global Bianchi identity demands to take into account the perturbation of the background matter as well.

Acknowledgments

DG is grateful to the Department of Physics of the National Central University (Taiwan) for hospitality and support during the Conference. He would like to thank J. Nester and C.-M. Chen for useful discussions. The work was supported in part by the RFBR grant 02-04-16949.

References

1. M. Abraham, *Theorie der Elektrizität*, vol. II, Springer, Leipzig, 1905.
2. P. Dirac, Proc. Roy. Soc. (London) **A167** (1938) 148.
3. D. Ivanenko and A. Sokolov, *Classical field theory (in Russian)*, Moscow, 1948; *Klassische Feldtheorie*, Berlin, 1953.
4. F. Rohrlich, *Classical charged particles*, Addison-Wesley, Reading, Mass. 1965; 2-nd edition: Redwood City, CA 1990).
5. C. Teitelboim, "Splitting of the Maxwell tensor: radiation reaction without advanced fields", Phys. Rev. **D1** (1970) 1572.
6. D. V. Gal'tsov and P. Spirin, "Radiation reaction reexamined: Bound momentum and Schott term," Gravitation and cosmology, No **1** (45) (2006) 1, hep-th/0405121.
7. B. S. DeWitt and R. W. Brehme, "Radiation Damping In A Gravitational Field," Annals Phys. **9** (1960) 220.
8. J. M. Hobbs, "A Vierbein Formalism of Radiation Damping," Annals Phys. **47** (1968) 141.
9. C. Morette-DeWitt and J. L. Ging, "Radiation Damping In A Gravitational Field," C. R. Acad. Sci. Paris **251** (1960) 1868.
10. Y. Mino, M. Sasaki and T. Tanaka, "Gravitational radiation reaction to a particle motion," Phys. Rev. D **55** (1997) 3457 [arXiv:gr-qc/9606018].
11. J.M. Sanchez and E. Poisson, "Extended body approach to the electromagnetic self-force in curved space-time", gr-qc/0512111.
12. S. Detweiler and B. F. Whiting, "Self-force via a Green's function decomposition," Phys. Rev. D **67** (2003) 024025 [arXiv:gr-qc/0202086].
13. T. C. Quinn and R. M. Wald, "An axiomatic approach to electromagnetic and gravitational radiation reaction of particles in curved spacetime," Phys. Rev. D **56** (1997) 3381 [arXiv:gr-qc/9610053].
14. D. W. Sciama,P. C. Waylen, and R. C. Gilman, "Generally Covariant Integral Formulation of Einstein's, Field Equations", Phys. Rev. **187, No.5** (1969) 1762.

Classical pseudotenors and positivity in small regions

Lau Loi So

Department of Physics, National Central University, Chungli 320, Taiwan

James M. Nester

Department of Physics and Institute of Astronomy, National Central University, Chungli 320, Taiwan

E-mail: nester@phy.ncu.edu.tw

Hsin Chen

Department of Physics, National Central University, Chungli 320, Taiwan

We have studied the famous classical pseudotensors in the small region limit, both inside matter and in vacuum. A recent work [Deser *et al.* 1999 *CQG* **16**, 2815] had found one combination of the Einstein and Landau-Lifshitz expressions which yields the Bel-Robinson tensor in vacuum. Using similar methods we found another independent combination of the Bergmann-Thomson, Papapetrou and Weinberg pseudotensors with the same desired property. Moreover we have constructed an infinite number of additional new holonomic pseudotensors satisfying this important positive energy requirement, all seem quite artificial. On the other hand we found that Møller's 1961 tetrad-teleparallel energy-momentum expression naturally has this Bel-Robinson property.

1. Introduction: quasilocal quantities for small regions

The localization of gravitational energy-momentum remains an important problem in GR. Using standard methods many famous researchers each found their own expression. None of these expressions is covariant, they are all reference frame dependent (referred to as "pseudotensors"). This feature can be understood in terms of the equivalence principle: gravity cannot be detected at a point, so it cannot have a point-wise defined energy-momentum density. Now there is another way to address the difficulty.

The new idea is quasilocal: energy-momentum is associated with a closed

2 surface surrounding a region [1]. A good quasilocal approach is in terms of the Hamiltonian [2]. Then the Hamiltonian boundary term determines the quasilocal quantities. In fact this approach includes all the traditional pseudotensors [3,4]. They are each generated by a superpotential which can serve as special type of Hamiltonian boundary term.

A good energy-momentum expression for gravitating systems should satisfy a variety of requirements, including giving the standard values for the total quantities for asymptotically flat space and reducing to the material energy-momentum in the appropriate limit. No entirely satisfactory expression has yet been identified. One of the most restrictive requirements is positivity. A general positivity proof is very difficult. One limit that is not so difficult is the small region limit. This has not previously been systematically investigated for all the classical pseudotensors expressions. In matter the expression should be dominated by the material energy-momentum tensor. In vacuum we require that its Taylor series expansion in Riemann normal coordinates have at the second order a positive multiple of the Bel-Robinson tensor. This will guarantee positive energy in small vacuum regions.

2. The classical pseudotensors

Recall the Einstein field equation $G_{\mu\nu} = R_{\mu\nu} - \frac{1}{2}g_{\mu\nu}R = \kappa T_{\mu\nu}$. Matter energy-momentum has a vanishing covariant divergence $\nabla_\mu T^\mu{}_\nu = 0$, but in curved spacetime this is not in the form of a conserved energy-momentum relation. But one can rewrite it in the form of a divergence

$$\partial_\mu \sqrt{-g}(T^\mu{}_\nu + t^\mu{}_\nu) = 0, \tag{1}$$

here $t^\mu{}_\nu$ is the gravitational energy-momentum pseudotensor. The energy-momentum complex $\mathcal{T}^\mu{}_\nu$ is then given as

$$\kappa \mathcal{T}^\mu{}_\nu = \kappa\sqrt{-g}(T^\mu{}_\nu + t^\mu{}_\nu) \equiv \partial_\lambda U_\nu{}^{[\mu\lambda]}, \tag{2}$$

where $U_\nu{}^{[\mu\lambda]}$ is called the superpotential.

3. Riemann normal coordinates and the adapted tetrads

For the total quantities of an isolated gravitating system the various expressions give the expected weak field asymptotic values. However they are quite different in the strong field region. To study the quasilocal quantities one can Taylor expand the Hamiltonian, including the divergence of its boundary term, in a small spatial region surrounding a point. The reference is the flat space geometry at this origin. Riemann normal coordinates (RNC) satisfy

$$g_{\mu\nu}(0) = \eta_{\mu\nu}, \quad \partial_\lambda g_{\mu\nu}(0) = 0, \quad \partial^2_{\mu\nu} g_{\alpha\beta}(0) = -\frac{1}{3}(R_{\alpha\mu\beta\nu} + R_{\alpha\nu\beta\mu})(0). \quad (3)$$

Later we will also need the associated adapted orthonormal frame (aka tetrad, vierbein) which satisfies $g_{\mu\nu} = \eta_{ab} e^a{}_\mu e^b{}_\nu$ and

$$e^a{}_\mu(0) = \delta^a_\mu, \quad \partial_\nu e^a{}_\nu(0) = 0, \quad \Gamma^a{}_{b\mu}(0) = 0, \quad \partial_\mu \Gamma^a{}_{b\nu}(0) = \frac{1}{2} R^a{}_{b\mu\nu}(0). \quad (4)$$

For the energy-momentum density we want at non-vacuum points the results to be dominated by the material energy-momentum. At vacuum points one wants a result proportional to the Bel-Robinson tensor. That will guarantee the proper positive energy property; i.e. the associated energy-momentum vector will then be future pointing and non-space like.

4. Quadratic curvature combinations

The Bel-Robinson tensor $B_{\alpha\beta\mu\nu}$ and the tensors $S_{\alpha\beta\mu\nu}$, $K_{\alpha\beta\mu\nu}$ and $T_{\alpha\beta\mu\nu}$ are defined as follows

$$B_{\alpha\beta\mu\nu} := R_{\alpha\lambda\mu\sigma} R_\beta{}^\lambda{}_\nu{}^\sigma + R_{\alpha\lambda\nu\sigma} R_\beta{}^\lambda{}_\mu{}^\sigma + 3T_{\alpha\beta\mu\nu}, \quad (5)$$

$$S_{\alpha\beta\mu\nu} := R_{\alpha\mu\lambda\sigma} R_{\beta\nu}{}^{\lambda\sigma} + R_{\alpha\nu\lambda\sigma} R_{\beta\mu}{}^{\lambda\sigma} - 6T_{\alpha\beta\mu\nu}, \quad (6)$$

$$K_{\alpha\beta\mu\nu} := R_{\alpha\lambda\mu\sigma} R_\beta{}^\lambda{}_\nu{}^\sigma + R_{\alpha\lambda\nu\sigma} R_\beta{}^\lambda{}_\mu{}^\sigma + 9T_{\alpha\beta\mu\nu}, \quad (7)$$

$$T_{\alpha\beta\mu\nu} := -\frac{1}{24} g_{\alpha\beta} g_{\mu\nu} R_{\lambda\sigma\xi\kappa} R^{\lambda\sigma\xi\kappa}. \quad (8)$$

A recent work [5] had found exactly one pseudotensor expression, a certain combination of the Einstein and Landau-Lifshitz expressions, which yields the Bel-Robinson tensor in vacuum. They argued that this combination is unique under their assumptions.

5. Classical holonomic pseudotensors

The well known classical superpotentials associated with the Einstein, Landau-Lifshitz, Bergmann-Thomson, Papapetrou, Weinberg and

Møller(1958) energy-momentum complexes are

$$
\begin{aligned}
{}_E U_\alpha{}^{[\mu\nu]} &= \sqrt{-g} g^{\beta\sigma} \Gamma^\tau{}_{\lambda\beta} \delta^{\lambda\nu\mu}_{\tau\sigma\alpha}, & (9)\\
{}_B U^{\alpha[\mu\nu]} &= \sqrt{-g} g^{\alpha\beta} g^{\pi\sigma} \Gamma^\tau{}_{\lambda\pi} \delta^{\lambda\nu\mu}_{\tau\sigma\beta} = \sqrt{-g}\, {}_L U^{\alpha[\mu\nu]}, & (10)\\
{}_P H^{[\mu\nu][\alpha\beta]} &= \sqrt{-g} g^{ma} g^{nd} \delta^{\mu\nu}_{ab} \delta^{\alpha\beta}_{mn}, & (11)\\
{}_W H^{[\mu\alpha][\nu\beta]} &= \sqrt{-\eta} \left(\eta^{mc} \eta^{nd} - \frac{1}{2} \eta^{mn} \eta^{cd} \right) g_{cd} \eta^{ab} \delta^{\alpha\mu}_{ma} \delta^{\nu\beta}_{nb}, & (12)\\
{}_{58} U_\alpha{}^{[\mu\nu]} &= \sqrt{-g} (\Gamma^{\nu\mu}{}_\alpha - \Gamma^{\mu\nu}{}_\alpha). & (13)
\end{aligned}
$$

The pseudotensors are obtained according to one of the prescriptions:

$$
\mathcal{T}_\alpha{}^\mu = \partial_\nu U_\alpha{}^{[\mu\nu]}, \quad \mathcal{T}^{\alpha\mu} = \partial_\nu U^{\alpha[\mu\nu]}, \quad \mathcal{T}^{\mu\nu} = \partial^2{}_{\alpha\beta} H^{[\mu\alpha][\nu\beta]}. \tag{14}
$$

Inside matter at the origin, the RNC expansion results are

$$
E_\alpha{}^\beta(0) = B_\alpha{}^\beta(0) = P_\alpha{}^\beta(0) = W_\alpha{}^\beta(0) = 2G_\alpha{}^\beta(0) = 2\kappa T_\alpha{}^\beta(0). \tag{15}
$$

This is as expected from the equivalence principle. However the Møller(1958) expression gives

$$
M_\alpha{}^\beta(0) = R_\alpha{}^\beta(0) = \kappa \left(T_\alpha{}^\beta - \frac{1}{2} \eta_\alpha{}^\beta T \right)(0) \neq 2\kappa T_\alpha{}^\beta(0). \tag{16}
$$

This result is not acceptable. From [5], the small vacuum ($G_{\mu\nu} = 0$) region non-vanishing terms are

$$
\begin{aligned}
E_\alpha{}^\beta &= -2\Gamma_{\lambda\sigma\alpha}\Gamma^{\beta\lambda\sigma} + \delta^\beta_\alpha \Gamma_{\lambda\sigma\tau}\Gamma^{\tau\lambda\sigma}, & (17)\\
L^{\alpha\beta} &= \Gamma^\alpha{}_{\lambda\sigma}(\Gamma^{\beta\lambda\sigma} - \Gamma^{\lambda\sigma\beta}) - \Gamma_{\lambda\sigma}{}^\alpha(\Gamma^{\beta\lambda\sigma} + \Gamma^{\sigma\lambda\beta}) + g^{\alpha\beta}\Gamma_{\lambda\sigma\tau}\Gamma^{\sigma\lambda\tau}. & (18)
\end{aligned}
$$

Using a similar technique we found, for the other pseudotensors (note Papapetrou and Weinberg depend explicitly on the Minkowski metric $\eta_{\alpha\beta}$),

$$
\begin{aligned}
P^{\alpha\beta} &= L^{\alpha\beta} + h^{\lambda\sigma}(\Gamma^{\alpha\beta}{}_{\lambda,\sigma} + \Gamma^{\beta\alpha}{}_{\lambda,\sigma}) - h^{\lambda\beta}{}_{,\sigma}(\Gamma^{\alpha\sigma}{}_\lambda + \Gamma^{\sigma\alpha}{}_\lambda), & (19)\\
W^{\alpha\beta} &= -2\Gamma_{\lambda\sigma}{}^\alpha\Gamma^{\lambda\sigma\beta} + g^{\alpha\beta}\Gamma_{\lambda\sigma\tau}\Gamma^{\lambda\sigma\tau}\\
&\quad - g^{\alpha\lambda} g^{\beta\pi} h^{\sigma\rho} \delta^{ck}_{\lambda\sigma} \delta^{\xi d}_{\pi\rho} (\Gamma_{dc\kappa,\xi} + \Gamma_{cd\kappa,\xi}), & (20)\\
M_\alpha{}^\beta &= \Gamma_{\lambda\sigma\alpha}\Gamma^{\lambda\sigma\beta} - \Gamma_{\lambda\sigma\alpha}\Gamma^{\sigma\lambda\beta} + (g^{\beta\sigma} g^{\psi\lambda} - g^{\beta\psi} g^{\lambda\sigma}) g_{\alpha\psi,\lambda\sigma}. & (21)
\end{aligned}
$$

Here $h_{\alpha\beta} := g_{\alpha\beta} - \eta_{\alpha\beta}$ and $h_{\alpha\beta} = -\frac{1}{3} R_{\alpha\lambda\beta\sigma} x^\lambda x^\sigma + O(x^3)$ in RNC. From [5] in vacuum we have

$$
\partial^2_{\mu\nu} E_{\alpha\beta}(0) = \frac{1}{9}(4B_{\alpha\beta\mu\nu} - S_{\alpha\beta\mu\nu}), \quad \partial^2_{\mu\nu} L_{\alpha\beta}(0) = \frac{1}{9}\left(7B_{\alpha\beta\mu\nu} + \frac{1}{2} S_{\alpha\beta\mu\nu}\right). \tag{22}
$$

In order to obtain the Bel-Robinson tensor, Deser *et al.* used a "by hand" combination

$$\partial^2_{\mu\nu}\left(\frac{1}{2}E_{\alpha\beta}+L_{\alpha\beta}\right)(0)=B_{\alpha\beta\mu\nu}, \tag{23}$$

Using similar methods we obtained for the other pseudotensors in vacuum at the origin

$$\partial^2_{\mu\nu}P_{\alpha\beta}(0)=\frac{2}{9}(4B_{\alpha\beta\mu\nu}-S_{\alpha\beta\mu\nu}-K_{\alpha\beta\mu\nu}), \tag{24}$$

$$\partial^2_{\mu\nu}W_{\alpha\beta}(0)=-\frac{2}{9}(B_{\alpha\beta\mu\nu}+2S_{\alpha\beta\mu\nu}+3K_{\alpha\beta\mu\nu}), \tag{25}$$

$$\partial^2_{\mu\nu}M_{\alpha\beta}(0)=\frac{1}{9}\left(2B_{\alpha\beta\mu\nu}-\frac{1}{2}S_{\alpha\beta\mu\nu}-K_{\alpha\beta\mu\nu}\right). \tag{26}$$

(Here we have included for completeness the result of the Møller(1958) expression, even though it does not have a good matter interior limit.) From this we find one more independent combination of the Landau-Lifshitz, Papapetrou and Weinberg pseudotensors with the same desired Bel-Robinson property. (Note: the earlier work cited above did not get this result, as they had excluded the explicit use of the Minkowski metric in the superpotentials they considered.) Inside matter and in vacuum at the origin, respectively, we find

$$\frac{1}{3}\left[2L_{\alpha\beta}+\frac{1}{2}(3P_{\alpha\beta}-W_{\alpha\beta})\right](0)=2G_{\alpha\beta}(0), \tag{27}$$

$$\frac{1}{3}\partial^2{}_{\mu\nu}\left[2L_{\alpha\beta}+\frac{1}{2}(3P_{\alpha\beta}-W_{\alpha\beta})\right](0)=B_{\alpha\beta\mu\nu}. \tag{28}$$

6. A large class of new pseudotensors

Moreover we have constructed an infinite number (a 3 parameter set) of new holonomic pseudotensors all satisfying this important Bel-Robinson/positivity property. The new general superpotential is

$$\mathcal{U}_\alpha{}^{[\mu\nu]}={}_EU_\alpha{}^{[\mu\nu]}+\left\{\begin{array}{l} c_1h^{\mu\pi}\Gamma_\alpha{}^\nu{}_\pi+c_2h^{\mu\pi}\Gamma^\nu{}_{\alpha\pi}+c_3h^{\mu\pi}\Gamma_\pi{}^\nu{}_\alpha \\ +c_4'\delta^\mu_\alpha h^{\pi\rho}\Gamma^\nu{}_{\pi\rho}+c_4''\delta^\mu_\alpha h^{\pi\rho}\Gamma_\pi{}^\nu{}_\rho+c_5h_{\alpha\pi}\Gamma^{\mu\nu\pi}-(\nu\leftrightarrow\mu)\end{array}\right\}, \tag{29}$$

where c_1 to c_5 are constants and $h_{\alpha\beta}:=g_{\alpha\beta}-\eta_{\alpha\beta}$. In Riemann normal coordinates $h^{\alpha\beta}=\frac{1}{3}R^\alpha{}_\xi{}^\beta{}_\kappa x^\xi x^\kappa+O(x^3)$. Actually, the leading superpotential ${}_EU_\alpha{}^{[\mu\nu]}$ in (29) is not necessary Freud's, it can be replaced by any other which offers a good spatially asymptotic and small region material limit. The resultant energy density inside matter at the origin is, as expected,

$$2\kappa\mathcal{E}_\alpha{}^\beta(0)=2G_\alpha{}^\beta(0)=2\kappa T_\alpha{}^\beta(0). \tag{30}$$

The RNC second derivatives in vacuum at the origin are

$$\partial^2_{\mu\nu}\mathcal{E}_{\alpha\beta}(0) = \frac{1}{9}\left\{\begin{array}{c}(4-2c_1+c_2+c_3-4c_4+3c_5)B_{\alpha\beta\mu\nu}\\ -\frac{1}{2}(2-c_1-4c_2+5c_3-2c_4-3c_5)S_{\alpha\beta\mu\nu}\\ +(c_1+c_2-2c_3-2c_4)K_{\alpha\beta\mu\nu}\end{array}\right\}. \tag{31}$$

where $c_4 = c_4' - \frac{1}{2}c_4''$. Note that we can choose

$$4-2c_1+c_2+c_3-4c_4+3c_5 > 0, \tag{32}$$

$$2-c_1-4c_2+5c_3-2c_4-3c_5 = 0, \tag{33}$$

$$c_1+c_2-2c_3-2c_4 = 0. \tag{34}$$

The solutions can be parameterized as follows,

$$c_1+c_2-2c_3 < 1, \tag{35}$$

$$(c_1,c_2,c_3,c_4,c_5) = \left(c_1,c_2,c_3,\frac{1}{2}(c_1+c_2+2c_3),\frac{1}{3}(2-2c_1-5c_2+7c_3)\right).$$

Then the second derivatives of the new pseudotensors in vacuum are

$$\partial^2_{\mu\nu}\mathcal{E}_{\alpha\beta}(0) = \frac{2}{3}(1-c_1-c_2+2c_3)B_{\alpha\beta\mu\nu}. \tag{36}$$

As there are three arbitrary constants that one can tune, obviously there exists an infinite number of solutions with any positive magnitude of $B_{\alpha\beta\mu\nu}$. They all appear to be highly artificial. It seems that there is no obstruction in going to higher order. From our analysis we infer that there are an infinite number of holonomic gravitational energy-momentum pseudotensor expressions which satisfy the highly desired small region Bel-Robinson/positive energy property.

7. Møller's 1961 tetrad-teleparallel energy-momentum tensor

On the other hand, unlike the aforementioned mathematically and physically contrived expressions, we found that Møller's 1961 teleparallel-tetrad energy-momentum expression naturally has the desired Bel-Robinson property. The superpotential has the same form as Freud's but the indices now refer to a tetrad:

$${}_{61}U_a{}^{[bc]} = \sqrt{-g}g^{df}\Gamma^i{}_{fe}\delta^{bce}_{ida}. \tag{37}$$

Expressed in terms of differential forms we have

$$m_a{}^b\eta_b = d(\Gamma^b{}_c \wedge \eta_b{}^c{}_a). \tag{38}$$

The RNC and adapted frame expansion results inside matter, and the vacuum second derivatives at the origin, respectively, are

$$m_a{}^b(0) = 2G_a{}^b(0) = 2\kappa T_a{}^b(0), \quad \partial^2_{\mu\nu} m_{ab}(0) = \frac{1}{2} B_{ab\mu\nu}. \tag{39}$$

Thus the desired Bel-Robinson property is naturally satisfied. An important consequence is that the gravitational energy according to this measure is positive, at least to this order. (We expected this positivity result since in fact Møller's 1961 expression has an associated positive energy proof [6].) Once again Møller's 1961 tensor stands out as one of the best descriptions for gravitational energy-momentum.

References

1. L. B. Szabados, Living Rev. Relativity, **7** (2004) 4; http://www.livingreviews.org/lrr-2004-4
2. C. M. Chen and J.M. Nester, *Class. Quantum Grav.* **16** (1999) 1279 [arXiv:gr-qc/9809020].
3. C. C. Chang, J. M. Nester and C. M. Chen, *Phys. Rev. Lett.* **83** (1999) 1897.
4. C. M. Chen and J. M. Nester, *Grav. & Cosmol.* **6** (2000) 257 [arXiv:gr-qc/0001088].
5. S. Deser, J. S. Franklin and D. Seminara, *Class. Quantum Grav.* **16** (1999) 2815.
6. J. M. Nester, *Int. J. Mod. Phys. A.* *4* (1989) 1755.

A Brief Review of Initial Data Engineering

Piotr T. Chruściel

AEI, Golm and LMPT, Tours*

E-mail: piotr@gargan.math.univ-tours.fr, URL: www.phys.univ-tours.fr/~piotr

James Isenberg†

University of Oregon

E-mail: jim@newton.uoregon.edu, URL: www.physics.uoregon.edu/~jim

Daniel Pollack‡

University of Washington

E-mail: pollack@math.washington.edu, URL: www.math.washington.edu/~pollack

We review the recently developed program for constructing and studying solutions of the Einstein constraint equations using gluing techniques. We discuss what we believe are sharp conditions sufficient for a pair of solutions to admit gluing via a connected sum or "wormhole", and describe how one carries out the gluing. We also discuss a number of useful applications.

1. Introduction

The initial value formulation is the most widely used procedure for constructing solutions of Einstein's gravitational filed equations, and the first step in carrying out such a construction is that of finding a set of initial data $(\Sigma^3, \gamma, K, \psi, \pi)$ which satisfies the Einstein constraint equations

$$R(\gamma) - |K|^2_\gamma - (\mathrm{tr}_\gamma K)^2 = 2\rho(\gamma, \psi, \pi) \tag{1.1}$$

$$D_i(K^{ij} - \mathrm{tr}_\gamma K \gamma^{ij}) = J(\gamma, \psi, \pi). \tag{1.2}$$

Here Σ^3 is a 3 dimensional manifold, γ is a Riemannian metric on Σ^3 with scalar curvature $R(\gamma)$, K is a symmetric tensor field, (ψ, π) represent any

*Visiting fellow.

†Partially supported by the NSF under Grant PHY-0099373

‡Partially supported by the NSF under Grant DMS-0305048

non gravitational fields which may be present, and ρ and J are the energy density and momentum density functions of these non gravitational fields.

Since the early 1970's, the predominately used procedure for constructing and studying sets of data which satisfy the constraints has been the conformal method [5,3] and the closely related conformal thin sandwich method [18]. These conformal techniques have been very useful both practically (for explicit construction of solutions) and theoretically (for proving theorems regarding properties of the solutions). They have not, however, been very successful in handling non constant mean curvature data sets. They are also somewhat limited in their ability to construct data for modeling prescribed physical systems.

As an alternative analytical tool, since 2000, "gluing techniques" have been developed for working with solutions of the constraints. One of the ideas in this context is to choose a pair of points p_1 and p_2 on a pair of known solutions $(\Sigma_1^3, \gamma_1, K_1, \psi_1, \pi_1)$ and $(\Sigma_2^3, \gamma_2, K_2, \psi_2, \pi_2)$ of the constraints, and construct a new solution on the connected sum manifold $\Sigma_1^3 \# \Sigma_2^3$ which agrees exactly with the original solutions outside a small neighborhood of the neck $S^2 \times I$ (for I an interval) now connecting the regions around the excised points p_1 and p_2.

Such a gluing operation cannot be carried out for *every* possible choice of pairs of data and pairs of points; however it can be done for generic such choices. We describe in Section 2 the explicit conditions on the data and the points which guarantee that gluing can be done, and we argue that these conditions are likely sharp. We then discuss in Section 3 some of the ideas and techniques used in carrying out the gluing and in showing that it can be carried out to completion in the appropriate cases. In Section 4 we review a number of the applications of gluing. Using it, we can attach black holes and wormholes to given spacetimes, we can prove that there are no topological restrictions on manifolds admitting asymptotically Euclidean or asymptotically hyperbolic solutions of the constraints, and we can show that there exist maximally developed vacuum solutions of Einstein's equations which contain a compact Cauchy surface but do not admit any constant mean curvature Cauchy surfaces. We make concluding remarks in Section 5.

Note that in this review paper, none of the proofs of the theorems we discuss are carried out in any detail; those appear in the succession of papers [14,15,7,8,13].

2. When Gluing Can be Done

Gluing techniques have been applied to solutions of a number of geometrically motivated PDE systems, so much so that they are now regarded as a standard tool in geometric analysis. In all cases, one has to impose a "non degeneracy" requirement which serves as a sufficient condition for successful gluing. This is true here as well for the Einstein constraint equations. While in our earlier results [14,15] we have stated alternatives, we now believe that the sharpest condition for gluing is based on the following

DEFINITIONHEADFONTDEFINITION 2.1.DEFINITIONFONT LET $(\Sigma^3, \gamma.K)$ BE A SET OF INITIAL DATA SATISFYING THE EINSTEIN VACUUM CONSTRAINT EQUATIONS, AND LET $p \in \Sigma^3$ AND LET U BE AN OPEN SET CONTAINING p. THE DATA HAS *No KIDs in* U IF THERE DO NOT EXIST NON TRIVIAL SOLUTIONS (N, Y) TO THE FORMAL ADJOINT OF THE LINEARIZED CONSTRAINT EQUATIONS:

$$0 = \begin{pmatrix} 2(\nabla_{(i}Y_{j)} - \nabla^l Y_l g_{ij} - K_{ij}N + \mathrm{TR}K\ N g_{ij}) \\ \\ \nabla^l Y_l K_{ij} - 2K^l{}_{(i}\nabla_{j)}Y_l + K^q{}_l \nabla_q Y^l g_{ij} - \Delta N g_{ij} + \nabla_i \nabla_j N \\ +(\nabla^p K_{lp} g_{ij} - \nabla_l K_{ij})Y^l - N\mathrm{Ric}\,(g)_{ij} \\ +2NK^l{}_i K_{jl} - 2N(\mathrm{TR}\ K)K_{ij} \end{pmatrix}, \quad (2.1)$$

IN U.

A similar definition holds for the Einstein-Maxwell, Einstein-Yang-Mills, Einstein-Vlasov, and other Einstein-matter systems. Geometrically, the No KIDs condition means that there are no Killing vectors defined on the domain of dependence of U in the spacetime development [17].

With the No KiDs definition in hand, we may now state the vacuum version of our main gluing result:

THEOREMHEADFONTTHEOREM 2.2.THEOREMFONT LET $(\Sigma_1^3, \gamma_1, K_1)$ AND $(\Sigma_2^3, \gamma_2, K_2)$ BE A PAIR OF SMOOTH INITIAL DATA SETS WHICH SATISFY THE VACUUM ($\rho = 0$ AND $J = 0$) CONSTRAINT EQUATIONS (1.1)-(1.2). LET $p_1 \in \Sigma_1^3$ AND $p_2 \in \Sigma_2^3$ BE A PAIR OF POINTS, WITH OPEN NEIGHBORHOODS $p_1 \in U_1$ AND $p_2 \in U_2$ IN WHICH THE NO KIDS CONDITION IS SATISFIED. THERE EXISTS A SMOOTH DATA SET $(\Sigma_1^3 \# \Sigma_2^3, \hat{\gamma}, \hat{K})$ WHICH SATISFIES THE EINSTEIN CONSTRAINT EQUATIONS EVERYWHERE, AND WHICH AGREES WITH (γ_1, K_1) AND (γ_2, K_2) AWAY FROM $U_1 \cup U_2$.

To see that some non degeneracy condition (of the nature of *No KIDs*) is indeed needed for gluing to be permitted, it is useful to consider the following example: Let $(\Sigma_1^3, \gamma_1, K_1)$ be any solution of the constraints which has Σ_1^3 compact and γ_1 non flat, and let $(\Sigma_2^3, \gamma_2, K_2) = (R^3, flat, 0)$. If one could glue these two initial data sets, then one would have an asymptotically Euclidean data set which is not data for Minkowski spacetime, and yet is identical to such data outside of a compact region. It would follow that the data would have mass zero, which would be a violation of the positive mass theorem [20,21,22]. We are this forced to conclude that the gluing of these particular data sets cannot be done. We note that the data $(\Sigma_2^3, \gamma_2, K_2) = (R^3, flat, 0)$ violates the No KIDs condition at every point.

How restrictive is the No KIDs condition? As shown by Beig, Chrusciel and Schoen [4], a generic initial data set (appropriately defined) satisfies the No KIDs condition almost everywhere, so the condition is fairly mild.

While we believe that a result very similar to Theorem 2.2 holds for Einstein-Maxwell and other Einstein-matter field theories, no such theorem has yet been proven. We have in fact established conformal ("non localized") gluing results of the nature discussed in Section 3 for many Einstein-matter theories (See [13]); however to complete the job and obtain results ("localized") of the nature of Theorem 2.2, we need also to show that the Corvino-Schoen [10,11,6] type procedures can be extended to non vacuum theories. While this has not yet been done, there do not appear to be any fundamental impediments to doing it.

3. How Gluing Works

Our gluing results have been developed and proven in three stages (with Rafe Mazzeo contributing to the first two). The first result, appearing in [14], shows that if we have a pair of initial data sets $(\Sigma_1^3, \gamma_1, K_1)$ and $(\Sigma_2^3, \gamma_2, K_2)$ which both have constant mean curvature ("CMC") of the same value, and which satisfy the non degeneracy condition that, if either Σ_1^3 or Σ_2^3 is a closed manifold, then the corresponding K_1 or K_2 may not be identically zero, and the corresponding geometry γ_1 or γ_2 does not have a conformal Killing field with a zero at the gluing points p_1 or p_2, then a gluing of the following sort can be carried out (which we call a "non localized gluing"). One can find a one parameter family $(\Sigma_1^3 \# \Sigma_2^3, \gamma_T, K_T)$ of initial data sets, all of which satisfy the constraints everywhere on $\Sigma_1^3 \# \Sigma_2^3$, with (γ_T, K_T) approaching arbitrarily close to $(\gamma_1 K_1)$ and (γ_2, K_2), away from the "neck" (or bridge) connecting Σ_1^3 to Σ_2^3, as $T \to \infty$. Note that this type of "non localized" gluing, involving changes in the data everywhere, approaching the original data only in a limit, is the traditional form of gluing theorem which has generally been proven for geometric PDE systems such as those corresponding to constant scalar curvature metrics.

In our second work [15], we show that the CMC condition need only be imposed locally near the points about which one wishes to glue: gluing can then be carried out regardless of the mean curvature of the data sets away from the resulting neck, so long as an additional non degeneracy condition (concerning the linearization of the equations which arise in solving the constraints via the conformal method) holds.

Finally in our third work [7,8] we obtain the result stated in Theorem 2.2. In particular, we show that gluing can be carried out with no restrictions whatsoever on the mean curvature. In addition, we show that the gluing can be done in such a way that the data changes only locally; away from a neighborhood of the gluing points, the data remains completely unchanged. We call this "localized gluing".

Since the proof (and constructions) of our localized gluing result (Theorem 2.2) rely on those of the non localized result, and since the non localized gluing theorem is of interest in its own right, we now sketch the ideas used to obtain both results.

The proof of the non localized gluing theorem relies primarily on the conformal method. It proceeds roughly as follows: We first apply to each of the given sets of data $(\Sigma_1^3, \gamma_1, K_1)$ and $(\Sigma_2^3, \gamma_2, K_2)$ a conformal transformation which is singular at the gluing point, and the identity away from a neighborhood of that point. Along with the transformations $\gamma \to \gamma_c = \psi^4\gamma$ of the metrics , one transforms the traceless part σ of K via the formula $\sigma \to \sigma_c = \psi^{-2}\sigma$, thereby guaranteeing that if $div_\gamma\sigma = 0$, then $div_{\gamma_c}\sigma_c = 0$. As a result of these transformations, the data $(\Sigma_1^3, \gamma_1, K_1)$ and $(\Sigma_2^3, \gamma_2, K_2)$ near the gluing points are each replaced by data on an infinite half tube whose geometry approaches that of a round $S^2 \times R^1$ cylinder.

Next, we connect the two cylinders at a coordinate parameter length $T/2$ along each, using cutoff functions to smoothly join the data fields from each side. We obtain $(\Sigma_1^3\#\Sigma_2^3, \hat{\gamma}_T, \hat{K}_T)$, which is no longer a solution of the constraints.

It follows from the construction of $\hat{\gamma}_T$ and $\hat{K}_T$ that $div_{\gamma_T}\sigma_T$ is non zero. The next step in the gluing construction is to find a vector field X_T such that $div_{\gamma_T}(LX_T) = -div_{\gamma_T}\sigma_T$, where L is the conformal Killing operator. For such a vector field, one verifies that $div_{\gamma_T}\tilde{\sigma}_T = 0$, where $\tilde{\sigma}_T := \sigma_T + LX_T$. In the course of finding X_T one also shows that, while LX_T is generally non zero everywhere on $\Sigma_1^3\#\Sigma_2^3$, it approaches zero away from the bridge joining Σ_1^3 and Σ_2^3 for large T.

There is one remaining step to carry out in proving our first gluing result: to solve the Lichnerowicz equation

$$\Delta_{\hat{\gamma}_T}\phi_T = \frac{1}{8}R_{\hat{\gamma}_T}\phi_T - \frac{1}{8}|\tilde{\sigma}|^2_{\hat{\gamma}_T}\phi^{-7} + \frac{1}{12}\tau^2\phi^5, \tag{3.1}$$

for the positive scalar ϕ_T, thus obtaining data $\bar{\gamma}_T = \phi_T^4\hat{\gamma}_T$ and $\bar{K}_T = \phi^{-2}\tilde{\sigma} + \frac{1}{3}\phi^4\hat{\gamma}_T\tau$ which satisfies the constraints everywhere on $\Sigma_1^3\#\Sigma_2^3$ for all T. (Here τ is the trace of K.) To prove that the solution ϕ_T exists, and further to prove that for large T, the data $(\bar{\gamma}, \bar{K})$ approaches the original data $(\gamma_1, K_1))$ and (γ_2, K_2) in appropriate regions, one needs to show that for an appropriate construction of a scalar ψ_T from the conformal blow up functions ψ_1 and ψ_2, together with cutoff functions, ψ_T is arbitrarily close to a solution of the Lichnerowicz equation for sufficiently large T.

We note that while this non localized gluing result is generally weaker than our later results, it does have the virtue that it allows Minkowski data to be glued (non locally) to other solutions of the constraints, since the hypotheses for our first result do not require a non degeneracy condition for data on R^3. There is no violation of the positive mass theorem, since after the non localized gluing is done, the data on $\Sigma^3\#R^3$ differs from Minkowski data in the asymptotically flat region, and in particular may have non zero mass.

As noted above, one of the key features of our work on the local gluing results of [7,8] is the removal altogether of any CMC requirement on the sets of data to be glued. We do this through the use of a result due to Bartnik [2], which says that for any choice of a set of initial data satisfying the constraints, for any real number τ, and for any point p, there is always a deformation of the data in a neighborhood of p (via the Einstein evolution equations) which is still a solution of the constraint equations, and which has mean curvature equal to the constant value τ throughout that neighborhood. Using this result, we show that we may glue any given pair of sets of initial data satisfying the constraints in a fixed (now CMC) neighborhood of the points about which we wish to glue, regardless of their mean curvatures, since we may first deform the data to CMC data in neighborhoods of the gluing points, and then proceed as in our first result adapted to a manifold with boundary (as a boundary value problem).

The further changes involved in going from our global results in [14,15] to Theorem 2.2 are two-fold. First, we replace our earlier non degeneracy condition by the No KIDs condition. Second, we introduce a non conformal deformation as a tool to replace non localized gluing by localized gluing, in which the data is completely unchanged away from the bridge connecting the gluing regions. Specifically, working with the constraints as an underdetermined system, we use techniques developed by Corvino and refined by Corvino-Schoen and Chruściel-Delay, to deform the data in an annular region around each end of the bridge in such a way that all of the gluing is done in the bridge and in its neighborhood, with no changes occurring away from this region. The details are found in [7,8].

Our discussion here has focussed on gluing solutions of the Einstein vacuum constraint equations. As shown in [13], non localized conformal gluing can readily be carried out for the Einstein-Maxwell, Einstein-Yang-Mills, Einstein-fluid, Einstein-Vlaxov, and other non vacuum field theories. To obtain non localized gluing for these field equations, it will be necessary to first extend the Corvino-Schoen technique to these non vacuum theories; this has not been done yet. We do note however that we have, in [8], established a non-vacuum version of Theorem 2.2 which allows for arbitrary non-gravitational fields satisfying the dominant energy condition. These results insure that the dominant energy condition is preserved under the gluing; we do not, however, claim to control any additional evolution equations (such as the Maxwell equations) which these additional fields may satisfy.

4. Applications of Gluing

Studies of the gluing of solutions of the Einstein constraint equations have always been strongly motivated by applications. Indeed, we initiated the whole program with the goal in mind of constructing "skew data sets", which are instrumental in showing that there are vacuum maximal globally hyperbolic spacetime solutions of the Einstein field equations which have no CMC Cauchy surfaces. The idea is this: We define a *skew data set* to be a solution (Σ^3, γ, K) of the constraint equations with $\Sigma^3 = \Lambda^3 \# \Lambda^3$ being a manifold which does *not* admit a metric with scalar curvature $R \geq 0$ [19], and with $\eta : \Lambda^3 \# \Lambda^3 \rightarrow \Lambda^3 \# \Lambda^3$ a reflection map ($\eta^2 = Id$) such that $\eta^* \gamma = \gamma$ (reflection symmetry) and $\eta^* K = -K$ (reflection skew symmetry). As indicated by Eardley and Witt (unpublished) and Bartnik [1] the maximal spacetime development of a skew symmetric set of data *cannot* contain a CMC Cauchy surface. This is because one verifies that if a Cauchy surface with data $(\bar{\gamma}, \bar{K})$ is contained in the development of skew symmetric data, then there is a Cauchy surface with data $(\bar{\gamma}, -\bar{K})$ in the development as well. Thus if $(\bar{\gamma}, \bar{K})$ has CMC τ, then there is a Cauchy surface with CMC $-\tau$ as well. It then follows from barrier theorems [1], that if the development of a set of skew symmetric data has a CMC Cauchy surface, then it must have a maximal ($trK = 0$) Cauchy surface as well. Now if this were true, it would follow from the constraints that the data on this maximal Cauchy surface would have $R = |K|^2 \geq 0$. This would violate our assumption regarding the geometries admitted by Σ^3. We conclude that the development admits no CMC Cauchy surfaces.

To prove that there are vacuum spacetimes with no CMC Cauchy data surfaces, it remains to show that we can construct skew data sets. But this can be done readily via using our local gluing techniques as follows (see [8] for details). We first use the conformal method to find a solution of the constraints (T^3, γ, K) which has no KIDs. Then noting that if (T^3, γ, K) solves the constraints, then so to does $(T^3, \gamma, -K)$. We proceed to glue $(T^3, \gamma, -K)$ to (T^3, γ, K) at equivalent points. A bit of analysis shows that this gluing produces skew symmetric data, as desired.

Most of our applications of gluing are more direct than this one. We may, for example, use gluing to produce initial data for spacetimes containing multiple black holes. We do this by choosing an arbitrary set of asymptotically Euclidean initial data, choosing a set of points $\{p_1, p_2, ...p_N\}$ on that data set, and then gluing (non locally) a copy of Euclidean space data (with K=0) to each of the points $\{...p_k...\}$. (This can also be done locally with a generic asymptotically flat solution which satisfies the no KIDs condition on every open set.) Clearly this procedure produces data with N minimal surfaces, or apparent horizons. As shown in [9], in fact, in certain situations, one can verify that N independent black holes develop.

We can also use gluing to add an arbitrary number of wormholes, at least for a short period, to a given spacetime. Indeed, given a set of constraint-satisfying initial data (Σ^3, γ, K) and a choice of a pair of open regions U and W in Σ^3, we can use gluing to find a new solution which is identical to (Σ^3, γ, K) outside of U and W, and which contains an arbitrary number of wormholes connecting U and W. To do this, we simply note that with small deformations, we can guarantee that U and W admit no KIDs; further, we note that while our gluing results have been stated for points on independent sets of data, in fact one readily shows using the same techniques that gluing can be carried out for two points on the same data set (see [14]). This tells us nothing about the long time future development of an initial data set with multiple wormholes.

We note one further application: verifying that there are no topological restrictions on constraint-satisfying initial data sets. To show this, we first recall that since any closed three manifold Σ^3 admits a constant negative scalar curvature metric $\hat{\gamma}$, one can always produce constraint-satisfying data on Σ^3 simply by choosing K to be pure trace of the right magnitude. This has long been known. Our new application is to show that for any closed three manifold Σ^3 we can always find asymptotically Euclidean as well as asymptotically hyperbolic solutions of the constraints on Σ^3 with a point removed. One finds these by gluing either an asymptotically flat or asymptotically hyperbolic solution of the constraints on R^3 to a small deformation of the simple solution $(\Sigma^3, \gamma, K = c\gamma)$. For the details of these applications we refer the interested reader to [14] (for the asymptotically hyperbolic case) and [15] (for the asymptotically Euclidean case).

5. Concluding Remarks

It is unlikely that the gluing results we have obtained for solutions of the Einstein constraint equations can be significantly strengthened, apart from allowing the presence of non gravitational fields (with coupled, additional evolution equations). (Note that while we have not discussed the issue here, as shown in our papers, all of our results do hold for general dimension.)

On the other hand, we believe that there might be a way to generalize gluing in the following sense: One might consider gluing along corresponding embedded submanifolds, of codimension at least three, of the initial data sets, rather than at corresponding points. Recent work of Mazzieri [16] with constant scalar curvature metrics suggests that this should indeed work. If so, gluing could prove useful in the study of the stability of black rings, and more generally, the topology of all "black objects" in higher dimensional spacetimes.

Whether or not this new version of gluing works, it is clear that gluing provides a powerful new tool for constructing and studying solutions of the Einstein constraint equations.

References

1. R. Bartnik, *Remarks on cosmological spacetimes and constant mean curvature surfaces,* Comm. Math. Phys. **117** (1988) no. 4, 615-624.
2. R. Bartnik, *Regularity of variational maximal surfaces*, Acta Math. **161** (1988), 145–181.
3. R. Bartnik and J. Isenberg, *The constraint equations*, 50 years of the Cauchy problem in general relativity (P.T. Chruściel and H. Friedrich, eds.), Birkhaeuser, Basel, 2004, gr-qc/0312047.
4. R. Beig, P.T. Chruściel, and R. Schoen, *KIDs are non-generic*, Ann. Henri Poincaré **6** (2005), no. 1, 155–194; gr-qc/0403042.
5. Y. Choquet-Bruhat and J. York, *The Cauchy Problem,* in General Relativity, A. Held ed., Plenum 1979.

6. P.T. Chruściel and E. Delay, *On mapping properties of the general relativistic constraints operator in weighted function spaces, with applications*, Mém. Soc. Math. de France. **94** (2003) 1–103. gr-qc/0301073.
7. P. T. Chruściel, J. Isenberg and D. Pollack, *Gluing initial data sets for general relativity,* Phys. Rev. Lett. **93**, 081101 (2004); ArXiv: gr-qc/0409047
8. P. T. Chruściel, J. Isenberg and D. Pollack, *Initial data engineering,* Comm. Math. Phys. **257** (2005) 1, 29–42; ArXiv: gr-qc/0403066.
9. P. T. Chruściel and R. Mazzeo, *On "many black hole" vacuum spacetimes*, Class. Quantum Grav. **20** (2003), 1–26.
10. J. Corvino, *Scalar curvature deformation and a gluing construction for the Einstein constraint equations,* Comm. Math. Phys. **214** (2000), no. 1, 137–189.
11. J. Corvino and R. Schoen, *On the asymptotics for the vacuum Einstein constraint equations*, J. Differential Geometry **73** (2006) 185–217; gr-qc/0301071.
12. J. Isenberg, *Constant mean curvature solutions of the Einstein constraint equations on closed manifolds,* Class. Quantum Grav. **12** (1995) 2249–2274.
13. J. Isenberg, D. Maxwell and D. Pollack, *A gluing constructions for non-vacuum solutions of the Einstein constraint equations,* ArXiv: gr-qc/0501083 Advances in Theoretical and Mathematical Physics, Vol. **9** No. 1, (2005) 129–172.
14. J. Isenberg, R. Mazzeo and D. Pollack, *Gluing and Wormholes for the Einstein constraint equations,* Comm. Math. Phys. **231** (2002) 3, 529–568; ArXiv: gr-qc/0109045.
15. J. Isenberg, R. Mazzeo and D. Pollack, *On the topology of vacuum spacetimes,* Ann. Henri Poincaré **4** (2003) 2, 369–383; ArXiv: gr-qc/0206034.
16. L. Mazzieri, *Generalized connected sum construction for constant scalar curvature metrics*, math.DG/0602471
17. V. Moncrief, *Spacetime symmetries and linearization stability of the Einstein equations I*, J. Math. Phys. **16** (1975), 493–498.
18. H. Pfeiffer and J. York , *Extrinsic curvature and the Einstein constraints*, Phys. Rev. D **67** (2003) 094022
19. R. Schoen and S.T. Yau, *On the structure of manifolds with positive scalar curvature*, Manuscripta Math. **28** (1979), no. 1-3, 159–183.
20. R. Schoen and S.T. Yau, *On the proof of the positive mass conjecture in general relativity*, Comm. Math. Phys. **65** (1979), no. 1, 45–76.
21. R. Schoen and S.T. Yau, *Proof of the positive mass theorem. II*, Comm. Math. Phys. **79** (1981), no. 2, 231–260.
22. E. Witten, *A new proof of the positive energy theorem*, Comm. Math. Phys. **80** (1981), no. 3, 381–402.

Abelian Decomposition of Einstein's Theory: Reformulation of General Relativity

Y. M. Cho and S. W. Kim

Center for Theoretical Physics and School of Physics, College of Natural Sciences, Seoul National University, Seoul 151-742, Korea

E-mail: ymcho@yongmin.snu.ac.kr, swkim@phya.snu.ac.kr

We propose a reformulation of general relativity by making the Abelian decomposition of Einstein's theory. Based on the view that Einstein's theory can be interpreted as a gauge theory of Lorentz group, we decompose the Einstein's gravitational connection into the restricted part made of the maximal Abelian subgroup of Lorentz group and the valence part which transforms covariantly under Lorentz group. With the decomposition we reconstruct Einstein's theory as a restricted theory of gravitation which has the valence part as the gravitational source.

1. Introduction

Einstein's theory of gavitation has been very successful describing the gravitational force. But perhaps a more important contribution of Einstein's theory is in theoretical physics in general, in particular in unified theory. It has been well known that an higher-dimensional Einstein's theory could provide a unification of all interaction [1,2,3]. In fact all modern unified theories, including the superstring, are based on Einstein's theory one way or another. This originates from the theoretical beauty of Einstein's theory. For this reason Einstein's theory has been the subject of intensive theoretical study.

Another important theory in theoretical physics which is closely related to Einstein's theory is the non-Abelian gauge theory. Actually the gauge theory itself can be viewed as an Einstein's theory originating from the extrinsic curvature of (4+n)-dimensional fiber bundle. Indeed in (4+n)-dimensional space-time Einstein's theory can reproduce the non-Abelian gauge theory when the n-dimensional internal space has an n-dimensional isometry [2,3]. This is known as the Kaluza-Klein miracle. Conversely Einstein's theory itself can be understooded as a non-Abelian gauge theory, in particular a gauge theory of Lorentz group [4]. This confirms that the two theories are closely related.

During last few decades our understanding of non-Abelian gauge theory has been extended very much. By now it has been well known that the non-Abelian gauge potential allows the Abelian decomposition [5,6]. It can be decomposed into the restricted potential which has an electric-magnetic duality and the valence potential which describes the gauge covariant valence gluon. A remarkable feature of this decomposition is that the restricted potential has the full non-Abelian gauge degrees of freedom, in particular the topological degrees of the gauge group G, in spite of the fact that it consistes of only the Abelian degrees of the maximal Abelian subgroup H of G. This means that we can construct a restricted gauge theory, a non-Abelian gauge theory made of only the resricted potential which has much less physical degrees of freedom. Furthermore, we can recover the full non-Abelian gauge theory simply by adding the valence part. And this decomposition plays a crucial role because the restricted part is known to be responsible for the confinement [7,8].

The purpose of this talk is to discuss a similar decomposition in Einstein's theory. Applying the Abelian decomposition to the gauge formalism of Einstein's theory, we show that we can construct a restricted theory of gravitation which has the full general invariance but has much less physical degrees. Moreover we show that we can recover the full Einstein's theory by adding the gauge covariant valence part of gravitational connection to the restricted theory.

Of course, Einstein's theory as a gauge theory of Lorentz group is different from the ordinary non-Abelian gauge theory. In gauge theory the fundamental field is the gauge potential, and in Einstein's theory the fundamental field is the metric. But in the gauge theory formulation the gauge potential of Lorentz group corresponds to the gravitational connection, not the metric. Also, in gauge theory the Yang-Mills Lagrangian is quadratic in field strength. But in gravitation the Einstein-Hilbert Lagrangian is made of the scalar curvature, which is linear in field strength [4]. Nevertheless we can still make the Abelian decomposition of the gravitational connection, and identify the restricted connection and the valence connection. With this we can separate the restricted part of gravitation, and recover the full Einstein's theory by adding the valence part to the restricted gravity.

2. Abelian decomposition: A Review

To understand how to make the desired Abelian decomposition of Einstein's theory, we have to understand the Abelian decomposition of gauge theory first. Consider $SU(2)$ QCD for simplicity. A natural way to make the Abelian decomposition is to introduce an isotriplet unit vector field $\hat{n}$ which selects the "Abelian" direction (i.e., the color charge direction) at each space-time point, and to decompose the potential into the restricted potential (called the Abelian projection) $\hat{A}_\mu$ which leaves $\hat{n}$ invariant and the valence potential $\vec{X}_\mu$ which forms a covariant vector field [5,6],

$$\vec{A}_\mu = A_\mu \hat{n} - \tfrac{1}{g}\hat{n} \times \partial_\mu \hat{n} + \bar{X}_\mu = \hat{A}_\mu + \bar{X}_\mu,$$
$$(\hat{n}^2 = 1, \quad \hat{n} \cdot \vec{X}_\mu = 0), \tag{1}$$

where $A_\mu = \hat{n} \cdot \vec{A}_\mu$ is the "electric" potential. Notice that the restricted potential is precisely the connection which leaves $\hat{n}$ invariant under the parallel transport,

$$\hat{D}_\mu \hat{n} = \partial_\mu \hat{n} + g\hat{A}_\mu \times \hat{n} = 0. \tag{2}$$

Under the infinitesimal gauge transformation

$$\delta \hat{n} = -\vec{\alpha} \times \hat{n}\,, \quad \delta \vec{A}_\mu = \frac{1}{g} D_\mu \vec{\alpha}, \tag{3}$$

one has

$$\delta A_\mu = \frac{1}{g}\hat{n} \cdot \partial_\mu \vec{\alpha}, \quad \delta \hat{A}_\mu = \frac{1}{g}\hat{D}_\mu \vec{\alpha},$$
$$\delta \bar{X}_\mu = -\vec{\alpha} \times \bar{X}_\mu. \tag{4}$$

This shows that $\hat{A}_\mu$ by itself describes an $SU(2)$ connection which enjoys the full $SU(2)$ gauge degrees of freedom. Furthermore $\vec{X}_\mu$ transforms covariantly under the gauge transformation. Most importantly, the decomposition is gauge-independent. Once the color direction $\hat{n}$ is selected, the decomposition follows independent of the choice of a gauge.

Our decomposition was first introduced long time ago in an attempt to demonstrate the monopole condensation in QCD [5,6]. But only recently the importance of the decomposition in clarifying the non-Abelian dynamics has become appreciated by many authors [9].

To understand the physical meaning of our decomposition notice that the restricted potential $\hat{A}_\mu$ actually has a dual structure. Indeed the field strength made of the restricted potential is decomposed as

$$\begin{aligned}\hat{F}_{\mu\nu} &= (F_{\mu\nu} + H_{\mu\nu})\hat{n}, \\ F_{\mu\nu} &= \partial_\mu A_\nu - \partial_\nu A_\mu, \\ H_{\mu\nu} &= -\frac{1}{g}\hat{n} \cdot (\partial_\mu \hat{n} \times \partial_\nu \hat{n}) = \partial_\mu \tilde{C}_\nu - \partial_\nu \tilde{C}_\mu,\end{aligned} \tag{5}$$

where $\tilde{C}_\mu$ is the "magnetic" potential [5,6]. Notice that we can always introduce the magnetic potential (at least locally section-wise), because $H_{\mu\nu}$ is closed

$$\partial_\mu \tilde{H}_{\mu\nu} = 0 \qquad (\tilde{H}_{\mu\nu} = \frac{1}{2}\epsilon_{\mu\nu\rho\sigma} H_{\rho\sigma}). \tag{6}$$

This allows us to identify the non-Abelian magnetic potential by

$$\vec{C}_\mu = -\frac{1}{g}\hat{n} \times \partial_\mu \hat{n}, \tag{7}$$

in terms of which the magnetic field is expressed as

$$\begin{aligned}\vec{H}_{\mu\nu} &= \partial_\mu \vec{C}_\nu - \partial_\nu \vec{C}_\mu + g\vec{C}_\mu \times \vec{C}_\nu \\ &= -g\vec{C}_\mu \times \vec{C}_\nu = -\frac{1}{g}\partial_\mu \hat{n} \times \partial_\nu \hat{n} = H_{\mu\nu}\hat{n}.\end{aligned} \tag{8}$$

As importantly $\hat{A}_\mu$, as an $SU(2)$ potential, retains all the essential topological characteristics of the original non-Abelian potential. This is because the topological field $\hat{n}$ can naturally describe the non-Abelian topology $\pi_2(S^2)$ and $\pi_3(S^2) \simeq \pi_3(S^3)$. Clearly the isolated singularities of $\hat{n}$ defines $\pi_2(S^2)$ which describes the non-Abelian monopoles. Indeed $\hat{A}_\mu$ with $A_\mu = 0$ and $\hat{n} = \hat{r}$ (or equivallently, $\vec{C}_\mu$ with $\hat{n} = \hat{r}$) describes precisely the Wu-Yang monopole [5]. Besides, with the S^3 compactification of R^3, $\hat{n}$ characterizes the Hopf invariant $\pi_3(S^2) \simeq \pi_3(S^3)$ which describes the topologically distinct vacua [10]. This tells that the restricted gauge theory made of $\hat{A}_\mu$ could describe the dual dynamics which should play an essential role in $SU(2)$ QCD [5,6,7,8].

With (1) we have

$$\vec{F}_{\mu\nu} = \hat{F}_{\mu\nu} + \hat{D}_\mu \vec{X}_\nu - \hat{D}_\nu \vec{X}_\mu + g\vec{X}_\mu \times \vec{X}_\nu, \tag{9}$$

so that the Yang-Mills Lagrangian is expressed as

$$\mathcal{L} = -\frac{1}{4}\vec{F}_{\mu\nu}^2$$
$$= -\frac{1}{4}\hat{F}_{\mu\nu}^2 - \frac{1}{4}(\hat{D}_\mu \bar{X}_\nu - \hat{D}_\nu \bar{X}_\mu)^2 - \frac{g}{2}\hat{F}_{\mu\nu} \cdot (\bar{X}_\mu \times \bar{X}_\nu) - \frac{g^2}{4}(\bar{X}_\mu \times \bar{X}_\nu)^2$$
$$+\lambda(\hat{n}^2 - 1) + \lambda_\mu \hat{n} \cdot \bar{X}_\mu, \qquad (10)$$

where λ and λ_μ are the Lagrangian multipliers. From the Lagrangian we have

$$\partial_\mu(F_{\mu\nu} + H_{\mu\nu} + X_{\mu\nu}) = -g\hat{n} \cdot \Big[\vec{X}_\mu \times (\hat{D}_\mu \vec{X}_\nu - \hat{D}_\nu \vec{X}_\mu)\Big],$$
$$\hat{D}_\mu(\hat{D}_\mu \vec{X}_\nu - \hat{D}_\nu \vec{X}_\mu) = g(F_{\mu\nu} + H_{\mu\nu} + X_{\mu\nu})\hat{n} \times \vec{X}_\mu. \qquad (11)$$

where

$$X_{\mu\nu} = g\hat{n} \cdot (\vec{X}_\mu \times \vec{X}_\nu). \qquad (12)$$

Notice that here $\hat{n}$ has no equation of motion even though the Lagrangian contains it explicitly. This implies that it is not a local degrees of freedom, but a topological degrees of freedom [5,7]. From this we conclude that the non-Abelian gauge theory can be viewed as a restricted gauge theory made of the restricted potential, which has an additional colored source made of the valence gluon.

Obviously the Lagrangian (10) is invariant under the active gauge transformation (3). But notice that the decomposition introduces another gauge symmetry that we call the passive gauge transformation [7,8],

$$\delta\hat{n} = 0, \qquad \delta\vec{A}_\mu = \frac{1}{g}D_\mu\vec{\alpha}, \qquad (13)$$

under which we have

$$\delta A_\mu = \frac{1}{g}\hat{n} \cdot D_\mu\vec{\alpha}, \qquad \delta\hat{A}_\mu = \frac{1}{g}(\hat{n} \cdot D_\mu\vec{\alpha})\hat{n},$$
$$\delta\vec{X}_\mu = \frac{1}{g}\Big[D_\mu\vec{\alpha} - (\hat{n} \cdot D_\mu\vec{\alpha})\hat{n}\Big]. \qquad (14)$$

This is because, for a given $\vec{A}_\mu$, one can have infinitely many different decomposition of (1), with different $\hat{A}_\mu$ and $\vec{X}_\mu$ by choosing different $\hat{n}$. Equivalently, for a fixed $\hat{n}$, one can have infinitely many different $\vec{A}_\mu$ which are gauge-equivalent to each other. So our decomposition automatically induce another type of gauge invariance which comes from different choices of decomposition. This extra gauge invariance plays a crucial role in quantizing the theory [7].

Another advantage of the decomposition (1) is that it can actually "Abelianize" (or more precisely "dualize") the non-Abelian dynamics [5,7]. To see this let $(\hat{n}_1,\ \hat{n}_2,\ \hat{n})$ be a right-handed orthonormal basis and let

$$\vec{X}_\mu = X^1_\mu\ \hat{n}_1 + X^2_\mu\ \hat{n}_2,$$
$$(X^1_\mu = \hat{n}_1 \cdot \vec{X}_\mu, \quad X^2_\mu = \hat{n}_2 \cdot \vec{X}_\mu)$$

and find

$$\hat{D}_\mu \vec{X}_\nu = \Big[\partial_\mu X^1_\nu - g(A_\mu + \tilde{C}_\mu) X^2_\nu\Big]\hat{n}_1 + \Big[\partial_\mu X^2_\nu + g(A_\mu + \tilde{C}_\mu) X^1_\nu\Big]\hat{n}_2 \quad (15)$$

So with

$$B_\mu = A_\mu + \widetilde{C}_\mu, \quad G_{\mu\nu} = \partial_\mu B_\nu - \partial_\nu B_\mu,$$
$$X_\mu = \frac{1}{\sqrt{2}}(X^1_\mu + iX^2_\mu), \quad (16)$$

one could express the Lagrangian explicitly in terms of the dual potential B_μ and the complex vector field X_μ,

$$\mathcal{L} = -\frac{1}{4}G^2_{\mu\nu} - \frac{1}{2}|\tilde{D}_\mu X_\nu - \tilde{D}_\nu X_\mu|^2 + igG_{\mu\nu}X^*_\mu X_\nu$$
$$-\frac{1}{2}g^2\Big[(X^*_\mu X_\mu)^2 - (X^*_\mu)^2(X_\nu)^2\Big], \quad (17)$$

where now

$$\tilde{D}_\mu = \partial_\mu + igB_\mu.$$

Clearly this describes an Abelian gauge theory coupled to the charged vector field X_μ. But the important point here is that the Abelian potential B_μ is given by the sum of the electric and magnetic potentials $A_\mu + \widetilde{C}_\mu$. In this form the equations of motion (11) is re-expressed as

$$\partial_\mu(G_{\mu\nu} + X_{\mu\nu}) = igX^*_\mu(\tilde{D}_\mu X_\nu - \tilde{D}_\nu X_\mu) - igX_\mu(\tilde{D}_\mu X_\nu - \tilde{D}_\nu X_\mu)^*,$$
$$\tilde{D}_\mu(\tilde{D}_\mu X_\nu - \tilde{D}_\nu X_\mu) = igX_\mu(G_{\mu\nu} + X_{\mu\nu}). \quad (18)$$

where now

$$X_{\mu\nu} = -ig(X^*_\mu X_\nu - X^*_\nu X_\mu).$$

This shows that one can indeed Abelianize the non-Abelian theory with our decomposition. The remarkable change in this "Abelian" formulation is that here the topological field $\hat{n}$ is replaced by the magnetic potential $\tilde{C}_\mu$.

One might ask whether this "Abelian" theory retains the original non-Abelian gauge symmetry, and if so, how the non-Abelian gauge symmetry is realized in this "Abelian" theory. To answer this notice that here we have never fixed the gauge to obtain this Abelian formalism, so that the original non-Abelian gauge symmetry must remain intact. To see this let

$$\begin{aligned}\vec{\alpha} &= \alpha_1\,\hat{n}_1 + \alpha_2\,\hat{n}_2 + \theta\,\hat{n},\\ \alpha &= \frac{1}{\sqrt{2}}(\alpha_1 + i\,\alpha_2),\\ \vec{C}_\mu &= -\frac{1}{g}\hat{n}\times\partial_\mu\hat{n} = -C_\mu^1\hat{n}_1 - C_\mu^2\hat{n}_2,\\ C_\mu &= \frac{1}{\sqrt{2}}(C_\mu^1 + i\,C_\mu^2).\end{aligned} \tag{19}$$

Then the Lagrangian (17) is invariant not only under the active gauge transformation (3) described by

$$\begin{aligned}\delta A_\mu &= \frac{1}{g}\partial_\mu\theta - i(C_\mu^*\alpha - C_\mu\alpha^*), \qquad \delta\tilde{C}_\mu = -\delta A_\mu,\\ \delta X_\mu &= 0,\end{aligned} \tag{20}$$

but also under the passive gauge transformation (13) described by

$$\begin{aligned}\delta A_\mu &= \frac{1}{g}\partial_\mu\theta - i(X_\mu^*\alpha - X_\mu\alpha^*), \qquad \delta\tilde{C}_\mu = 0,\\ \delta X_\mu &= \tfrac{1}{g}\tilde{D}_\mu\alpha - i\theta X_\mu.\end{aligned} \tag{21}$$

This tells that the "Abelian" theory not only retains the original non-Abelian gauge symmetry, but actually has an enlarged (both the active and passive) gauge symmetries. And we emphasize that this Abelianization is not the "naive" Abelianization of the $SU(2)$ gauge theory which one obtains by fixing the gauge. Our Abelianization is a gauge independent Abelianization. Besides, here the Abelian gauge group is actually made of $U(1)_e \otimes U(1)_m$, so that the theory becomes a dual gauge theory [5,7]. This is evident from (20) and (21).

3. Decomposition of Gravitation

Now, we apply the above decomposition to Einstein's theory, regarding Einstein's theory as a gauge theory of Lorentz group. To do this we introduce a coordinate basis

$$[\,\partial_\mu,\ \partial_\nu\,] = 0, \quad (\mu,\ \nu = 0,1,2,3)$$

and an orthonormal basis

$$[\ \xi_a,\ \xi_b\ ] = c_{ab}{}^c\ \xi_c\ . \tag{22}$$

Let J_{ab} be the generators of Lorentz group which acts on the orthonormal frame,

$$[\ J_{ab},\ J_{cd}\ ] = \eta_{bc}J_{ad} - \eta_{ac}J_{bd} + \eta_{ad}J_{bc} - \eta_{bd}J_{ac}\ ,$$

where $\eta_{ab} = diag(-1,1,1,1)$ is the Minkowski metric. Instead of $a,\ b,...$, we can use $A,\ B,... = 1,2,3,4,5,6$

$$[\ J_A,\ J_B\ ]\ =\ f_{AB}{}^C\ J_C\ .$$

Now, with

$$J_{01,02,03} = K_{1,2,3} = J_{4,5,6},\quad J_{23,31,12} = L_{1,2,3} = J_{1,2,3},$$

the Lorentz algebra is written as

$$\begin{aligned}
[\ L_i,\ L_j\ ] &= \epsilon_{ijk}L_k, \\
[\ L_i,\ K_j\ ] &= \epsilon_{ijk}K_k, \\
[\ K_i,\ K_j\ ] &= -\ \epsilon_{ijk}L_k,
\end{aligned} \tag{23}$$

where $i,j = 1,2,3$.

As we have pointed out, we can regard Einstein's theory as a gauge theory of Lorentz group. In this view the gravitational connection corresponds to the gauge potential of Lorentz group. To make the desired decomposition we have to choose the gauge covariant sextet vector fields which form adjoint representation of Lorentz group which describe the desired magnetic isometry. To see what types of isometry is possible, we notice that Lorentz group has two maximal Abelian subgroups, A_2 made of L_3 and K_3 and B_2 made of $(L_1 + K_2)/\sqrt{2}$ and $(L_2 - K_1)/\sqrt{2}$ [11]. So we have two possible Abelian decomposition of the gravitational connection, and in both cases the magnetic isometry is described by two, not one, sextet vector fields of Lorentz group. To see this let us denote one of the isometry vector field l_{ab} by $\vec{l}$

$$\vec{l} = \begin{pmatrix} \vec{n} \\ \vec{m} \end{pmatrix}, \tag{24}$$

where $n_i = \epsilon_{ijk} l_{jk}/2$ is the rotation part and $m_i = l_{0i}$ is the boost part of $\vec{l}$. This l_{ab} automatically gives us the dual vector $\tilde{l}_{ab} = \epsilon_{abcd} l_{cd}/2$ or $\vec{\tilde{l}}$,

$$\vec{\tilde{l}} = \begin{pmatrix} \vec{m} \\ -\vec{n} \end{pmatrix}, \tag{25}$$

which also become an isometry. It is useful to charactrize the isometry by Lorentz invariants. Since Lorentz group has two Casimir invariants, $\vec{l}$ has two Lorentz invariants α and β,

$$\alpha \;=\; \vec{l}\cdot\vec{l} \;=\; \vec{n}^2 - \vec{m}^2, \quad \beta \;=\; \vec{l}\cdot\vec{\tilde{l}} \;=\; \vec{n}\cdot\vec{m}. \tag{26}$$

With this the corresponding Lorentz invariants of $\vec{\tilde{l}}$ becomes $-\alpha$ and β. Moreover, we can normalize α and β to be ± 1 or 0 without loss of generality. So $(\alpha,\ \beta)$ provides a gauge invariant description of the isometry vector fields.

Now we discuss the two cases separately.

A. **Case 1**: In this case the isometry is A_2 made of L_3 and K_3, and we have two sextet vector fields which describes the isometry. Let $\vec{l}$ and $\vec{\tilde{l}}$ be the vector fields which correspond to L_3 and K_3. We normalize them by $\vec{l}\cdot\vec{l} \;=\; -\vec{\tilde{l}}\cdot\vec{\tilde{l}} \;=\; 1$ and $\vec{l}\cdot\vec{\tilde{l}} \;=\; 0$. With this $(\alpha,\ \beta)$ of $\vec{l}$ and $\vec{\tilde{l}}$ are given by $(1,0)$ and $(-1,0)$. In this case the full connection for Lorentz group is given by

$$\begin{aligned} \vec{A}_\mu \;&=\; A^1_\mu\,\vec{l} - A^2_\mu\,\vec{\tilde{l}} - \vec{l}\times\partial_\mu\vec{l} + \vec{X}_\mu, \\ A^1_\mu = \vec{l}\cdot\vec{A}_\mu\ ,\quad & A^2_\mu = \vec{\tilde{l}}\cdot\vec{A}_\mu, \quad \vec{l}\cdot\vec{X}_\mu = \vec{\tilde{l}}\cdot\vec{X}_\mu = 0. \end{aligned} \tag{27}$$

And the corresponding field strength (the curvature tensor) $\vec{F}_{\mu\nu}$ is

$$\vec{F}_{\mu\nu} = (F^1_{\mu\nu} + H^1_{\mu\nu} + X^1_{\mu\nu})\vec{l} - (F^2_{\mu\nu} + H^2_{\mu\nu} + X^2_{\mu\nu})\vec{\tilde{l}} + \hat{D}_\mu\vec{X}_\nu - \hat{D}_\nu\vec{X}_\mu, \tag{28}$$

where

$$\begin{aligned} F^i_{\mu\nu} \;&=\; \partial_\mu A^i_\nu - \partial_\nu A^i_\mu, (i \;=\; 1,2) \\ H^1_{\mu\nu} \;&=\; -\vec{l}\cdot(\partial_\mu\vec{l}\times\partial_\nu\vec{l}), \quad H^2_{\mu\nu} = \; -\vec{\tilde{l}}\cdot(\partial_\mu\vec{l}\times\partial_\nu\vec{l}) \\ X^1_{\mu\nu} \;&=\; \vec{l}\cdot(\vec{X}_\mu\times\vec{X}_\nu), \qquad X^2_{\mu\nu} \;=\; \vec{\tilde{l}}\cdot(\vec{X}_\mu\times\vec{X}_\nu). \end{aligned}$$

B. **Case 2**: This is when the isometry is B_2 made of $(L_1+K_2)/\sqrt{2}$ and $(L_2-K_1)/\sqrt{2}$. Let $\vec{l}_+$ and $\vec{\tilde{l}}_-$ be the vector fields which correspond to $(L_1+K_2)/\sqrt{2}$ and $(L_2-K_1)/\sqrt{2}$. In this case $(\alpha,\ \beta)$ of $\vec{l}_+$ and $\vec{\tilde{l}}_+$ are given by (0,0). Let's introduce 6 independent basis vectors in Lorentz group manifold as follows,

$$\begin{aligned}
l_+ &= \begin{pmatrix} \vec{n} \\ \vec{m} \end{pmatrix} \; : \; \frac{L_1+K_2}{\sqrt{2}}, \quad \widetilde{l}_+ = \begin{pmatrix} \vec{m} \\ -\vec{n} \end{pmatrix} \; : \; \frac{L_2-K_1}{\sqrt{2}}, \\
l_- &= \frac{1}{\vec{n}^2+\vec{m}^2}\begin{pmatrix} \vec{n} \\ -\vec{m} \end{pmatrix}, \quad \widetilde{l}_- = \frac{1}{\vec{n}^2+\vec{m}^2}\begin{pmatrix} -\vec{m} \\ -\vec{n} \end{pmatrix}, \\
l &= \widetilde{l}_+ \times l_- = -\frac{2}{\vec{n}^2+\vec{m}^2}\begin{pmatrix} \vec{n}\times\vec{m} \\ 0 \end{pmatrix}, \\
\widetilde{l} &= -l_+ \times l_- = \frac{2}{\vec{n}^2+\vec{m}^2}\begin{pmatrix} 0 \\ \vec{n}\times\vec{m} \end{pmatrix}.
\end{aligned} \tag{29}$$

With this we obtain the full connection for Lorentz group,

$$\begin{aligned}
\vec{A}_\mu &= A^1_\mu \, l_+ - A^2_\mu \, \widetilde{l}_+ - l_- \times \partial_\mu l_+ + \vec{X}_\mu \\
A^1_\mu &= l_- \cdot \vec{A}_\mu \,, \quad A^2_\mu = \widetilde{l}_- \cdot \vec{A}_\mu \,, \quad l_- \cdot \vec{X}_\mu = \widetilde{l}_- \cdot \vec{X}_\mu = 0 \,.
\end{aligned} \tag{30}$$

The corresponding field strength is given by

$$\begin{aligned}
\vec{F}_{\mu\nu} = {} & (F^1_{\mu\nu} + H^1_{\mu\nu} + X^1_{\mu\nu})l - (F^2_{\mu\nu} + H^2_{\mu\nu} + X^2_{\mu\nu})\widetilde{l} \\
& + Y^1_{\mu\nu} \, l_- + Y^2_{\mu\nu} \, \widetilde{l}_- + Y^3_{\mu\nu} \, l - Y^4_{\mu\nu} \, \widetilde{l},
\end{aligned} \tag{31}$$

where

$$\begin{aligned}
F^i_{\mu\nu} &= \partial_\mu A^i_\nu - \partial_\nu A^i_\mu, \; (i = 1,2) \\
H^1_{\mu\nu} &= -l_- \cdot (\partial_\mu l_- \times \partial_\nu l_+ + \partial_\mu l_+ \times \partial_\nu l_-), \\
H^2_{\mu\nu} &= -\widetilde{l}_- \cdot (\partial_\mu l_- \times \partial_\nu l_+ + \partial_\mu l_+ \times \partial_\nu l_-), \\
X^1_{\mu\nu} &= \vec{X}_\mu \cdot \hat{D}_\nu l_- - \vec{X}_\nu \cdot \hat{D}_\mu l_-, \\
X^2_{\mu\nu} &= \vec{X}_\mu \cdot \hat{D}_\nu \widetilde{l}_- - \vec{X}_\nu \cdot \hat{D}_\mu \widetilde{l}_- \\
Y^1_{\mu\nu} &= l_+ \cdot (\hat{D}_\mu \vec{X}_\nu - \hat{D}_\nu \vec{X}_\mu + \vec{X}_\mu \times \vec{X}_\nu), \\
Y^2_{\mu\nu} &= \widetilde{l}_+ \cdot (\hat{D}_\mu \vec{X}_\nu - \hat{D}_\nu \vec{X}_\mu + \vec{X}_\mu \times \vec{X}_\nu) \\
Y^3_{\mu\nu} &= l \cdot (\hat{D}_\mu \vec{X}_\nu - \hat{D}_\nu \vec{X}_\mu), \\
Y^4_{\mu\nu} &= \widetilde{l} \cdot (\hat{D}_\mu \vec{X}_\nu - \hat{D}_\nu \vec{X}_\mu).
\end{aligned} \tag{32}$$

From the above results one can decompose the scalar curvature, and the corresponding Einstein-Hilbert action.

4. Decomposition of Einstein's theory

To decompose the Einstein-Hilbert action we introduce the following notations,

$$
\begin{aligned}
\vec{P}_{ab} &= -\vec{J}^{ab} - \vec{l}\, l^{ab} + \vec{\tilde{l}}\, \tilde{l}^{ab}\,, \qquad : \text{projection operator} \\
l^{ab} &= -\vec{l}\cdot \vec{J}^{ab}\,, \quad \tilde{l}^{ab} = -\frac{1}{2}\, l^{ab} \vec{J}_{ab}\,, \\
\vec{A}\cdot\vec{J} &= \frac{1}{2} A^{ab} J_{ab}\,, \\
J^1_{23} &= (J_{23})_{23} = -1\,, \quad J^4_{01} = (J_{01})_{01} = 1\,, \qquad [J,J] \sim 4\eta J\,, \\
(\vec{A}\times\vec{B})\cdot\vec{J}_{ab} &= [A,B]_{ab} = \frac{1}{2}(A_{ac}\,B_{bc} - A_{bc}\,B_{ac}).
\end{aligned}
\tag{33}
$$

Now, since the Einstein-Hilbert action is described by the metric we have to express the above decomposition of the gravitational connection in terms of the metric. To do this we use the first order formalism of Einstein theory. Consider the fiest case (Case 1). In this case the Einstein-Hilbert action can be expressed as

$$
\begin{aligned}
S[e^\mu_a,\, A^1_\mu,\, A^1_\mu,\, \vec{X}_\mu] &= \int d^4x\, \{e\, e^\mu_a\, e^\nu_b\, \vec{F}_{\mu\nu}\cdot\vec{J}^{ab} \\
&\quad + \lambda^1(\vec{l}^2 - 1) + \lambda^2(\vec{l}\cdot\vec{\tilde{l}}) + \lambda^3_\mu(\vec{l}\cdot\vec{X}^\mu) + \lambda^4_\mu(\vec{\tilde{l}}\cdot\vec{X}^\mu)\}, \\
\vec{F}_{\mu\nu}\cdot\vec{J}^{ab} &= (F+H+X)^1_{\mu\nu} l^{ab} - (F+H+X)^2_{\mu\nu}\tilde{l}^{ab} + (\hat{D}_\mu\vec{X}_\nu - \hat{D}_\nu\vec{X}_\mu)\cdot\vec{P}^{ab},
\end{aligned}
\tag{34}
$$

where e^μ_a is the tetrad and $\lambda's$ are the Lagrange multipliers. From this we get the following equations of motion

$$
\begin{aligned}
\delta A^1_\nu &: \partial_\mu(e\, e^\mu_a\, e^\nu_b\, \vec{l}\cdot\vec{J}^{ab}) = e\, e^\mu_a\, e^\nu_b\, (\vec{X}_\mu\times\vec{l})\cdot\vec{J}^{ab} \\
\delta A^2_\nu &: \partial_\mu(e\, e^\mu_a\, e^\nu_b\, \vec{\tilde{l}}\cdot\vec{J}^{ab}) = e\, e^\mu_a\, e^\nu_b\, (\vec{X}_\mu\times\vec{\tilde{l}})\cdot\vec{J}^{ab} \\
\delta\vec{X}_\nu &: \partial_\mu(e\, e^\mu_a\, e^\nu_b\, \vec{P}^{ab}) = e\, e^\mu_a\, e^\nu_b\, \{(\vec{l}\times\vec{X}_\mu)l^{ab} - (\vec{\tilde{l}}\times\vec{X}_\mu)\tilde{l}^{ab} + \vec{P}^{ab}\times\hat{A}_\mu\} \\
\delta e^\mu_a &: R^a_\mu - \frac{1}{2}\, e^a_\mu\, R = 0,
\end{aligned}
\tag{35}
$$

where

$$
\vec{P}_{ab} = -\vec{J}^{ab} - \vec{l}\, l^{ab} + \vec{\tilde{l}}\,\tilde{l}^{ab}\,, \quad l^{ab} = -\vec{l}\cdot\vec{J}^{ab}\,, \quad X^{ab}_\mu = -\vec{X}_\mu\cdot\vec{J}^{ab}.
$$

Now, using the fact that $(\hat{D}_\mu \vec{l}) \cdot \vec{J}^{ab} = 0$, we can combine the first three equations into a single equation,

$$
\begin{aligned}
e^c_\mu \partial_a \, e^\mu_b \; - \; \delta^c_b \, (e^d_\mu \, \partial_a \, e^\mu_d \; - \; \partial_\mu \, e^\mu_a) \; &= \; A_{ab}{}^c \; - \; \delta^c_b \, A_{ad}{}^d \\
\rightarrow -2\vec{A}_a \cdot \vec{J}_{bc} \; &= \; e^\nu_a \, \partial_b \, e_{c\nu} + e^\nu_c \, \partial_a \, e_{b\nu} + e^\nu_c \, \partial_b \, e_{a\nu} \\
&\quad - e^\nu_a \, \partial_c \, e_{b\nu} - e^\nu_b \, \partial_a \, e_{c\nu} - e^\nu_b \, \partial_c \, e_{a\nu}. \qquad (36)
\end{aligned}
$$

Notice that we can easily reproduce the same equations varying the full connection $\vec{A}_\nu$:

$$
\begin{aligned}
&\partial_\mu (e \, e^\mu_a \, e^\nu_b \, \vec{J}^{ab}) = \; e \, e^\mu_a \, e^\nu_b \, \vec{J}^{ab} \times \vec{A}_\mu \\
&\rightarrow e^c_\mu \partial_a \, e^\mu_b - \; \delta^c_b \, (e^d_\mu \, \partial_a \, e^\mu_d \; - \; \partial_\mu \, e^\mu_a) = \; A_{ab}{}^c \; - \; \delta^c_b \, A_{ad}{}^d. \qquad (37)
\end{aligned}
$$

So finally with

$$
S[e^\mu_a, \, \vec{A}_\mu] \; = \; \int d^4x \, e \, e^\mu_a \, e^\nu_b \vec{F}_{\mu\nu} \cdot \vec{J}^{ab}, \qquad (38)
$$

we have

$$
\begin{aligned}
\delta e^\mu_a \; &\longrightarrow \; R^a_\mu \; - \frac{1}{2} \, e^a_\mu \, R = \; 0 \\
\delta \vec{A}_\mu \; &\longrightarrow \; \vec{A}_a \cdot \vec{J}_{bc} = \; - \, \omega_{abc}, \qquad (39)
\end{aligned}
$$

where

$$
\begin{aligned}
\omega_{abc} \; &= \; \frac{1}{2} (e^\nu_a \, \partial_b \, e_{c\nu} + e^\nu_c \, \partial_a \, e_{b\nu} + e^\nu_c \, \partial_b \, e_{a\nu} \\
&\quad - e^\nu_a \, \partial_c \, e_{b\nu} - e^\nu_b \, \partial_a \, e_{c\nu} - e^\nu_b \, \partial_c \, e_{a\nu}) \qquad (40)
\end{aligned}
$$

Notice that the first equation is nothing but the Einstein'seqution and the second equation is the equation of the spin connection.

we can repear the same procedure for the second case (Case 2) to obtain the desired decomposition.

5. Restricted gravity

So far we kept the gauge covariant part of the connection $\vec{X}_\mu$ to be general. But as we have seen in the $SU(2)$ gauge theory, we can obtain the restricted gravity excluding the covariant part of the connection. Again consider the first case, and notice that

$$
\begin{aligned}
A^1_\mu \; &= \; \omega_{\mu ab} \, l^{ab} \; , \\
A^2_\mu \; &= \; \omega_{\mu ab} \, \tilde{l}^{ab} \; , \\
\vec{X}_\mu \; &= \; \omega_{\mu ab} \, \vec{P}_{ab} + \; \vec{l} \times \partial_\mu \vec{l}, \qquad (41)
\end{aligned}
$$

from which we have

$$\omega_{\mu ab} = A_\mu^1\, l_{ab} - A_\mu^2\, \widetilde{l}_{ab} + l_{ca}\partial_\mu l_b^c - l_{cb}\partial_\mu l_a^c + X_{\mu ab}. \qquad (42)$$

Now, with $\vec{X}_\mu = 0$, we have the equation for the restricted Einstein's theory

$$\begin{aligned} \hat{\nabla}_a l_{bc} &= 0 , \\ A_\mu^1 &= \omega_{\mu ab}(\hat{e}_\mu^a)\, l^{ab} , \\ A_\mu^2 &= \omega_{\mu ab}(\hat{e}_\mu^a)\, \widetilde{l}^{ab} , \\ \hat{R}_\mu^a - \frac{1}{2}\, \hat{e}_\mu^a\, \hat{R} &= 0, \end{aligned} \qquad (43)$$

where $\hat{\nabla}_a$ is the covariant derivative with respect to the restricted connection $\omega(\hat{e})$. Notice that the first equation determines the restricted tetrad $\hat{e}_\mu^a$. For the general case, we have

$$\nabla_a l_{bc} = X_{ab}{}^d\, l_{dc} - X_{ac}{}^d\, l_{db}, \qquad (44)$$

which clearly tells that the first equation determines the restricted tetrad.

We can obtain the same result starting with the restricted Einstein-Hilbert action,

$$\begin{aligned} S[\hat{e}_a^\mu,\, A_\mu^1,\, A_\mu^2] &= \int d^4x\, \{\hat{e}\, \hat{e}_a^\mu\, \hat{e}_b^\nu\, \vec{J}^{ab} \cdot \hat{F}_{\mu\nu} + \lambda^1(\vec{l}^2 - 1) + \lambda^2(\vec{l} \cdot \vec{\widetilde{l}})\}, \\ \hat{F}_{\mu\nu} \cdot \vec{J}^{ab} &= (F + H)_{\mu\nu}^1 l^{ab} - (F + H)_{\mu\nu}^2 \widetilde{l}^{ab}. \end{aligned} \qquad (45)$$

The restricted action gives the following equations

$$\begin{aligned} \delta A_\nu^1 &: \ \partial_\mu(\hat{e}\, \hat{e}_a^\mu\, \hat{e}_b^\nu\, \vec{l} \cdot \vec{J}^{ab}) = 0 \\ \delta A_\nu^2 &: \ \partial_\mu(\hat{e}\, \hat{e}_a^\mu\, \hat{e}_b^\nu\, \vec{\widetilde{l}} \cdot \vec{J}^{ab}) = 0 \\ \delta \hat{e}_a^\mu &: \ \hat{R}_\mu^a - \frac{1}{2}\, \hat{e}_\mu^a\, \hat{R} = 0. \end{aligned} \qquad (46)$$

From this, we have the restricted Einstein equation

$$\begin{aligned} \hat{\nabla}^a l_{ab} &= 0 , \\ \hat{\nabla}^a \widetilde{l}_{ab} &= 0 , \\ \hat{R}_\mu^a - \frac{1}{2}\, \hat{e}_\mu^a\, \hat{R} &= 0. \end{aligned} \qquad (47)$$

We can check that the first two equations can be replaced by

$$\begin{aligned} \hat{\nabla}_a l_{bc} &= 0 , \\ A_\mu^1 &= \omega_{\mu ab}(\hat{e}_\mu^a)\, l^{ab} , \\ A_\mu^2 &= \omega_{\mu ab}(\hat{e}_\mu^a)\, \widetilde{l}^{ab} . \end{aligned} \qquad (48)$$

These equations have clear meaning that they give the relation between restricted metric and restricted connection.

Notice that with

$$\vec{E}_{\mu\nu} = e\ e_a^\mu\ e_b^\nu\ \vec{J}_{ab} = E^1_{\mu\nu}\ \vec{l}\ -\ E^2_{\mu\nu}\ \tilde{\vec{l}}\ +\ \vec{E}_{\perp\mu\nu}\ . \tag{49}$$

we can express the full action by

$$S[e_a^\mu,\ \vec{A}_\mu] = \int d^4x\ \vec{E}^{\mu\nu}\cdot\vec{F}_{\mu\nu}. \tag{50}$$

This gives us the following equations

$$\begin{aligned} \delta\vec{A}_\nu\ &:\ \vec{D}_\mu\vec{E}^{\mu\nu}\ = 0 \\ \delta e_a^\mu\ &:\ \vec{E}^{\rho\mu}\cdot\vec{F}_{\rho\nu}\ =\ 0, \end{aligned} \tag{51}$$

which can be split into

$$\begin{aligned} \partial_\mu E^{1\mu\nu}\ &=\ (\vec{E}_\perp^{\mu\nu}\times\vec{X}_\mu)\cdot\vec{l} \\ \partial_\mu E^{2\mu\nu}\ &=\ (\vec{E}_\perp^{\mu\nu}\times\vec{X}_\mu)\cdot\tilde{\vec{l}} \\ \partial_\mu(\vec{l}\times\vec{E}_\perp^{\mu\nu})\ &=\ \vec{E}_\perp^{\mu\nu}A^1_\mu - \tilde{\vec{E}}_\perp^{\mu\nu}A^2_\mu - E^{1\mu\nu}\vec{X}_\mu + E^{2\mu\nu}\tilde{\vec{X}}_\mu \\ E^{1\rho\mu}\ &(F+H+X)^1_{\rho\nu} - E^{2\rho\mu}\ (F+H+X)^2_{\rho\nu} \\ &= \vec{E}^{\perp\rho\mu}\cdot(\hat{D}_\rho\vec{X}_\nu - \hat{D}_\nu\vec{X}_\rho). \end{aligned} \tag{52}$$

And with $\vec{E}_{\mu\nu}$ the restricted action is written as

$$\begin{aligned} S[e_a^\mu,\ A^1_\mu,\ A^2_\mu] = \int d^4x\ \{&E^{1\mu\nu}(F+H)^1_{\mu\nu} - E^{2\mu\nu}(F+H)^2_{\mu\nu} \\ &+\lambda^1(\vec{l}^2-1)+\lambda^2(\vec{l}\cdot\tilde{\vec{l}})\}, \end{aligned} \tag{53}$$

from which we have

$$\begin{aligned} \partial_\mu E^{1\mu\nu}\ &=\ 0 \\ \partial_\mu E^{2\mu\nu}\ &=\ 0 \\ E^{1\rho\mu}\ (F+H)^1_{\rho\nu} - E^{2\rho\mu}\ (F+H)^2_{\rho\nu}\ &=\ 0\ . \end{aligned} \tag{54}$$

Similar procedure for the second case gives us the restricted gravity which has B_2 isometry. This tells that we can construct a restricted theory of gravitation which is generally invariant but which has less physical degree.

6. Discussions

In this talk we have outline how we can decompose the Einstein's theory of gravitation to an Abelian part and and gauge covariant part which plays the role of the gravitational source. A common difficulty in non-Abelian gauge theory and gravitation is the highly non-linear self interaction. In gauge theory it has been shown that one can simplify this non-linear interaction by separating the gauge covariant valence part from the potential. Here we have shown how one can separate the covariant part of the gravitational connection which plays the role of gravitational source without compromising the general invariance.

Perhaps more importantly we have shown that one can actually construct a restricted theory of gravitation made of the restricted part of connection which has the full general invariance. We hope that this generally invariant decomposition of general relativity will enhance our understanding of gravitation.

From theoretical point of view, however, the above decomposition of gravitation differs from the Abelian decomposition of non-Abelian gauge theory in one important respect. In gauge theory the fundamental ingredient is the gauge potential, and the decomposition of the potential provides a neat decomposition of the theory. But in gravitation the fundamental field is not the potential (the connection) but the metric. And the decomposition of the connection gives us the the decomposition of the metric only indirectly, through the equation of motion, as we have seen.

Nevertheless the above decomposition of Einstein's theory has many interesting applications. The details of the subject with interesting applications will be discussed separately [12].

ACKNOWLEDGEMENT

The work is supported in part by the ABRL Program of Korea Science and Enginering Foundation (Grant R14-2003-012-01002-0).

References

1. T. Kaluza, Sitzber. Preuss. Akad. Wiss. 966 (1921); O. Klein, Z. Physik 37, 895 (1926).
2. Y. M. Cho, J. Math. Phys. **16**, 2029 (1975).
3. Y. M. Cho and P. G. O. Freund, Phys. Rev. **D12**, 1711 (1975); Y. M. Cho and Pongsoo Zhang, Phys. Rev. **D12**, 3789 (1975).

4. Y. M. Cho, Phys. Rev. **D14**, 3335 (1976); T. W. B. Kibble, J. Math. Phys. **2**, 212 (1961).
5. Y. M. Cho, Phys. Rev. **D21**, 1080 (1980); Phys. Rev **D62**, 074009 (2000).
6. Y. M. Cho, Phys. Rev. Lett. **46**, 302 (1981); Phys. Rev. **D23**, 2415 (1981); W. S. Bae, Y. M. Cho, and S. W. Kim, Phys. Rev. **D65**, 025005 (2002).
7. Y. M. Cho, H. W. Lee, and D. G. Pak, Phys. Lett. **B 525**, 347 (2002); Y. M. Cho and D. G. Pak, Phys. Rev. **D65**, 074027 (2002).
8. Y. M. Cho, D. G. Pak, and M. Walker, JHEP **05**, 073 (2004);Y. M. Cho and M. L. Walker, Mod. Phys. Lett. **A19**, 2707 (2004).
9. L. Faddeev and A. Niemi, Phys. Rev. Lett. **82**, 1624 (1999); Phys. Lett. **B449**, 214 (1999); S. Shabanov, Phys. Lett. **B458**, 322 (1999); **B463**, 263 (1999); H. Gies, Phys. Rev. **D63**, 125023 (2001); R. Zucchini, Int. J. Geom. Meth. Mod. Phys. **1**, 813 (2004).
10. A. Belavin, A. Polyakov, A. Schwartz, and Y. Tyupkin, Phys. Lett. **B59**, 85 (1975); Y. M. Cho, Phys. Lett. **B81**, 25 (1979).
11. Y. M. Cho, J. Math. Phys. **20**, 2605 (1979).
12. Y. M. Cho and S. W. Kim, to be published.

Quasi-local energy flux

James M. Nester
Department of Physics and Institute of Astronomy,
National Central University, Chungli 320, Taiwan
E-mail: nester@phy.ncu.edu.tw

Chiang-Mei Chen
Department of Physics, National Central University, Chungli 320, Taiwan
E-mail: cmchen@phy.ncu.edu.tw

Roh-Suan Tung
Center for Astrophysics, Shanghai Normal University,
100 Guilin Road, Shanghai 200234, China
Email: tung@shnu.edu.cn

The Hamiltonian includes a boundary term which determines the quasi-local values and the boundary conditions. Using our covariant Hamiltonian formalism we found four particular quasi-local energy-momentum boundary term expressions. Here we show how a fundamental Hamiltonian identity naturally leads to the associated quasi-local energy flux expressions. For electromagnetism one of the four is distinguished by gauge invariance; it gives the familiar energy density and Poynting flux. For Einstein's general relativity two choices correspond to quasi-local expressions which asymptotically give the ADM energy, the Trautman-Bondi energy and, moreover, an associated energy flux. Again there is a distinguished expression: the one which is covariant.

1. Introduction

For gravitating systems the localization of energy-momentum is still an outstanding fundamental problem. Unlike matter and the other interaction fields, the gravitational field has no proper energy-momentum density. This can be understood as a consequence of Einstein's *equivalence principle*: gravity cannot be detected at a point. The energy-momentum of gravitating systems (hence for all physical systems) is inherently *non-local*. The modern idea is *quasi-local*: i.e. associated with a closed 2-surface bounding a region.[1]

We have addressed this issue via a Hamiltonian approach.[2,3,4,5,6,7,8,9,10] The Hamiltonian, which generates displacements of a spatial region, necessarily includes an integral over the closed 2-surface bounding the region. This boundary term can take on many forms while playing the same two key roles. A particular boundary term expression gives the quasi-local energy-momentum values and, via the *the boundary variational principle* (the requirement that the boundary term in the variation of the Hamiltonian vanish) each particular expression is associated with specific boundary conditions.

This formulation works well for finite regions and for infinite spatial regions with suitable field fall offs. However for boundaries which approach null infinity the natural field fall offs do not satisfy the boundary variation principle—the Hamiltonian is *not functionally differentiable* on the phase space of fields with radiative fall offs. This apparent calamity, along with a natural differential identity, led us to a new expression for *energy-flux*.[9]

Here we develop enough formalism to derive this result, illustrate its application in the familiar case of Maxwell's electrodynamics, and apply it to Einstein gravity, obtaining thereby a preferred expression for gravitational quasi-local energy momentum and the associated quasi-local energy flux.

2. First order Lagrangian

The *first order* Lagrangian for an f-form field φ (which may carry suppressed indices) and its conjugate momentum p is

$$\mathcal{L} = d\varphi \wedge p - \Lambda(\varphi, p). \tag{1}$$

The variation,

$$\delta\mathcal{L} = d(\delta\varphi \wedge p) + \delta\varphi \wedge \frac{\delta\mathcal{L}}{\delta\varphi} + \frac{\delta\mathcal{L}}{\delta p} \wedge \delta p, \tag{2}$$

implicitly determines a pair of first order equations (here $\varsigma := (-1)^f$):

$$\frac{\delta\mathcal{L}}{\delta p} := d\varphi - \partial_p\Lambda = 0, \quad \frac{\delta\mathcal{L}}{\delta\varphi} := -\varsigma dp - \partial_\varphi\Lambda = 0. \tag{3}$$

As an example, the first order Lagrangian 4-form for the (source free) U(1) gauge field one-form A of Maxwell electrodynamics is

$$\mathcal{L}_{\mathrm{EM}} = dA \wedge H - \frac{1}{2}{}^\star H \wedge H. \tag{4}$$

Variation leads to the pair of first order equations

$$dH = 0, \qquad dA - {}^\star H = 0, \tag{5}$$

equivalent to the vacuum Maxwell equations.

3. Translation invariance

Infinitesimal diffeomorphism invariance, from (2) with $\delta \to \pounds_N$, the Lie derivative, requires

$$di_N\mathcal{L} \equiv \pounds_N\mathcal{L} \equiv d(\pounds_N\varphi \wedge p) + \pounds_N\varphi \wedge \frac{\delta\mathcal{L}}{\delta\varphi} + \frac{\delta\mathcal{L}}{\delta p} \wedge \pounds_N p. \tag{6}$$

This means that $\mathcal{L}$ is a 4-form which depends on position only through the fields φ, p (so the fields necessarily include dynamic geometry, i.e. gravity).

From (6) it follows that the "translational current" (3-form),

$$\mathcal{H}(N) := \pounds_N\varphi \wedge p - i_N\mathcal{L}, \tag{7}$$

satisfies

$$-d\mathcal{H}(N) \equiv \pounds_N\varphi \wedge \frac{\delta\mathcal{L}}{\delta\varphi} + \frac{\delta\mathcal{L}}{\delta p} \wedge \pounds_N p. \tag{8}$$

This is a conservation law, as the right hand side vanishes "on shell" (i.e., when the field equations are satisfied). Note that, as with other Noether conserved currents, $\mathcal{H}(N)$ is not unique: it can be modified to an alternate conserved current expression by adding the differential of any 2-form.

With geometric gravity included one, moreover, has *local* diffeomorphism invariance, and hence an associated differential identity. A short explicit calculation using (1) reveals that $\mathcal{H}(N)$ always has the form

$$\mathcal{H}(N) = N^\mu\mathcal{H}_\mu + d\mathcal{B}(N). \tag{9}$$

Consequently $d(N^\mu\mathcal{H}_\mu + d\mathcal{B}(N)) \equiv dN^\mu \wedge \mathcal{H}_\mu + N^\mu d\mathcal{H}_\mu$ is, from (8), proportional to the field equations, thus $\mathcal{H}_\mu$ in particular vanishes "on shell". Hence for gravitating systems the Noether translational "charge"—energy-momentum—is *quasi-local*; it is given by the integral of the boundary term, $\mathcal{B}(N)$ over the 2-dimensional boundary of the region. But this boundary term, as we noted, can be *completely* modified. The Hamiltonian approach tames this enormous ambiguity.

4. Hamiltonian approach

Energy can be identified as the value of the Hamiltonian associated with a timelike displacement vector field N. The Hamiltonian $H(N)$ is given by an integral of a suitable Hamiltonian 3-form (density) $\mathcal{H}(N)$ over a 3-dimensional (spacelike) region Σ.

From the first order Lagrangian one can construct the Hamiltonian 3-form by projecting along a "timelike" displacement vector field:

$$i_N\mathcal{L} = \pounds_N\varphi \wedge p - \mathcal{H}(N). \tag{10}$$

The resultant Hamiltonian 3-form is just the previously identified Noether translational current. Since $\mathcal{H}_\mu$ vanishes "on shell", the *quasi-local energy* is determined only by the boundary integral

$$E(N) = \int_\Sigma \mathcal{H}(N) = \int_\Sigma N^\mu \mathcal{H}_\mu + \oint_{\partial\Sigma} \mathcal{B}(N) = \oint_{\partial\Sigma} \mathcal{B}(N). \tag{11}$$

The two parts of the Hamiltonian have distinct functions. The 3-form part $N^\mu\mathcal{H}_\mu$ generates the equations of motion. The boundary term $\mathcal{B}(N)$ is the key, it has a dual role.

The Hamiltonian boundary term with different choices of displacement N determines the values of the various quasi-local quantities: *energy-momentum* from a suitable spacetime translation, and angular momentum (center-of-mass) from a rotation (boost). However our Noether analysis has revealed that $\mathcal{B}(N)$ can be adjusted, thereby changing the conserved values. Fortunately the variational principle contains an additional (largely overlooked) feature which distinguishes all of these choices.

5. Boundary Conditions

According to the *boundary variation principle* the boundary term in the variation should vanish.[11] The different Hamiltonian boundary terms are each associated with distinct *boundary conditions.* As in thermodynamics or electrostatics there are various "energies" which correspond to how the system interacts with the outside through its boundary.

In general (in particular for gravity) it is *necessary* (in order to guarantee functional differentiability of the Hamiltonian on the phase space with the desired boundary conditions) to adjust the boundary term, $\mathcal{B}(N) = i_N\varphi\wedge p$, which is naturally inherited from the Lagrangian. Moreover a reference configuration, $\bar{\varphi}$ and $\bar{p}$, (which determines the ground state) is needed at least for gravity (where the ground sate of the metric is the non-vanishing Minkowski metric) in order to allow the desired phase space asymptotics.[12,13,14]

6. Quasi-local Expressions

From (2) and (10) short calculations give

$$\delta i_N \mathcal{L} \equiv \pounds_N(\delta\varphi \wedge p) - \delta\varphi \wedge \pounds_N p + \pounds_N \varphi \wedge \delta p - \delta\mathcal{H}(N), \tag{12}$$

$$i_N \delta\mathcal{L} \equiv \pounds_N(\delta\varphi \wedge p) - d i_N(\delta\varphi \wedge p) + i_N \left[\delta\varphi \wedge \frac{\delta\mathcal{L}}{\delta\varphi} + \frac{\delta\mathcal{L}}{\delta p} \wedge \delta p\right]. \tag{13}$$

The last term in (13) vanishes on shell. Equating the two relations then gives an identity involving the variation of the Hamiltonian (7):

$$\delta\mathcal{H}(N) = -\delta\varphi \wedge \pounds_N p + \pounds_N \varphi \wedge \delta p + d i_N(\delta\varphi \wedge p). \tag{14}$$

When integrated the total differential term becomes a boundary term that vanishes if $\delta\varphi$ vanishes on the boundary; then the remaining terms give the Hamiltonian field equations.

One can modify the boundary term in the Hamiltonian. We found two boundary choices (here we use $\Delta\varphi := \varphi - \bar{\varphi}$, $\Delta p := p - \bar{p}$),

$$\mathcal{B}_\varphi = i_N \varphi \wedge \Delta p - \varsigma \Delta\varphi \wedge i_N \bar{p} \quad \Longrightarrow \quad i_N(\delta\varphi \wedge \Delta p), \tag{15}$$

$$\mathcal{B}_p = i_N \bar{\varphi} \wedge \Delta p - \varsigma \Delta\varphi \wedge i_N p \quad \Longrightarrow \quad -i_N(\Delta\varphi \wedge \delta p), \tag{16}$$

which lead to the indicated *covariant* Dirichlet and Neumann type boundary conditions. Moreover we found two other physical interesting choices:

$$\mathcal{B}_{\rm dyn} = i_N \bar{\varphi} \wedge \Delta p - \varsigma \Delta\varphi \wedge i_N \bar{p} \quad \Longrightarrow \quad \varsigma \delta\varphi \wedge i_N \Delta p - i_N \Delta\varphi \wedge \delta p, \tag{17}$$

$$\mathcal{B}_{\rm con} = i_N \varphi \wedge \Delta p - \varsigma \Delta\varphi \wedge i_N p \quad \Longrightarrow \quad i_N \delta\varphi \wedge \Delta p - \varsigma \Delta\varphi \wedge i_N \delta p. \tag{18}$$

leading to the indicated boundary terms in the variation of the Hamiltonian.

At spatial infinity the above Hamiltonians are well defined on the phase space of fields with the standard fall offs for the indicated parity terms[12,13,14]

$$\Delta\varphi \simeq O^+(r^{-1}) + O^-(r^{-2}), \qquad \Delta p \simeq O^-(r^{-2}) + O^+(r^{-3}). \tag{19}$$

This is not the case near null infinity; with radiative fall offs[15] $\Delta p \simeq d\varphi \simeq O(1/r)$ so the above Hamiltonians *are not* functionally differentiable (the variational derivatives which determine the field equations *are not well defined*). This apparent calamity is actually providential. The non-vanishing of the boundary term in the Hamiltonian variation indicates that the system is not closed; it gives an expression for the *energy flux.*[9]

7. Quasi-local Energy Flux

In classical mechanics from $\delta H = \dot{q}^k \delta p_k - \dot{p}_k \delta q^k$ with $\delta = d/dt$ one obtains $\dot{H} \equiv 0$. Likewise for the time evolution of the Hamiltonian (7), one can insert $\pounds_N$ for δ in (14) and arrive, via symplectic[16,17] cancelations, at a simple relation: $\pounds_N \mathcal{H}(N) = d i_N(\pounds_N \varphi \wedge p)$. Similarly the evolution of the Hamiltonians with our improved boundary terms (15–18) are

$$\pounds_N \mathcal{H}_\varphi = d i_N(\pounds_N \varphi \wedge \Delta p), \tag{20}$$

$$\pounds_N \mathcal{H}_p = -d i_N(\Delta\varphi \wedge \pounds_N p), \tag{21}$$

$$\pounds_N \mathcal{H}_{\rm dyn} = d(\varsigma \pounds_N \varphi \wedge i_N \Delta p - i_N \Delta\varphi \wedge \pounds_N p), \tag{22}$$

$$\pounds_N \mathcal{H}_{\rm con} = d(i_N \pounds_N \varphi \wedge \Delta p - \varsigma \Delta\varphi \wedge i_N \pounds_N p). \tag{23}$$

When integrated these relations yield the energy flux expressions.

8. Applications

For (vacuum) Maxwell electrodynamics (4), (5) the natural background choice is $\bar{A} = 0$ and $\bar{H} = 0$. The Hamiltonian 3-form reduces to the usual energy density, $\frac{1}{2}(E^2 + B^2)$, plus a gauge transformation generating term, $-i_N A dH$, involving the scalar potential multiplying the Gauss constraint.

Our four boundary term choices are

$$\mathcal{B}_A = i_N A \wedge H, \qquad \mathcal{B}_H = A \wedge i_N H,$$
$$\mathcal{B}_{\rm dyn} = 0, \qquad \mathcal{B}_{\rm con} = i_N A \wedge H + A \wedge i_N H. \tag{24}$$

The "dynamical" boundary choice gives a *distinguished* energy expression, the only one which is *gauge invariant*. (Bonus: $\mathcal{H}_{\rm dyn}$ is *positive definite*). The time derivative of this "dynamical" Hamiltonian density is

$$\pounds_N \mathcal{H}_{\rm dyn} = -d\left(i_N dA \wedge i_N H\right) = -d\left(E_i B_j \, dx^i \wedge dx^j\right), \tag{25}$$

the flux of the usual Poynting vector.

For Einstein's (vacuum) gravity theory a first order Lagrangian is

$$\mathcal{L}_{\rm GR} = R^\alpha{}_\beta \wedge \eta_\alpha{}^\beta, \tag{26}$$

where $\Gamma^\alpha{}_\beta$ is the connection 1-form; $R^\alpha{}_\beta := d\Gamma^\alpha{}_\beta + \Gamma^\alpha{}_\gamma \wedge \Gamma^\gamma{}_\beta$ is the curvature 2-form and $\eta^{\alpha\cdots} := *(\vartheta^\alpha \wedge \cdots)$, with ϑ^α the orthonormal co-frame.

Detailed calculations[8,9] show that two different boundary condition choices ($\mathcal{B}_\vartheta$, $\mathcal{B}_{\rm con}$) correspond to quasi-local expressions which asymptotically give the ADM energy, the Trautman-Bondi energy[18] and the Bondi flux: (both outgoing and incoming). Again one expression stands out:

$$\mathcal{B}_\vartheta = \Delta\Gamma^{\alpha\beta} \wedge i_N \eta_{\alpha\beta} + \bar{D}_\beta \bar{N}^\alpha \Delta\eta_\alpha{}^\beta. \tag{27}$$

It is distinguished by a very desirable and appropriate property: it corresponds to imposing boundary conditions on a *covariant* object. The associated energy flux expression is

$$\pounds_N \mathcal{H}_\vartheta = di_N \left(\Delta\Gamma^{\alpha\beta} \wedge \pounds_N \eta_{\alpha\beta} \right) . \tag{28}$$

Acknowledgments

The work of C. M. C. and J. M. N. was supported by the National Science Council of the R.O.C. under the grants NSC 94-2112-M-008-012 and NSC 94-2112-M-008-038; the work of R. S. T. was supported by the National Nature Science Foundation of China under the grants 10375081 and 10375087.

References

1. L. B. Szabados, *Living Rev. Relativity* 7, 4 (2004); www.livingreviews.org/lrr-2004-4.
2. C.-M. Chen, J. M. Nester, and R.-S. Tung, *Phys. Lett. A* **203**, 5–11 (1995).
3. C.-M. Chen and J. M. Nester, *Class. Quantum Grav.* **16**, 1279–1304 (1999).
4. C.-C. Chang, J. M. Nester and C.-M. Chen, *Phys. Rev. Lett.* **83**, 1897 (1999).
5. C.-C. Chang, J. M. Nester, and C.-M. Chen, in *Gravitation and Astrophysics* ed Liao Liu, Jun Luo, X.-Z. Li, and J.P. Hsu (World Scientific, Singapore) pp 163-73; gr-qc/9912058.
6. C.-M. Chen and J. M. Nester, *Gravitation & Cosmology* **6**, 257–70 (2000).
7. J. M. Nester, *Class. Quantum Grav.* **21**, S261–S280 (2004).
8. X. Wu, C.-M. Chen and J. M. Nester, *Phys. Rev. D* **71**, 124010 (2005).
9. C.-M. Chen, J. M. Nester and R.-S. Tung, *Phys. Rev.* **D72**, 104020 (2005).
10. J. M. Nester, C.-M Chen and R.-S. Tung, "The Hamiltonian boundary term", in these proceedings.
11. C. Lanczos, *The variational principles of mechanics* (University of Toronto Press, Toronto, 1949).
12. T. Regge and C. Teitelboim, *Annals Phys.* **88**, 286–318 (1974).
13. R. Beig and N. Ó Murchadha, *Ann. Phys.* **174**, 463–98 (1987).
14. L. B. Szabados, *Class. Quantum Grav.* **20**, 2627–61 (2003).
15. P. T. Chrusciel, J. Jezierski, and J. Kijowski, *A Hamiltonian framework for field theories in the radiating regime* (Springer, 2001).
16. J. Kijowski and W. M. Tulczyjew, *A Symplectic Framework for Field Theories*, (Lecture Notes in Physics **107**, Springer-Verlag, Berlin, 1979).
17. J. Kijowski, *Gen. Rel. Grav.* **29**, 307–343 (1997).
18. P. T. Chruściel, J. Jezierski and M. A. H. MacCallum, *Phys. Rev. D* **58**, 084001 (1998).

The Hamiltonian boundary term

James M. Nester and Chiang-Mei Chen
Department of Physics, National Central University, Chungli 320, Taiwan
E-mail: nester@phy.ncu.edu.tw, cmchen@phy.ncu.edu.tw

Roh-Suan Tung
Center for Astrophysics, Shanghai Normal University,
100 Guilin Road, Shanghai 200234, China
Email: tung@shnu.edu.cn

The Hamiltonian for a gravitating region necessarily includes a boundary term. Its role is to determine the quasi-local values and, via the vanishing of the boundary term in the variation of the Hamiltonian (required for a well defined Hamiltonian), the boundary conditions. With this boundary variation principle, the Hamiltonian formalism tames the ambiguities inherent in the traditional approaches to the (quasi-)localization of energy-momentum. Using our covariant Hamiltonian formalism, we identified special boundary terms associated with particular physical boundary conditions. In the radiating regime the Hamiltonian fails to be "well defined". This led to new energy flux expressions. We here report on new results for homogeneous cosmologies, the small region limit, and positivity. Although there are many possibilities for the boundary term (reflecting the many choices for the boundary conditions and reference), we found that one expression is distinguished.

1. Introduction: energy and its localization

Ninety years ago Einstein presented the field equations for his covariant gravitation theory, general relativity, after having, for some two years, turned away from requiring general covariance. How large a role did the inherent non-localizability of gravitational energy play in his struggles? We know that energy-momentum conservation was central to his approach, and that he had, by rearranging the gravitational field equation, already identified an expression for the density of gravitational energy-momentum, which was (inevitably) non-covariant. With such a density, it was not unreasonable to consider, as he did, field equations which were not generally covariant, and to expect that energy-momentum conservation could be used to distinguish the preferred reference frame.[1]

To describe the physics associated with the 25 November covariant field equations, later investigators proposed other gravitational energy-momentum expressions; all were reference frame dependent. There are two ambiguities: (i) which energy-momentum expression should one use, and (ii) which reference frame should be used to get the real physical energy-momentum? An influential textbook concluded that *"Anyone who looks for a magic formula for "local gravitational energy-momentum" is looking for the right answer to the wrong question. Unhappily, enormous time and effort were devoted in the past to trying to "answer this question" before investigators realized the futility of the enterprise."*[2] Physically this "non-localizability" can be understood as a consequence of Einstein's equivalence principle, one implication of which is that gravity cannot be detected at a point and hence, unlike all other physical fields, cannot have a proper local energy-momentum density. Energy localization thus reveals a certain kind of incompatibility between the two fundamental pillars of Einstein's theory: *the equivalence principle* and *the principle of general covariance.*

For gravitating systems (i.e. all physical systems) energy-momentum localization remains an outstanding fundamental problem. One expects from conservation and the *local* interaction with sources some kind of meaningful local description, yet investigations found only reference frame dependent quantities. These *pseudotensors* have been obtained both from the Lagrangian by using Noether symmetry (note: all Noether currents can be modified without loosing the conservation property) or by splitting the field equations (with a similar ambiguity).[3]

We have argued that the apparent ambiguities fundamentally inherent in the localization of energy-momentum are given clear physical and geometric meanings by looking to the Hamiltonian and, in particular, to the Hamiltonian boundary term.[3,4,5,6,7,8,9]

Arnowitt, Deser and Misner, using a 3+1 approach, had proposed certain 2-sphere-at-spatial-infinity integral expressions for energy-momentum in connection with their Hamiltonian formulation.[10] In a seminal work, Regge and Teitelboim argued that these ADM expressions were an *essential* boundary term part of the Hamiltonian, which was necessary to render it functionally differentiable.[11] Since then certain important improvements in the formulation were made by Beig and Ó Murchadha[12] and Szabados.[13]

In our work, we have used symplectic ideas[14] to develop a covariant Hamiltonian formulation applicable to both finite and infinite regions. The main objective was to identify the proper form and role of the Hamiltonian boundary term. Using this formulation, we have tamed the ambiguities inherent in energy localization and, moreover, identified a distinguished quasi-local energy-momentum and its associated energy-flux expression.

2. The Hamiltonian approach: quasi-local quantities

Energy can be identified as the value of the Hamiltonian associated with a timelike displacement vector field N. The Hamiltonian is given by an integral over a 3-dimensional (spacelike) region of a suitable Hamiltonian 3-form (density). This density takes the form

$$\mathcal{H}(N) = N^\mu \mathcal{H}_\mu + d\mathcal{B}(N). \tag{1}$$

The Hamiltonian thus includes an integral over the boundary of the region:

$$H(N, \Sigma) = \int_\Sigma \mathcal{H}(N) = \int_\Sigma N^\mu \mathcal{H}_\mu + \oint_{\partial\Sigma} \mathcal{B}(N). \tag{2}$$

The two parts of the Hamiltonian have distinct functions. The 3-form part $N^\mu \mathcal{H}_\mu$ generates the equations of motion; for geometric gravity theories it has vanishing value. The Hamiltonian boundary term $\mathcal{B}(N)$ plays two key roles: (i) quasi-local values,[15] and (ii) boundary conditions.

For gravitating systems, the quasi-local energy-momentum (and angular momentum/center-of -mass) values are given by the integral of the boundary term, $\mathcal{B}(N)$. However, this boundary term can be modified, fortunately the formulation includes a principle which tames this freedom. According to the boundary variation principle the boundary term in the variation tells us what to hold fixed on the boundary. Different Hamiltonian boundary terms are associated with distinct boundary conditions. Thus for a physical system there are various "energies" which correspond to how the system interacts with the outside through its boundary.

The idea is well known in some other physical situations. In thermodynamics one uses different energies: internal, Helmholtz, enthalpy, and Gibbs, depending on which variables one "controls" on the "boundary" of the system. A familiar field theory example is (vacuum) electrodynamics. One Hamiltonian (related to the *symmetric* energy-momentum tensor) is

$$H_s = \int \left[\frac{1}{2}(E^2 + B^2) + \phi \vec{\nabla} \cdot \vec{E}\right] d^3x, \qquad \delta H_s \sim \oint \phi\, \delta(\vec{E} \cdot \vec{n}) dS. \tag{3}$$

The indicated boundary term is the electric part of its variation; in order for it to vanish one should fix the normal component of the electric field, i.e. the surface charge. An alternative Hamiltonian choice (related to the *canonical* energy-momentum tensor) differs only by a spatial boundary term:

$$H_c = \int \left[\frac{1}{2}(E^2 + B^2) - \vec{E}\cdot\vec{\nabla}\phi\right] d^3x, \qquad \delta H_c \sim -\oint \delta\phi\,(\vec{E}\cdot\vec{n})dS. \quad (4)$$

Now the indicated electric boundary term in the variation will vanish if one fixes the electric potential on the boundary. Both choices correspond to situations where the energy can be associated with the work done in physical processes which can actually be realized, e.g. by inserting/removing a dielectric in a parallel plate capacitor with or without a battery connected. Yet the choice H_s is favored, as it is gauge invariant and positive definite.

3. The variational boundary term and energy flux

The idea applies to any field. From a first-order Lagrangian 4-form

$$\mathcal{L} = d\varphi \wedge p - \Lambda(\varphi, p). \quad (5)$$

one constructs the Hamiltonian 3-form

$$\mathcal{H}(N) := \pounds_N\varphi \wedge p - i_N\mathcal{L}, \quad (6)$$

which satisfies the key *variational identity*

$$\delta\mathcal{H}(N) = -\delta\varphi \wedge \pounds_N p + \pounds_N\varphi \wedge \delta p + d i_N(\delta\varphi \wedge p). \quad (7)$$

Upon integration this identity gives the variation of the Hamiltonian: δH. The total differential becomes a boundary term, vanishing if $\delta\varphi$ vanishes on the boundary. Then the Hamiltonian is functionally differentiable and its variation yields the Hamiltonian equations.

Substitute $\delta \to \pounds_N$ into the identity. Symplectic cancelations then yield

$$\pounds_N\mathcal{H}(N) = d i_N(\pounds_N\varphi \wedge p), \quad (8)$$

the integral of which gives a natural *energy flux* relation.

The Hamiltonian 3-form (6), has the form (1). However the natural boundary term inherited from the Lagrangian, $\mathcal{B}(N) := i_N\vartheta \wedge p$, can, and often should, be modified. We found four particular types of improved boundary terms and their associated energy flux relations.[9,16]

4. A distinguished expression

For Einstein's gravity theory, using the orthonormal co-frame ϑ^α and the metric compatible connection one-form $\Gamma^\alpha{}_\beta$ as independent variables, we found four expressions corresponding to different types of boundary conditions. For some purposes, e.g., spatial asymptotics, they are all acceptable.

However for certain applications, in particular for *energy flux*, one particular boundary expression (fixing the metric on the boundary) stands out:

$$\mathcal{B}_\vartheta(N) = \Delta\Gamma^\alpha{}_\beta \wedge i_N \eta_\alpha{}^\beta + \bar{D}_\beta \bar{N}^\alpha \, \Delta\eta_\alpha{}^\beta , \tag{9}$$

where $\eta^{\alpha\beta} := *(\vartheta^\alpha \wedge \vartheta^\beta)$ and $\Delta\Gamma := \Gamma - \bar{\Gamma}$, $\Delta\eta := \eta - \bar{\eta}$, with the barred quantities being necessary reference (i.e. vacuum or "ground state") values. The integral of the associated energy flux expression

$$\pounds_N \mathcal{H}_\vartheta = d i_N \left(\Delta\Gamma^{\alpha\beta} \wedge \pounds_N \eta_{\alpha\beta}\right) . \tag{10}$$

directly gives the Bondi flux.[9]

5. Quasi-local values for small regions

Take the divergence of the Hamiltonian boundary term and Taylor expand. One should get to zeroth order the material energy-momentum. In vacuum one hopes to get a positive multiple of the Bel-Robinson tensor to second order (in particular this will guarantee positive energy in the small).[15]

Just one of our Hamiltonian boundary term expressions satisfies this important *ultra-local positivity* requirement; it is the same expression (9) as was distinguished by the Bondi flux.

Note: none of the classical holonomic pseudotensors satisfies this vacuum Bel-Robinson requirement but Møller's (1961) tetrad-teleparallel expression (to which our favored expression reduces in this limit) does.[17]

6. Quasi-local positivity

Under certain conditions our favored quasi-local expression (9) gives positive quasi-local energy. This can be shown in two ways. We can adapt to quasi-local regions the positivity proof[18] which uses certain frame gauge conditions[19] to get a non-negative Hamiltonian density. One could instead use the Shi-Tam proof[20] for the Brown-York expression[21] (our energy reduces to the Brown-York energy for certain reference/displacement choices).

7. Quasi-local values for large regions

Consider homogeneous cosmologies. Let the metric have the Bianchi form

$$ds^2 = -dt^2 + g_{ij}(t)\sigma^i\sigma^j, \qquad d\sigma^i = \frac{1}{2}C^i{}_{jk}\sigma^j \wedge \sigma^k, \tag{11}$$

where the $C^i{}_{jk}$ are certain specific constants. From our favored quasi-local expression (9), with *homogeneous* boundary conditions and reference, for *all regions* we find the energy to be given by

$$E(V) = A_k A_l g^{kl}(t) V(t) \geq 0, \tag{12}$$

where $A_k := C^i{}_{ik}$. More specifically we obtain *zero energy* for Class A models: $A_k = 0$ (Bianchi Types I, II, VI_0, VII_0, VIII, IX) and *positive energy* for Class B models: $A_k \neq 0$ (Bianchi Types III, IV, V, VI_h, VII_h), results which are in line with the requirement that closed universes have vanishing energy, since all Class A models can be closed and Class B cannot.

For homogeneous and isotropic (FRW) cosmologies the metric has the form $ds^2 = -dt^2 + a^2(t)dl^2$, where

$$dl^2 = \frac{1}{1-kr^2}dr^2 + r^2 d\Omega^2 = \frac{1}{(1+kR^2/4)^2}(dR^2 + R^2 d\Omega^2). \tag{13}$$

We take the boundary conditions and reference to be those natural to these explicitly isotropic-about-a-particular point representations. Then our expression for the energy within a given radius is

$$E_k = ar(1-\sqrt{1-kr^2}) = k\frac{a}{2}R^3\left[1+kR^2/4\right]^{-2}. \tag{14}$$

Note that $E_0 = 0$ and $E_{-1} < 0$, *contrary* to a common quasi-local criterion.

Certain Bianchi models can be isotropic, in particular isotropic Bianchi I, V (and VII_h), IX are, respectively, isometric to FRW $k = 0, -1, +1$. Except for the $k = 0$ case, the Bianchi and FRW representations give different values for the same physical situation. This illustrates the fact that our quasi-local energy boundary expressions naturally give a different result for different reference, boundary condition, and displacement choices. We note that another common quasi-local desiderata is *violated* by our expressions: $E = 0$ need not be Minkowski space.

8. Discussion

We have been investigating the proper form and role of the Hamiltonian boundary term. We found that it gives conserved quasi-local quantities and, via the boundary variational principle, boundary conditions. The ambiguities inherent in the quasi-localization of energy-momentum for gravitating systems have been tamed. From our investigations a particular choice of Hamiltonian boundary term stands out.

Our expressions have been shown to give good values asymptotically at spatial and future null infinity for energy-momentum and angular momentum/center-of-mass, values which are under appropriate conditions compatible with widely recognized works.[2,10,11,12,13,21]

We found that energy flux distinguishes one particular expression. The same expression has positive energy proofs and gives good values for both large (homogeneous cosmology) and small regions. The most outstanding issue is how to choose the necessary reference values on the boundary.

Acknowledgements

The work of C. M. C. and J. M. N. was supported by the National Science Council of the R.O.C. under the grants NSC 94-2112-M-008-012, NSC 94-2112-M-008-038; the work of R. S. T. was supported by the National Nature Science Foundation of China under the grants 10375081 and 10375087.

References

1. J. Norton "How Einstein Found His Field Equations: 1912-1915," in D. Howard and J. Stachel (eds.), *Einstein and the History of General Relativity: Einstein Studies Vol. I*, Boston: Birkhauser, 1989, pp 101-159.
2. C. W. Misner, K. Thorne and J. A. Wheeler, *Gravitation* (Freeman, 1973). p 467.
3. J. M. Nester, *Class. Quantum Grav.* **21**, S261–S280 (2004).
4. C.-M. Chen, J. M. Nester, and R.-S. Tung, *Phys. Lett.* A **203**, 5–11 (1995).
5. C. C. Chang, J. M. Nester and C.-M. Chen, *Phys. Rev. Lett.* **83**, 1897 (1999).
6. C.-M. Chen and J. M. Nester, *Class. Quantum Grav.* **16**, 1279–1304 (1999).
7. C.-C. Chang, J. M. Nester, and C.-M. Chen, in *Gravitation and Astrophysics* ed Liao Liu, Jun Luo, X.-Z. Li, J. P. Hsu (World Scientific, Singapore) p 163.
8. C.-M. Chen and J. M. Nester, *Gravitation & Cosmology* **6**, 257–70 (2000).
9. C.-M. Chen, J. M. Nester and R. S. Tung, *Phys. Rev. D* **72**, 104020 (2005).
10. R. Arnowitt, S. Deser, C. W. Misner, in: *Gravitation: An Introduction to Current Research*, ed L. Witten (Wiley, New York, 1962).
11. T. Regge and C. Teitelboim, *Annals Phys.* **88**, 286–318 (1974).
12. R. Beig and N. Ó Murchadha, *Ann. Phys.* **174**, 463–98 (1987).
13. L. B. Szabados, *Class. Quantum Grav.* **20**, 2627–61 (2003).
14. J. Kijowski and W. M. Tulczyjew, *A Symplectic Framework for Field Theories*, (Lecture Notes in Physics **107**, Springer-Verlag, Berlin, 1979). J. Kijowski, *Gen. Rel. Grav.* **29**, 307–343 (1997).
15. L. B. Szabados, *Living Rev. Relativity* **7**, 4 (2004); www.livingreviews.org/lrr-2004-4.
16. C.-M. Chen, J. M. Nester and R.-S. Tung, "Quasi-local energy flux", in these proceedings.
17. L. L. So, J. M. Nester, and H. Chen, "Classical pseudotensors and positivity in small regions", in this volume.
18. J. M. Nester, *Int. J. Mod. Phys. A* **4**,1755 (1989);*Phys. Lett.* **139A**, 112 (1989).
19. J. M. Nester, *J. Math. Phys. 30*, 624 (1989); *J. Math. Phys.* **33**, 910 (1992).
20. Y. Shi and L.-F. Tam, *J. Diff. Geometry* **62**, 79–125 (2002).
21. J. D. Brown and J. W. York, Jr., *Phys. Rev. D* **47**, 1407–19 (1993).

DYNAMICAL UNTRAPPED HYPERSURFACES

ROH-SUAN TUNG

Center for Astrophysics, Shanghai Normal University,
100 Guilin Road, Shanghai 200234, China
E-mail: tung@shnu.edu.cn

A smooth, three-dimensional submanifold of spacetime is a *dynamical untrapped hypersurface* if it can be foliated by a family of closed 2-manifolds S such that each foliation is an untrapped (mean convex) surface. If further on each leaf S, the dual expansion vector is tangent to the dynamical untrapped hypersurface, then it is called a *stationary untrapped hypersurface*. Note that for an untrapped surface S, the dual expansion vector is unique, always timelike and it is the direction for zero expansion. Thus for stationary untrapped hypersurfaces, the dual expansion vector plays the role which the stationary Killing vector plays for stationary black holes. We show that Hamiltonian for the spatial hypersurface associated with the timelike vector extended by the dual expansion vector for each leaf is well-defined for stationary untrapped hypersurface, and thus provides a definition of total energy-momentum for the region with the boundary S. For dynamical untrapped hypersurface, there does not exist a Hamiltonian in general, but a gravitational radiation energy flux can be obtained.

1. Introduction: Untrapped Hypersurfaces

Recently with the concepts of dynamical and trapping horizons, expressions of fluxes of energy carried by gravitational waves across the horizons were obtained.[1,2,3] This leads to a question whether we can generalize the concepts to certain situations of general evolving spacetime regions outside the horizons such that we can obtain expressions of energy conservation carried by gravitational waves across hypersurfaces outside the horizons.

Given the geometry of 2-surface S embedded in a 4-dimensional spacetime M. Let $(e_0)^a, (e_1)^a$ be a set of timelike and spacelike unit normals to S and $(e_A)^a = [(e_2)^a, (e_3)^a]$ be a set of local frames tangent to S. On S the trace of the extrinsic curvatures with respect to $(e_0)^a$ and $(e_1)^a$ directions are given by $k(0) = -\Gamma^{0A}{}_A$ and $k(1) = -\Gamma^{1A}{}_A$. The expansion vector (mean curvature vector) H^a and the dual expansion vector (dual mean-curvature vector) $H^a_\perp$ of the 2-surface are given by

$$H^a_\perp = (k(1)(e_0)^a - k(0)(e_1)^a) \tag{1}$$

$$H^a = (k(1)(e_1)^a - k(0)(e_0)^a) \tag{2}$$

The norm of the expansion vector is the mean curvature $|H| = \sqrt{H^2} = \sqrt{k(1)^2 - k(0)^2}$. For an untrapped (mean convex) surface, $k(1)^2 > k(0)^2$, there is a set of unique special timelike and spacelike unit normals on the 2-surfaces, given by $(\hat{e}_0)^a = H^a_\perp/|H|$ and $(\hat{e}_1)^a = H^a/|H|$. These special normals are independent of the choice of normal frames for the 2-surface ($(\hat{e}_0)^a$ and $(\hat{e}_1)^a$ are invariant under arbitrary boost transformation of $(e_0)^a$ and $(e_1)^a$). Thus, they depend only on the 2-surface.[4,5]

Dynamical Untrapped Hypersurfaces

A smooth, three-dimensional submanifold Δ of spacetime is said to be a *Dynamical Untrapped Hypersurface* if it can be foliated by a family of closed 2-manifolds S such that each foliation is an untrapped (mean convex) surface.

Note that for each leaf S of a Dynamical Untrapped Hypersurface, the dual expansion vector $H^a_\perp$ is always timelike. Moreover, the trace extrinsic curvature is zero along the direction of the dual expansion vector, i.e. $k(H^a_\perp) = 0$, thus we have

$$\pounds_{H^a_\perp} \sigma_{ab} = 0. \tag{3}$$

where σ_{ab} is the intrinsic metric of S.

Stationary Untrapped Hypersurfaces

If on each leaf of the Dynamical Untrapped Hypersurface, the dual expansion vector $H^a_\perp$ is tangent to the dynamical untrapped hypersurface, then it is called a *Stationary Untrapped Hypersurface* Δ_S.

The dual expansion vector plays the role for Stationary Untrapped Hypersurfaces, which the stationary Killing vector plays for stationary black holes.

2. Hamiltonian and Conserved Quantities

A systematic way to study conserved quantities is through Hamiltonian method.[7] To obtain quasilocal quantities, we shall first extend the Hamiltonian principle for spatial infinity studied by Regge and Teitelboim [8] to finite regions.

For a general diffeomorphism-invariant field theory in four dimensions with a Lagrangian 4-form $\mathcal{L}(\varphi)$, where φ denotes an arbitrary collection of dynamical fields, the equations of motion, are obtained by computing the first variation of the Lagrangian.

$$\delta\mathcal{L} = d(\delta\varphi \wedge p) + \delta\varphi \wedge \frac{\delta\mathcal{L}}{\delta\varphi} + \frac{\delta\mathcal{L}}{\delta p} \wedge \delta p. \tag{4}$$

For any diffeomorphism generated by a smooth vector field ξ^a, one can define a translational current 3-form $\mathcal{H}(\xi)$ [11] (also known as Noether current 3-form, in the second order formulation) by

$$\mathcal{H}(\xi) := \pounds_\xi\varphi \wedge p - i_\xi\mathcal{L} \tag{5}$$

where $\pounds_\xi$ denotes the Lie derivative and i_ξ is the inner product. On shell, the translational current is closed, and can be written in terms of a 2-form $\mathcal{Q}(\xi)$ (the Noether charge) as $\mathcal{H}(\xi) = d\mathcal{Q}(\xi)$. The variation of the translational current 3-form is given by (here we assume $\delta\xi^a = 0$),

$$\begin{aligned}\delta\mathcal{H}(\xi) \equiv\ & \pounds_\xi\varphi \wedge \delta p - \delta\varphi \wedge \pounds_\xi p \\ & + di_\xi(\delta\varphi \wedge p) - i_\xi\left[\delta\varphi \wedge \frac{\delta\mathcal{L}}{\delta\varphi} + \frac{\delta\mathcal{L}}{\delta p} \wedge \delta p\right].\end{aligned} \tag{6}$$

The last term vanishes "on shell". Thus

$$\delta\mathcal{H}(\xi) = \omega(\varphi, \delta\varphi, \pounds_\xi\varphi) + di_\xi(\delta\varphi \wedge p), \tag{7}$$

where ω is the symplectic current 3-form defined by

$$\omega(\varphi, \delta_1\varphi, \delta_2\varphi) = \delta_2\varphi \wedge \delta_1 p - \delta_1\varphi \wedge \delta_2 p. \tag{8}$$

Its integral over a 3-surface Σ defines the presymplectic form Ω. If the presymplectic form is a total variation

$$\Omega(\varphi, \delta\varphi, \pounds_\xi\varphi) \equiv \int_\Sigma \omega(\varphi, \delta\varphi, \pounds_\xi\varphi) = \delta H(\xi) \tag{9}$$

for some function $H(\xi)$ on the field space, then $H(\xi)$ is conserved along ξ^a, i.e. $\pounds_\xi H(\xi) = 0$. The function $H(\xi)$ is called the Hamiltonian conjugate to ξ^a [6,10,11].

In order to have a well-defined Hamiltonian, there must exist a 2-form $\mathcal{B}$ such that at the boundary we have $i_\xi(\delta\varphi \wedge p) = \delta\mathcal{B}$. Then since on shell $\mathcal{H}(\xi) = d\mathcal{Q}(\xi)$, $H(\xi) = \oint \mathcal{Q}(\xi) - \mathcal{B}(\xi)$ is our desired Hamiltonian.

3. Boundary conditions and quasilocal mass for Stationary Untrapped Hypersurfaces

Using Hamiltonian method, we shall show that for evolution along the region bounded by a Stationary Untrapped Hypersurface, the Hamiltonian (associated with a spacelike hypersurface spanning S) is well-defined. Thus the total energy-momentum for the region can be defined and is conserved.

A related question was in fact proposed long ago in the study of *quasilocal mass* [9], i.e. to find a suitable definition of total energy-momentum and angular-momentum, surrounded by a spacelike 2-surface S, with S^2 topology, in 4-dimensional spacetime M. The construction is quasilocal in the sense that it refers only to the geometry of S (intrinsic metric, first fundamental form) and the extrinsic curvatures (second fundamental form) for its embedding in M. Instead of looking for a general formula for quasilocal mass, here we focus on quasilocal mass for regions bounded by Stationary Untrapped Hypersurfaces. Instead of looking at perturbation δ, we look at a Lie derivatives along a Stationary Untrapped Hypersurface $\delta = \pounds_{H^a_\perp}$. For Stationary Untrapped Hypersurface, the following boundary conditions are satisfied:

Boundary Condition I. The timelike flow ξ is fixed by the dual mean curvature vector $H^a_\perp$,

$$\delta\xi^a = \delta H^a_\perp = 0, \tag{10}$$

i.e. the untrapped surface evolves along the dual mean curvature vector direction.

Boundary Condition II. The intrinsic metric of the 2-surface is fixed, thus the area of each cross section does not change along the dual expansion vector direction, i.e.,

$$\delta\sigma_{ab} = 0. \tag{11}$$

The boundary conditions *I.* and *II.* are satisfied for $\delta = \pounds_{H^a_\perp}$. For General Relativity the Hamiltonian exist with these boundary conditions satisfied. Let us start with the Hilbert action for General Relativity with Lagrangian 4-form given by

$$\mathcal{L} = R^{ab} \wedge *(\vartheta_a \wedge \vartheta_b) \tag{12}$$

where $R^{ab} = d\Gamma^{ab} + \Gamma^a{}_c \wedge \Gamma^{cb}$ is the curvature 2-form, $*(\vartheta^a \wedge \vartheta^b) = \frac{1}{2}\epsilon^{ab}{}_{cd}\vartheta^c \wedge \vartheta^d$, and $g = \eta_{ab}\,\vartheta^a \otimes_s \vartheta^b$ is the metric, where $\eta_{ab} = \mathrm{diag}(-1,1,1,1)$ and ϑ^a is the orthonormal frame 1-form field.

Using the boundary conditions (10) and (11), the Hamiltonian is well-defined and is given by [5,6]:

$$H(\xi) = \oint_S (i_\xi \vartheta^c)\, \Gamma^{ab} \wedge \vartheta^d \epsilon_{abcd} + f(\sigma) dS. \tag{13}$$

where $f(\sigma)$ is a function constructed by just the metric σ_{ab} of the 2-surface S, so that $\pounds_{H^a_\perp} f(\sigma) = 0$. By using an identity on the 2-surface S for any vector field V^a:

$$V^c\, \Gamma^{ab} \wedge \epsilon_{abcd} \vartheta^d = 2\left(V^0 k(1) + V^1 k(0) + V^A \zeta_A\right) dS$$

where $\zeta_A = -\Gamma^{01}{}_A$ is the twist and $dS = \epsilon_{01CD} \vartheta^C \wedge \vartheta^D$, this yields

$$H(\xi) = \oint_S [i_\xi \vartheta^0 k(1) + i_\xi \vartheta^1 k(0) + (i_\xi \vartheta^A \zeta_A) + f(\sigma)] dS. \tag{14}$$

On the Stationary Untrapped Hypersurface, let $\xi = h(\sigma) H_\perp$, the Hamiltonian is given by

$$H(\xi) = h(\sigma) \oint_S \left(g(\sigma) - H^2\right) dS \tag{15}$$

where $H^2 = k(1)^2 - k(0)^2$ and $g(\sigma) = f(\sigma)/h(\sigma)$. The free functions on the 2-surface can be chosen such that the expression gives ADM mass at spatial infinity and irriducible mass at the apparent horizon. This can be done by letting $g(\sigma)$ to be the 2 dimensional scalar curvature $\mathcal{R}$ of S and let $h(\sigma)$ to be $(1/16\pi)\sqrt{A/16\pi}$ (where $A = \oint_S dS$ is the area of S) then

$$H(\xi) = \frac{1}{16\pi}\sqrt{\frac{A}{16\pi}} \oint_S \left(\mathcal{R} - H^2\right) dS. \tag{16}$$

By Gauss-Bonnet theorem, the expression reduced to

$$H(\xi) = \sqrt{\frac{A}{16\pi}} \left(1 - \frac{1}{16\pi} \oint_S H^2 dS\right) \tag{17}$$

which is precisely the Hawking mass. At marginal surfaces, $H = 0$, it is the irreducible mass.

Thus when an untrapped surface evolves along the direction of the dual mean curvature vector, the Hamiltonian is well-defined and thus provides a definition of total energy-momentum for this stationary untrapped hypersurface.

When the dual expansion vector is null, the expansion vector is also a null vector. The boundary conditions for Stationary Untrapped Hypersurfaces reduce to Non-expanding Horizons[1].

4. Energy flux for Dynamical Untrapped Hypersurfaces

For evolution along the region bounded by a Dynamical Untrapped Hypersurface. Let ξ be a vector normal to a cross section and tangent to the Dynamical Untrapped Hypersurface, since $\xi \neq H_\perp$ in general, there does not in general exist a Hamiltonian, but a gravitational radiation energy flux can be obtained using Hamiltonian methods.[14,15,16] Consider a perturbation process along $\xi = aH_\perp + bH$, where a and b are constants, similar to the expression obtained in the study of Dynamical Horizons[1,2] and Trapping Horizons[3], the quasilocal energy flux can be obtained.[17]

Acknowledgments

This work was supported by the National Natural Science Foundation of China under the grants 10375081 and 10375087, by the Shanghai Pujiang Program 05PJ14081 and by the Shanghai Education Foundation Shuguang Award.

References

1. A. Ashtekar and B. Krishnan, *Phys. Rev. Lett.* ***89*** (2002) 261101.
2. A. Ashtekar and B. Krishnan, *Living Review in Relativity* *7* (2004) 10.
3. S. A. Hayward, *Phys. Rev. Lett.* ***93*** (2004) 251101.
4. L. B. Szabados, *Class. Quantum Grav.* ***11*** (1994) 1833-1846.
5. S. C. Anco and R. S. Tung, *J. Math. Phys.* *43* (2002) 3984-4019.
6. S. C. Anco and R. S. Tung, *J. Math. Phys.* *43* (2002) 5531-5566.
7. J. M. Nester, C. M. Chen and R. S. Tung, "The Hamiltonian boundary term", in the proceedings (pp. 396-402).
8. T. Regge and C. Teitelboim, *Annals of Physics* ***88*** (1974) 286-318.
9. L. B. Szabados, *Living Review in Relativity* *7* (2004) 4.
10. J. Lee and R. M. Wald, *J. Math. Phys.* *31* (1990) 725-743.
11. J. M. Nester, *Mod. Phys. Lett. A* ***6*** (1991) 2655-2661.
12. S. W. Hawking, *J. Math. Phys.* ***9*** (1968) 598-604.
13. Y. Choquet-Bruhat, in Gravitational Radiation, edited by N. Deruelle and T. Piran (North-Holland, Amsterdam, 1983), p. 399.
14. C. M. Chen, J. M. Nester and R. S. Tung, *Phys. Rev. D* **72**, 104020 (2005).
15. C. M. Chen, J. M. Nester and R. S. Tung, "Quasi-local energy flux", in the proceedings (pp.389-395).
16. R. M. Wald and A. Zoupas, *Phys. Rev. D* ***61*** (2000) 084027.
17. R. S. Tung, "Energy Conservation in Dynamical Untrapped Hypersurfaces", preprint.

The VII Asia-Pacific International Conference on Gravitation and Astrophysics

on the Occasion of the 90th Year of General Relativity

November 23-26, 2005
National Central University, Taiwan

International Organizing Committee:

Australia	D. Blair
Brazil	M. Novello
Canada	D. Page, V. Frolov
China	Y.-J. Wang, L. Liu, J. Luo, K. S. Cheng
France	J.-M. Alimi, T. Damour, R. Triay
Germany	H. Dehnen
Greece	S. Cotsakis
India	V. Sahni
Italy	V. de Sabbata
Japan	Y. Fujii, K. Kuroda, M. Kenmoku, K. Sato, H. Kodama, M. Sasaki
Korea	S. -W. Kim, C. W. Kim, Y. M. Cho, C. H. Lee, J. Yang, H. K. Lee
Mexico	A. Garcia
Russia	V. N. Melnikov, A. P. Efremov, G. S. Bisnovaty-Kogan, A. A. Starobinsky, D. V. Gal'tsov, V. N. Lukash, I. D. Novikov, A. B. Balakin, V. K. Milyukov, P. I. Pronin, V. N. Rudenko
Taiwan	J. M. Nester, W.-T. Ni
USA	E. Fischbach, F. Everitt, H. J. Paik, J. P. Hsu

Local Organizing Committee:

Chairman

James M. Nester (Central U.)
nester@phy.ncu.edu.tw

Scientific Secretary

Chiang-Mei Chen (Central U.)
cmchen@phy.ncu.edu.tw

Hing-Tong Cho (Tamkang U.)
Pei-Ming Ho (Taiwan U.)
Hsien-Chung Kao (Taiwan Normal U.)
Wen-Fang Kao (Chiao-Tung U.)
Chung-I Kuo (Soochow U.)
Da-Shin Lee (Dong Hwa U.)
Kin-Wang Ng (Academia Sinica)
Wei-Tou Ni (Tsing-Hua U.)
Chopin Soo (Cheng-Kung U.)
Roh-Suan Tung (Shanghai Normal U.)
Hwei-Jang Yo (Cheng-Kung U.)

List of Speakers

	Name	Affiliation	Nationality
1	Chen, Chiang-Mei	Central U	Taiwan
2	Chen, Wenfeng	NCTS, Taiwan	Canada
3	Cho, Yong-Min *	Seoul National U	Korea
4	Fan, Jun-Hui	Guangzhou U	China
5	Fan, Shu-Hua	HUST, Wuhan	China
6	Ford, Larry *	Tufts U	USA
7	Frolov, Valeri P. *	U of Alberta	Russia
8	Gal'tsov, Dmitry V. *	Moscow State U	Russia
9	Gu, Je-An	Taiwan U	Taiwan
10	Harada, Tomohiro	Kyoto U.	Japan
11	Hayward, Sean	Penn State U	UK
12	Hsiang, Jen-Tsung	Dong-Hwa U	Taiwan
13	Hsu, Jong-Ping *	U Mass Dartmouth	Taiwan
14	Hu, Bei-Lok *	U of Maryland	USA
15	Huang, Chao-Guang *	CAS, Beijing	China
16	Isenberg, James *	U of Oregon	USA
17	Ivanov, Michael	Belarus State U	Belarus
18	Kenmoku, Masakatsu *	Nara Women U.	Japan
19	Keum, Yong-Yeon	Taiwan U	Korea
20	Kim, Sang-Pyo	Kunsan National U	Korea
21	Kim, Sung-Won *	Ewha Women's U	Korea
22	Lau, Yun-Kau *	CAS, Beijing	China
23	Lee, Seokcheon	Academia Sinica, TW	Korea
24	Lin, Shih-Yuin	Academia Sinica, TW	Taiwan
25	Lindblom, Lee *	Caltech	USA
26	Ling, Yi	Nanchang U	China
27	Liu, Liao *	Beijing Normal U	China

28	Luo, Jun *	HUST, Wuhan	China
29	Maeda, Hideki	Waseda U	Japan
30	Maeda, Kei-ichi *	Waseda U.	Japan
31	Mann, Robert *	U of Waterloo	Canada
32	Melinkov, Vitaly N. *	Grav Metrology Ctr	Russia
33	Misner, Charles *	U of Maryland	USA
34	Miyamoto, Umpei	Waseda U.	Japan
35	Morikawa, Masahiro	Ochanomizu U	Japan
36	Narita, Makoto	Central U	Japan
37	Nester, James M. *	Central U	USA
38	Page, Don N. *	U of Alberta	Canada
39	Paik, Ho-Jung	U of Maryland	Korea
40	So, Lau-Loi	Central U	Taiwan
41	Soo, Chopin	Cheng Kung U	Taiwan
42	Starobinsky, Alexi A *.	Landau Inst.	Russia
43	Tian, Yong	Army Armor School	Taiwan
44	Triay, Roland *	Provence U	France
45	Tung, Roh-Suan *	Shanghai Normal U	Taiwan
46	Wu, Xue-Jun	Nanjing Normal U	China
47	Xu, Chong-Ming *	Nanjing Normal U	China
48	Xu, Ren-Xin	Peking U	China
49	Yo, Hwei-Jang	Cheng Kung U	Taiwan
50	Yoon, Jong-Hyuk *	Konkuk U	Korea
51	Yu, Hoi-Lai	Academia Sinica, TW	Taiwan
52	Yu, Hong-Wei	Hunan Normal U	China
53	Yuan, Ye-Fei	USTC, Hefei	China
54	Zhu, Zong-Hong	Beijing Normal U	China
	* Invited		

List of Participants

	Name	Affiliation	Nationality	e-mail
1	Chan, Chuan-Tsung	NCTS, Taiwan	Taiwan	ctchan@phys.cts.nthu.edu.tw
2	Chang, Chia-Hung	Taiwan Normal U	Taiwan	chchang@phy.ntnu.edu.tw
3	Charng, Yeo-Yie	Academia Sinica, Taiwan	Taiwan	charng@phys.sinica.edu.tw
4	Chen, Chiang-Mei	Central U	Taiwan	cmchen@phy.ncu.edu.tw
5	Chen, Hsin	Central U	Taiwan	eppp8@yahoo.com.tw
6	Chen, Wenfeng	NCTS, Taiwan	Canada	wfchen@phys.cts.nthu.edu.tw
7	Chiu, Mu-Chen	Central U	Taiwan	chiumuchen@gmail.com
8	Cho, Hing Tong	Tamkang U	Taiwan	htcho@mail.tku.edu.tw
9	Cho, Yong-Min	Seoul National U	Korea	ymcho@yongmin.snu.ac.kr
10	Chou, Chung-Hsien	Academia Sinica	Taiwan	chouch@phys.sinica.edu.tw
11	Fan, Jun-Hui	Guangzhou U	China	Jhfan_cn@yahoo.com.cn
12	Fan, Shu-Hua	HUST, Wuhan	China	
13	Ford, Larry	Tufts U	USA	ford@cosmos.phy.tufts.edu
14	Frolov, Valeri P.	U of Alberta	Russia	frolov@phys.ualberta.ca
15	Gal'tsov, Dmitry V.	Moscow State U	Russia	gdmv04@mail.ru
16	Gu, Je-An	Taiwan U	Taiwan	jagu@phys.ntu.edu.tw
17	Harada, Tomohiro	Kyoto U.	Japan	harada@scphys.kyoto-u.ac.jp
18	Hayward, Sean	Penn State U	UK	Sean_a_hayward@yahoo.co.uk
19	Ho, Fei-Hung	Central U	Taiwan	flyparadox@gmail.com
20	Ho, Pei-Ming	Taiwan U	Taiwan	pmho@ntu.edu.tw
21	Hsiang, Jen-Tsung	Dong-Hwa U	Taiwan	cosmology@gmail.com
22	Hsu, Jong-Ping	U Mass Dartmouth	Taiwan	jhsu@umassd.edu
23	Hu, Bei-Lok	U of Maryland	USA	hub@phys.umd.edu
24	Huang, Chao-Guang	CAS, Beijing	China	huangcg@ihep.ac.cn
25	Huang, Ying-Ming	Central U	Taiwan	acutedog@hotmail.com
26	Isenberg, James	U of Oregon	USA	jim@newton.uoregon.edu
27	Ivanov, Michael	Belarus State U	Belarus	ivanovma@gw.bsuir.unibel.by
28	Kenmoku, Masakatsu	Nara Women U.	Japan	Kenmoku@asuka.phys.narawu.ac.jp

29	Keum, Yong-Yeon	Taiwan U	Korea	yykeum@phys.ntu.edu.tw
30	Kim, Sang-Pyo	Kunsan National U	Korea	sangkim@kunsan.ac.kr
31	Kim, Sung-Won	Ewha Women's U	Korea	sungwon@ewha.ac.kr
32	Kong, Otto	Central U	Taiwan	otto@phy.ncu.edu.tw
33	Kung, Dan-Sheng	Central U	Taiwan	etbigwin@yahoo.com.tw
34	Kuo, Chung-I	Soochow U	Taiwan	cikuo@scu.edu.tw
35	Lau, Yun-Kau	CAS, Beijing	China	lau@amss.ac.cn
36	Lee, Da-Shin	Dong-Hwa U	Taiwan	dslee@mail.ndhu.edu.tw
37	Lee, Seokcheon	Academia Sinica, Taiwan	Korea	skylee@phys.sinica.edu.tw
38	Lee, Wolung	Taiwan Normal U	Taiwan	wolung@gmail.com
39	Li, Hsian-Nan	Academia Sinica, Taiwan	Taiwan	hnli@phys.sinica.edu.tw
40	Li, Xiao-Jing	Central U	Taiwan	
41	Lin, Shih-Yuin	Academia Sinica, Taiwan	Taiwan	sylin@phys.sinica.edu.tw
42	Lindblom, Lee	Caltech	USA	lindblom@tapir.caltech.edu
43	Ling, Yi	Nanchang U	China	yling@ncu.edu.cn
44	Liu, Jian-Liang	Central U	Taiwan	tendauliang@yahoo.com.tw
45	Liu, Liao	Beijing Normal U	China	liuliao1928@yahoo.com.cn
46	Luo, Jun	HUST, Wuhan	China	junluo@mail.hust.edu.cn
47	Luo, Zhi-Quan	Xihua Normal U	China	lzq_sc@tom.com
48	Maeda, Hideki	Waseda U	Japan	hideki@gravity.phys.waseda.ac.jp
49	Maeda, Kei-ichi	Waseda U.	Japan	maeda@gravity.phys.waseda.ac.jp
50	Mann, Robert	U of Waterloo	Canada	rmann@avatar.uwaterloo.ca
51	Melinkov, Vitaly N.	Grav Metrology Ctr	Russia	melnikov@phys.msu.ru
52	Misner, Charles	U of Maryland	USA	misner@physics.umd.edu
53	Miyamoto, Umpei	Waseda U.	Japan	umpei@gravity.phys.waseda.ac.jp
54	Morikawa, Masahiro	Ochanomizu U	Japan	hiro@phys.ocha.ac.jp
55	Narita, Makoto	Central U	Japan	narita@phy.ncu.edu.tw
56	Nester, James M.	Central U	USA	nester@phy.ncu.edu.tw
57	Ng, Kin-Wang	Academia Sinica, Taiwan	Taiwan	nkw@phys.sinica.edu.tw

58	Page, Don N.	U of Alberta	Canada	don@phys.ualberta.ca
59	Paik, Ho-Jung	U of Maryland	Korea	hpaik@physics.umd.edu
60	Pu, Hung-Yi	Tsing Hwa U	Taiwan	hypu@asiaa.sinica.edu.tw
61	Shie, Kun-Feng	Central U	Taiwan	x93222038@yahoo.com.tw
62	Shiokawa, Kazutomu	U of Maryland	Japan	kshiok@physics.umd.edu
63	So, Lau-Loi	Central U	Taiwan	s0242010@cc.ncu.edu.tw
64	Soo, Chopin	Cheng Kung U	Taiwan	cpsoo@mail.ncku.edu.tw
65	Starobinsky, Alexi A.	Landau Inst.	Russia	alstar@landau.ac.ru
66	Tian, Yong	Army Armor School	Taiwan	yonngtian@gmail.com
67	Triay, Roland	Provence U	France	triay@cpt.univ-mrs.fr
68	Tung, Roh-Suan	Shanghai Normal U	Taiwan	tung@shnu.edu.cn
69	Wang, I-Chin	Cheng Kung U	Taiwan	wzboson@yahoo.com.tw
70	Wang, John	NCTS, Taiwan	USA	jwang@phys.cts.nthu.edu.tw
71	Wen, Wen-Yu	Taiwan Normal U	Taiwan	wenw@umich.edu
72	Wu, Chun-Hsien	Academia Sinica	Taiwan	chunwu@phys.sinica.edu.tw
73	Wu, Min-Fan	Central U	Taiwan	wuminfan@yahoo.com.tw
74	Wu, Shang-Yu	Taiwan U	Taiwan	R92244001@ntu.edu.tw
75	Wu, Xue-Jun	Nanjing Normal U	China	
76	Xu, Chong-Ming	Nanjing Normal U	China	cmxu@njnu.edu.cn
77	Xu, Ren-Xin	Peking U	China	r.x.xu@pku.edu.cn
78	Yo, Hwei-Jang	Cheng Kung U	Taiwan	hjyo@phys.ncku.edu.tw
79	Yong, Poh-Hoong	Central U	Taiwan	u1202046@cc.ncu.edu.tw
80	Yoon, Jong-Hyuk	Konkuk U	Korea	yoonjh@konkuk..ac.kr
81	Yu, Hoi-Lai	Academia Sinica, Taiwan	Taiwan	hlyu@phys.sinica.edu.tw
82	Yu, Hong-Wei	Hunan Normal U	China	hwyu@hunnu.edu.cn
83	Yuan, Ye-Fei	USTC, Hefei	China	yfyuan@ustc.edu.cn
84	Zhu, Zong-Hong	Beijing Normal U	China	zhuzh@bnu.edu.cn
85	Zou, Shou-Jyun	Central U	Taiwan	sgzou@hotmail.com

VII Asia-Pacific International Conference on Gravitation and Astrophysics
23-26 November 2005 National Central University, Taiwan

FRAGIL